CAMBRIDGE TRACTS IN MATHEMATICS

General Editors

151 Frobenius manifolds and moduli spaces for singularities

Frobenius manifolds and moduli spaces for singularities

CLAUS HERTLING

CAMBRIDGE
UNIVERSITY PRESS

PUBLISHED BY THE PRESS SYNDICATE OF THE UNIVERSITY OF CAMBRIDGE
The Pitt Building, Trumpington Street, Cambridge, United Kingdom

CAMBRIDGE UNIVERSITY PRESS
The Edinburgh Building, Cambridge CB2 2RU, UK
40 West 20th Street, New York, NY 10011-4211, USA
477 Williamstown Road, Port Melbourne, VIC 3207, Australia
Ruiz de Alarcón 13, 28014 Madrid, Spain
Dock House, The Waterfront, Cape Town 8001, South Africa

http://www.cambridge.org

First published 2002

Printed in the United Kingdom at the University Press, Cambridge

Typeface Times 10/13 pt *System* LaTeX 2_ε [TB]

A catalogue record for this book is available from the British Library

ISBN 0 521 81296 8 hardback

Contents

Preface

Frobenius manifolds are complex manifolds with a rich structure on the holomorphic tangent bundle, a multiplication and a metric which harmonize in the most natural way. They were defined by Dubrovin in 1991, motivated by the work of Witten, Dijkgraaf, E. Verlinde, and H. Verlinde on topological field theory. Originally coming from physics, Frobenius manifolds now turn up in very different areas of mathematics, giving unexpected relations between them, in quantum cohomology, singularity theory, integrable systems, symplectic geometry, and others. The isomorphy of certain Frobenius manifolds in quantum cohomology and in singularity theory is one version of mirror symmetry.

This book is devoted to the relations between Frobenius manifolds and singularity theory. It consists of two parts.

In part 1 F-manifolds are studied, manifolds with a multiplication on the tangent bundle with a natural integrability condition. They were introduced in [**HM**][**Man2**, I§5]. Frobenius manifolds are F-manifolds. Studying F-manifolds, one is led directly to discriminants, a classical subject of singularity theory, and to Lagrange maps and their singularities. Our development of the general structure of F-manifolds is at the same time an introduction to discriminants and Lagrange maps. As an application, we use some work of Givental to prove a conjecture of Dubrovin about Frobenius manifolds and Coxeter groups.

In part 2 we take up the construction of Frobenius manifolds in singularity theory. Already in 1983 K. Saito and M. Saito had found that the base space of a semiuniversal unfolding of an isolated hypersurface singularity can be equipped with the structure of a Frobenius manifold. Their construction involves the Gauß-Manin connection, polarized mixed Hodge structures, K. Saito's higher residue pairings, and his primitive forms. It was hardly accessible for nonspecialists. We give a more elementary detailed account of the construction, explain all ingredients, and develop or cite all necessary results.

We give a number of applications. The deepest one is the construction of global moduli spaces for isolated hypersurface singularities.

The construction of K. Saito and M. Saito is related to a recent construction of Frobenius manifolds via oscillating integrals by Sabbah and Barannikov. We comment upon that.

Background and other books. The reader should know the basic concepts of complex analytic geometry, including coherent sheaves and flatness (cf. for example [**Fi**]). All notions from symplectic geometry which are used can be found in [**AGV1**, chapter 18]. An excellent basic reference on flat connections and vector bundles (and much more) is the forthcoming book [**Sab4**]. It also gives a viewpoint on Frobenius manifolds which complements ours. Two fundamental books on Frobenius manifolds are [**Du3**] and [**Man2**]. Our treatment of singularities and their Gauß-Manin connection is essentially self-contained and gives precisely what is needed, but it is quite compact. Some books which expound several aspects in much more detail are [**AGV1**][**AGV2**][**Ku**][**Lo2**].

Acknowledgements. This book grew out of my habilitation. I would like to thank many people. E. Brieskorn was my teacher in singularity theory and defined in 1970 the wonderful object H_0'', which is now called the Brieskorn lattice. Yu. Manin introduced me to Frobenius manifolds. The common paper [**HM**] was the starting point for part 1. His papers and those of B. Dubrovin and C. Sabbah, and discussions with them were very fruitful. G.-M. Greuel and G. Pfister sharpened my view of moduli problems. M. Schulze and M. Rosellen made useful comments.

Of course, this book builds on the work of many people in singularity theory; Arnold, Givental, Looijenga, Malgrange, K. Saito, M. Saito, Scherk, O.P. Shcherbak, Slodowy, Steenbrink, Teissier, Varchenko, Wall, and many others. A good part of the book was written during a stay at the mathematics department of the University Paul Sabatier in Toulouse. I thank the department and especially J.-F. Mattei for their hospitality.

Bonn, July 2001 Claus Hertling

Part 1

Multiplication on the tangent bundle

Chapter 1

Introduction to part 1

An F-manifold is a complex manifold M such that each holomorphic tangent space $T_tM, t \in M$, is a commutative and associative algebra with unit element, and the multiplication varies in a specific way with the point $t \in M$. More precisely, it is a triple $(M, \circ, e)$ where $\circ$ is an $\mathcal{O}_M$-bilinear commutative and associative multiplication on the holomorphic tangent sheaf $\mathcal{T}_M$, e is a global unit field, and the multiplication satisfies the integrability condition

$$\mathrm{Lie}_{X \circ Y}(\circ) = X \circ \mathrm{Lie}_Y(\circ) + Y \circ \mathrm{Lie}_X(\circ) \tag{1.1}$$

for any two local vector fields X and Y in $\mathcal{T}_M$. This notion was first defined in [**HM**][**Man2**, I§5], motivated by Frobenius manifolds. Frobenius manifolds are F-manifolds.

Part 1 of this book is devoted to the local structure of F-manifolds. It turns out to be closely related to singularity theory and symplectic geometry. Discriminants and Lagrange maps play a key role.

In the short section 1.1 of this introduction the reader can experience some of the geometry of F-manifolds. We sketch a construction of 2-dimensional F-manifolds which shows how F-manifolds turn up 'in nature' and how they are related to discriminants. In section 1.2 we offer a fast track through the main notions and results of chapters 2 to 5.

In chapters 2 to 4 the general structure of F-manifolds is developed. In chapter 5 the most important classes of F-manifolds are discussed.

In chapter 2 F-manifolds are defined and some basic properties are established. One property shows that F-manifolds decompose locally in a nice way. Another one describes the relation to connections, metrics, and the potentiality condition of Frobenius manifolds.

In chapter 3 the relation to symplectic geometry and especially to Lagrange maps is discussed. This allows use to be made of Givental's paper [**Gi2**] on singular Lagrange varieties and their Lagrange maps.

Chapter 4 presents several notions and results, which are mostly motivated by corresponding notions and results in singularity theory. Most important are the discriminants and their geometry.

In chapter 5 F-manifolds from hypersurface singularities, boundary singularities, and Coxeter groups are discussed. In the case of Coxeter groups we extend some results of Givental [**Gi2**] and use them to prove a conjecture of Dubrovin about their Frobenius manifolds.

The reader should have the following background. There should be familiarity with the basic concepts of complex analytic geometry, including coherent sheaves and flatness. One reference is [**Fi**]. There should also be awareness of those notions from symplectic geometry which are treated in [**AGV1**, chapter 18] (canonical 1-form on the cotangent bundle, Lagrange fibration, Lagrange map, generating function). We recommend this reference. In chapter 5 some acquaintance with singularity theory makes the reading easier, but it is not necessary. Good references are [**AGV1**] and [**Lo2**].

1.1 First examples

To give the reader an idea of what F-manifolds look like and how they arise naturally, a construction of 2-dimensional F-manifolds is sketched. A systematic treatment is given in sections 4.1 and 4.2.

Let W be a finite Coxeter group of type $I_2(m)$, $m \geq 2$, acting on $\mathbb{R}^2$ and (by $\mathbb{C}$-linear extension) on $\mathbb{C}^2$. Then the ring $\mathbb{C}[x_1, x_2]^W \subset \mathbb{C}[x_1, x_2]$ of W-invariant polynomials is $\mathbb{C}[x_1, x_2]^W \cong \mathbb{C}[t_1, t_2]$ with 2 homogeneous generators t_1 and t_2 of degrees m and 2. Therefore the quotient space $\mathbb{C}^2/W =: M$ is isomorphic to $\mathbb{C}^2$ as an affine algebraic variety, and the vector field $e := \frac{\partial}{\partial t_1}$ is unique up to multiplication by a constant. The image in M of the union of the complexified reflection hyperplanes is the discriminant $\mathcal{D}$. We choose t_1 and t_2 such that it is given as $\mathcal{D} = \{t \in M \mid t_1^2 - \frac{4}{m^2} t_2^m = 0\}$.

For a point $t \in M$ with $t_2 \neq 0$, the pair $(e, \mathcal{D})$ gives rise to a multiplication on $T_t M$ in the following way, which is illustrated in figure 1.1.

The e-orbit through the point t intersects the discriminant at 2 points. We shift the tangent hyperplanes of $\mathcal{D}$ at these points with the flow of e to $T_t M$. We find that they are transverse to one another and to e. Therefore there are 2 unique vectors e_1 and e_2 in $T_t M$ which are tangent to these lines and satisfy $e = e_1 + e_2$. We define a multiplication on $T_t M$ by $e_i \circ e_j := \delta_{ij} e_i$. It is obviously commutative and associative, and e is the unit vector.

If we write this multiplication in terms of the coordinate fields $e := \frac{\partial}{\partial t_1}$ and $\frac{\partial}{\partial t_2}$, after some calculation we find $\frac{\partial}{\partial t_2} \circ \frac{\partial}{\partial t_2} = t_2^{m-2} \cdot \frac{\partial}{\partial t_1}$, and e is the

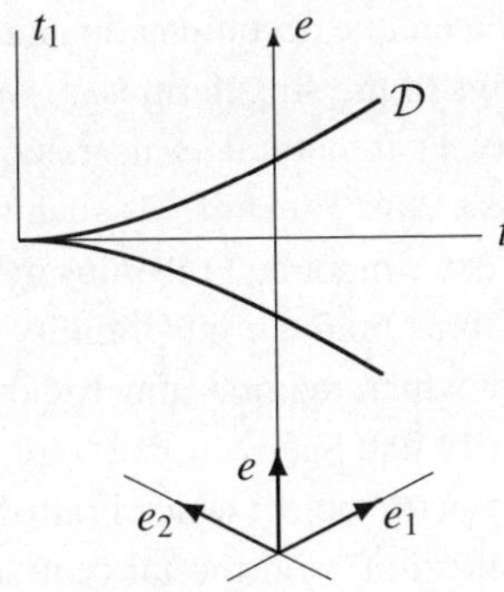

Figure 1.1

unit field. Therefore the multiplication extends holomorphically to the whole tangent bundle TM. One can show that it satisfies (1.1). The orbit space M is an F-manifold.

This construction of an F-manifold from a discriminant $\mathcal{D}$ and a transversal vector field e extends to higher dimensions (Corollary 4.6) and yields F-manifolds in many other cases, for example for all finite Coxeter groups (section 5.3).

1.2 Fast track through the results

The most notable (germs of) F-manifolds with many typical and some special properties are the base spaces of semiuniversal unfoldings of isolated hypersurface singularities and of boundary singularities (sections 5.1 and 5.2). Here the tangent space at each parameter is canonically isomorphic to the sum of the Jacobi algebras of the singularities above this parameter. Many of the general results on F-manifolds have been known in another guise in the hypersurface singularity case and all should be compared with it.

One reason why the integrability condition (1.1) is natural is the following: Let (M, p) be the germ of an F-manifold $(M, \circ, e)$. The algebra T_pM decomposes uniquely into a sum of (irreducible) local algebras which annihilate one another (Lemma 2.1). Now the integrability condition (1.1) ensures that this infinitesimal decomposition extends to a unique decomposition of the germ (M, p) into a product of germs of F-manifolds (Theorem 2.11).

If the multiplication at T_pM is semisimple, that is, if T_pM decomposes into 1-dimensional algebras, then this provides canonical coordinates $u_1, \ldots, u_n$ on (M, p) with $\frac{\partial}{\partial u_i} \circ \frac{\partial}{\partial u_j} = \delta_{ij} \frac{\partial}{\partial u_i}$. In fact, at points with semisimple multiplication the integrability condition (1.1) is equivalent to the existence of such canonical coordinates. In the hypersurface case, the decomposition of the germ (M, p)

for some parameter p is the unique decomposition into a product of base spaces of semiuniversal unfoldings of the singularities above p.

Another reason why (1.1) is natural is its relation to the potentiality of Frobenius manifolds. There exist F-manifolds such that not all tangent spaces are Frobenius algebras. They cannot be Frobenius manifolds. But if all tangent spaces are Frobenius algebras then the integrability condition (1.1) is related to a version of potentiality which requires a metric on M that is multiplication invariant, but not necessarily flat. See section 2.5 for details.

The most important geometric object which is attributed to an n-dimensional manifold M with multiplication $\circ$ on the tangent sheaf $\mathcal{T}_M$ and unit field e (with or without (1.1)) is the analytic spectrum $L := \text{Specan}(\mathcal{T}_M) \subset T^*M$ (see section 2.2). The projection $\pi : L \to M$ is flat and finite of degree n. The fibre $\pi^{-1}(p) \subset L$ above $p \in M$ consists of the set $\text{Hom}_{alg}(T_pM, \mathbb{C})$ of algebra homomorphisms from T_pM to $\mathbb{C}$; they correspond 1-1 to the irreducible subalgebras of $(T_pM, \circ)$ (see Lemma 2.1). The multiplication on $\mathcal{T}_M$ can be recovered from L, because the map

$$\mathbf{a} : \mathcal{T}_M \to \pi_* \mathcal{O}_L, \quad X \mapsto \alpha(\widetilde{X})|_L \tag{1.2}$$

is an isomorphism of $\mathcal{O}_M$-algebras; here $\widetilde{X}$ is any lift of X to T^*M and α is the canonical 1–form on T^*M. The values of the function $\mathbf{a}(X)$ on $\pi^{-1}(p)$ are the eigenvalues of $X\circ : T_pM \to T_pM$.

The analytic spectrum L is a reduced variety if and only if the multiplication is generically semisimple. Then the manifold with multiplication $(M, \circ, e)$ is called *massive*. Now, a third reason why the integrability condition (1.1) is natural is this: Suppose that $(M, \circ, e)$ is a manifold with generically semisimple multiplication. Then $L \subset T^*M$ is a Lagrange variety if and only if $(M, \circ, e)$ is a massive F-manifold (Theorem 3.2).

The main body of part 1 is devoted to the study of germs of massive F-manifolds at points where the multiplication is not semisimple.

We will make use of the theory of singular Lagrange varieties and their Lagrange maps, which has been worked out by Givental in [**Gi2**]. In fact, the notion of an irreducible germ (with respect to the above decomposition) of a massive F-manifold is equivalent to Givental's notion of a miniversal germ of a flat Lagrange map (Theorem 3.16). Via this equivalence Givental's paper contains many results on massive F-manifolds and will be extremely useful.

Locally the canonical 1–form α on T^*M can be integrated on the analytic spectrum L of a massive F-manifold $(M, \circ, e)$ to a *generating function* $F : L \to \mathbb{C}$ which is continuous on L and holomorphic on L_{reg}. It depends on a property of L, which is weaker than normality or maximality of the complex structure of L, whether F is holomorphic on L (see section 3.2).

If F is holomorphic on L then it corresponds via (1.2) to an *Euler field* $E = \mathbf{a}^{-1}(F)$ of weight 1, that is, a vector field on M with $\text{Lie}_E(\circ) = \circ$ (Theorem 3.3).

In any case, a generating function $F : L \to \mathbb{C}$ gives rise to a Lyashko–Looijenga map $\Lambda : M \to \mathbb{C}^n$ (see sections 3.3 and 3.5) and a discriminant $\mathcal{D} = \pi(F^{-1}(0)) \subset M$.

If F is holomorphic and an Euler field $E = \mathbf{a}^{-1}(F)$ exists then the discriminant $\mathcal{D}$ is the hypersurface of points where the multiplication with E is not invertible. Then it is a free divisor with logarithmic fields $\text{Der}_M(\log \mathcal{D}) = E \circ \mathcal{T}_M$ (Theorem 4.9). This generalizes a result of K. Saito for the hypersurface case.

From the unit field e and a discriminant $\mathcal{D} \subset M$ one can reconstruct everything. One can read off the multiplication on TM in a very nice elementary way (Corollary 4.6 and section 1.1): The e-orbit of a generic point $p \in M$ intersects $\mathcal{D}$ at n points. One shifts the n tangent hyperplanes with the flow of e to T_pM. Then there exist unique vectors $e_1(p), \ldots, e_n(p) \in T_pM$ such that $\sum_{i=1}^n e_i(p) = e(p)$ and $\sum_{i=1}^n \mathbb{C} \cdot e_i(p) = T_pM$ and such that the subspaces $\sum_{i \neq j} \mathbb{C} \cdot e_i(p)$, $j = 1, \ldots, n$, are the shifted hyperplanes. The multiplication on T_pM is given by $e_i(p) \circ e_j(p) = \delta_{ij} e_i(p)$.

In the case of hypersurface singularities and boundary singularities, the classical discriminant in the base space of a semiuniversal unfolding is such a discriminant. The critical set C in the total space of the unfolding is canonically isomorphic to the analytic spectrum L; this isomorphism identifies the map $\mathbf{a}$ in (1.2) with a Kodaira–Spencer map $\mathbf{a}_C : \mathcal{T}_M \to (\pi_C)_* \mathcal{O}_C$ and a generating function $F : L \to \mathbb{C}$ with the restriction of the unfolding function to the critical set C. This Kodaira–Spencer map $\mathbf{a}_C$ is the source of the multiplication on $\mathcal{T}_M$ in the hypersurface singularity case. The multiplication on $\mathcal{T}_M$ had first been defined in this way by K. Saito.

Critical set and analytic spectrum are smooth in the hypersurface singularity case. By the work of Arnold and Hörmander on Lagrange maps and singularities an excellent correspondence holds (Theorem 5.6): each irreducible germ of a massive F-manifold with smooth analytic spectrum comes from an isolated hypersurface singularity, and this singularity is unique up to stable right equivalence.

By the work of Nguyen huu Duc and Nguyen tien Dai the same correspondence holds for boundary singularities and irreducible germs of massive F-manifolds whose analytic spectrum is isomorphic to $(\mathbb{C}^{n-1}, 0) \times (\{(x, y) \in \mathbb{C}^2 \mid xy = 0\}, 0)$ with ordered components (Theorem 5.14).

The complex orbit space $M := \mathbb{C}^n / W \cong \mathbb{C}^n$ of a finite irreducible Coxeter group W carries an (up to some rescaling) canonical structure of a massive F-manifold: A generating system $P_1, \ldots, P_n$ of W-invariant homogeneous polynomials induces coordinates $t_1, \ldots, t_n$ on M. Precisely one polynomial, e.g. P_1, has highest degree. The field $\frac{\partial}{\partial t_1}$ is up to a scalar independent of any choices. This field $e := \frac{\partial}{\partial t_1}$ as the unit field and the classical discriminant $\mathcal{D} \subset M$, the image of the reflection hyperplanes, determine in the elementary way described above the structure of a massive F-manifold. This follows from [**Du2**][**Du3**, Lecture 4] as well as from [**Gi2**, Theorem 14].

Dubrovin established the structure of a Frobenius manifold on the complex orbit space $M = \mathbb{C}^n / W$, with this multiplication, with K. Saito's flat metric on M, and with a canonical Euler field with positive weights (see Theorem 5.23). At the same place he conjectured that these Frobenius manifolds and their products are (up to some well-understood rescalings) the only massive Frobenius manifolds with an Euler field with positive weights. We will prove this conjecture (Theorem 5.25).

Crucial for the proof is Givental's result [**Gi2**, Theorem 14]. It characterizes the germs $(M, 0)$ of F-manifolds of irreducible Coxeter groups by geometric properties (see Theorem 5.21). We obtain from it the following intermediate result (Theorem 5.20): An irreducible germ (M, p) of a *simple* F-manifold such that $T_p M$ is a Frobenius algebra is isomorphic to the germ at 0 of the F-manifold of an irreducible Coxeter group.

A massive F-manifold $(M, \circ, e)$ is called *simple* if the germs (M, p), $p \in M$, of F-manifolds are contained in finitely many isomorphism classes. Theorem 5.20 complements in a nice way the relation of irreducible Coxeter groups to the simple hypersurface singularities A_n, D_n, E_n and the simple boundary singularities B_n, C_n, F_4.

In dimensions 1 and 2, up to isomorphism all the irreducible germs of massive F-manifolds come from the irreducible Coxeter groups A_1 and $I_2(m)$ $(m \geq 3)$ with $I_2(3) = A_2,\ I_2(4) = B_2,\ I_2(5) =: H_2,\ I_2(6) = G_2$. But already in dimension 3 the classification is vast (see section 5.5).

Chapter 2

Definition and first properties of F-manifolds

An F-manifold is a manifold with a multiplication on the tangent bundle which satisfies a certain integrability condition. It is defined in section 2.3. Sections 2.4 and 2.5 give two reasons why this is a good notion. In section 2.4 it is shown that germs of F-manifolds decompose in a nice way. In section 2.5 the relation to connections and metrics is discussed. It turns out that the integrability condition is part of the potentiality condition for Frobenius manifolds. Therefore Frobenius manifolds are F-manifolds.

Section 2.1 is a self-contained elementary account of the structure of finite dimensional algebras in general (e.g. the tangent spaces of an F-manifold) and Frobenius algebras in particular. Section 2.2 discusses vector bundles with multiplication. There the caustic and the analytic spectrum are defined, two notions which are important for F-manifolds.

2.1 Finite-dimensional algebras

In this section $(Q, \circ, e)$ is a $\mathbb{C}$-algebra of finite dimension (≥ 1) with commutative and associative multiplication and with unit e. The next lemma gives precise information on the decomposition of Q into irreducible algebras. The statements are well known and elementary. They can be deduced directly in the given order or from more general results in commutative algebra (Q is an Artin algebra). Algebra homomorphisms are always supposed to map the unit to the unit.

Lemma 2.1 *Let $(Q, \circ, e)$ be as above. As the endomorphisms $x\circ : Q \to Q$, $x \in Q$, commute, there is a unique simultaneous decomposition $Q = \bigoplus_{k=1}^{l} Q_k$ into generalized eigenspaces Q_k (with $\dim_{\mathbb{C}} Q_k \geq 1$). Define $e_k \in Q_k$ by $e = \sum_{k=1}^{l} e_k$. Then*

(i) One has $Q_j \circ Q_k = 0$ for $j \neq k$; also $e_k \neq 0$ and $e_j \circ e_k = \delta_{jk} e_k$; the element e_k is the unit of the algebra $Q_k = e_k \circ Q$.

(ii) The function $\lambda_k : Q \to \mathbb{C}$ *which associates to* $x \in Q$ *the eigenvalue of* $x\circ$ *on* Q_k *is an algebra homomorphism;* $\lambda_j \neq \lambda_k$ *for* $j \neq k$.

(iii) The algebra $(Q_k, \circ, e_k)$ *is an irreducible and a local algebra with maximal ideal* $\mathbf{m}_k = Q_k \cap \ker(\lambda_k)$.

(iv) The subsets $\ker(\lambda_k) = \mathbf{m}_k \oplus \bigoplus_{j\neq k} Q_j$, $k = 1, \ldots, l$, *are the maximal ideals of the algebra* Q*; the complement* $Q - \bigcup_k \ker(\lambda_k)$ *is the group of invertible elements of* Q.

(v) The set $\{\lambda_1, \ldots, \lambda_l\} = \mathrm{Hom}_{\mathbb{C}-alg}(Q, \mathbb{C})$.

(vi) The localization $Q_{\ker(\lambda_k)}$ *is isomorphic to* Q_k.

We call this decomposition the *eigenspace decomposition* of $(Q, \circ, e)$. The set $L := \{\lambda_1, \ldots, \lambda_l\} \subset Q^*$ carries a natural complex structure $\mathcal{O}_L$ such that $\mathcal{O}_L(L) \cong Q$ and $\mathcal{O}_{L,\lambda_k} \cong Q_k$. More details on this will be given in section 2.2.

The algebra (or its multiplication) is called *semisimple* if Q decomposes into 1-dimensional subspaces, $Q \cong \bigoplus_{k=1}^{\dim Q} Q_k = \bigoplus_{k=1}^{\dim Q} \mathbb{C} \cdot e_k$.

An irreducible algebra $Q = \mathbb{C} \cdot e \oplus \mathbf{m}$ with maximal ideal $\mathbf{m}$ is a *Gorenstein ring* if the socle $\mathrm{Ann}_Q(\mathbf{m})$ has dimension 1.

An algebra $Q = \bigoplus_{k=1}^{l} Q_k$ is a *Frobenius algebra* if each irreducible subalgebra is a Gorenstein ring (cf. for example [**Kun**]).

The next (also well known) lemma gives equivalent conditions and additional information. Note that this classical definition of a Frobenius algebra is slightly weaker than Dubrovin's: he calls an algebra $(Q, \circ, e)$ together with a *fixed* bilinear form g as in Lemma 2.2 (a) (iii) a Frobenius algebra.

Lemma 2.2 *(a) The following conditions are equivalent.*

(i) The algebra $(Q, \circ, e)$ *is a Frobenius algebra.*

(ii) As a Q-module $\mathrm{Hom}(Q, \mathbb{C}) \cong Q$.

(iii) There exists a bilinear form $g : Q \times Q \to \mathbb{C}$ *which is symmetric, non-degenerate and multiplication invariant, i.e.* $g(a \circ b, c) = g(a, b \circ c)$.

(b) Let $Q = \bigoplus_{k=1}^{l} Q_k$ *be a Frobenius algebra and* $Q_k = \mathbb{C} \cdot e_k \oplus \mathbf{m}_k$. *The generators of* $\mathrm{Hom}(Q, \mathbb{C})$ *as a Q-module are the linear forms* $f : Q \to \mathbb{C}$ *with* $f(\mathrm{Ann}_{Q_k}(\mathbf{m}_k)) = \mathbb{C}$ *for all* k.

One obtains a 1-1 correspondence between these linear forms and the bilinear forms g *as in (a) (iii) by putting* $g(x, y) := f(x \circ y)$.

Proof: (a) Any of the conditions (i), (ii), (iii) in (a) is satisfied for Q if and only if it is satisfied for each irreducible subalgebra Q_k. One checks this with $Q_j \circ Q_k = 0$ for $j \neq k$. So we may suppose that Q is irreducible.

(i) $\iff$ (ii) A linear form $f \in \text{Hom}(Q, \mathbb{C})$ generates $\text{Hom}(Q, \mathbb{C})$ as a Q-module if and only if the linear form $(x \mapsto f(y \circ x))$ is nontrivial for any $y \in Q - \{0\}$, that is, if and only if $f(y \circ Q) = \mathbb{C}$ for any $y \in Q - \{0\}$.

The socle $\text{Ann}_Q(\mathbf{m})$ is the set of the common eigenvectors of all endomorphisms $x\circ : Q \to Q, x \in Q$. If $\dim \text{Ann}_Q(\mathbf{m}) \geq 2$ then for any linear form f an element $y \in (\ker f \cap \text{Ann}_Q(\mathbf{m})) - \{0\}$ satisfies $y \circ Q = \mathbb{C} \cdot y$ and $f(y \circ Q) = 0$; so f does not generate $\text{Hom}(Q, \mathbb{C})$. If $\dim \text{Ann}_Q(\mathbf{m}) = 1$ then it is contained in any nontrivial ideal, because any such ideal contains a common eigenvector of all endomorphisms. The set $y \circ Q$ for $y \in Q - \{0\}$ is an ideal. So, then a linear form f with $f(\text{Ann}_Q(\mathbf{m})) = \mathbb{C}$ generates $\text{Hom}(Q, \mathbb{C})$ as a Q-module.

(i) $\Rightarrow$ (iii) Choose any linear form f with $f(\text{Ann}_Q(\mathbf{m})) = \mathbb{C}$ and define g by $g(x, y) := f(x \circ y)$. It remains to show that g is nondegenerate. But for any $x \in Q - \{0\}$ there exists a $y \in Q$ with $\mathbb{C} \cdot x \circ y = \text{Ann}_Q(\mathbf{m})$, because $\text{Ann}_Q(\mathbf{m})$ is contained in the ideal $x \circ Q$.

(iii) $\Rightarrow$ (i) The equalities $g(\mathbf{m}, \text{Ann}_Q(\mathbf{m})) = g(e, \mathbf{m} \circ \text{Ann}_Q(\mathbf{m})) = g(e, 0) = 0$ imply $\dim \text{Ann}_Q(\mathbf{m}) = 1$.

(b) Starting with a bilinear form g, the corresponding linear form f is given by $f(x) = g(x, e)$. The rest is clear from the preceding discussion. □

The semisimple algebra $Q \cong \bigoplus_{k=1}^{\dim Q} \mathbb{C} \cdot e_k$ is a Frobenius algebra.

A classical result is that the complete intersections $\mathcal{O}_{\mathbb{C}^m,0}/(f_1, \ldots, f_m)$ are Gorenstein. But there are other Gorenstein algebras, e.g. $\mathbb{C}\{x, y, z\}/(x^2, y^2, xz, yz, xy - z^2)$ is Gorenstein, but not a complete intersection.

Finally, in the next section vector bundles with multiplication will be considered. Condition (ii) of Lemma 2.2 (a) shows that there the points whose fibres are Frobenius algebras form an open set in the base.

2.2 Vector bundles with multiplication

Now we consider a holomorphic vector bundle $Q \to M$ on a complex manifold M with multiplication on the fibres: The sheaf $\mathcal{Q}$ of holomorphic sections of the bundle $Q \to M$ is equipped with an $\mathcal{O}_M$-bilinear commutative and associative multiplication $\circ$ and with a global unit section e.

The set $\bigcup_{p \in M} \text{Hom}_{\mathbb{C}-alg}(Q(p), \mathbb{C})$ of algebra homomorphisms from the single fibres $Q(p)$ to $\mathbb{C}$ (which map the unit to $1 \in \mathbb{C}$) is a subset of the dual bundle Q^* and has a natural complex structure. It is the analytic spectrum $\text{Specan}(\mathcal{Q})$. We sketch the definition ([**Hou**, ch. 3], also [**Fi**, 1.14]):

The $\mathcal{O}_M$-sheaf $\text{Sym}_{\mathcal{O}_M} \mathcal{Q}$ can be identified with the $\mathcal{O}_M$-sheaf of holomorphic functions on Q^* which are polynomial in the fibres. The canonical $\mathcal{O}_M$-algebra homomorphism $\text{Sym}_{\mathcal{O}_M} \mathcal{Q} \to \mathcal{Q}$ which maps the multiplication in $\text{Sym}_{\mathcal{O}_M} \mathcal{Q}$

to the multiplication $\circ$ in $\mathcal{Q}$ is surjective. The kernel generates an ideal $\mathcal{I}$ in $\mathcal{O}_{Q^*}$.

One can describe the ideal locally explicitly: suppose $U \subset M$ is open and $\delta_1, \ldots, \delta_n \in \mathcal{Q}(U)$ is a base of sections of the restriction $Q|_U \to U$ with $\delta_1 = e$ and $\delta_i \circ \delta_j = \sum_k a_{ij}^k \delta_k$; denote by $y_1, \ldots, y_n$ the fibrewise linear functions on $Q^*|U$ which are induced by $\delta_1, \ldots, \delta_n$; then the ideal $\mathcal{I}$ is generated in $Q^*|U$ by

$$y_1 - 1 \quad \text{and} \quad y_i y_j - \sum_k a_{ij}^k y_k. \tag{2.1}$$

The support of $\mathcal{O}_{Q^*}/\mathcal{I}$ with the restriction of $\mathcal{O}_{Q^*}/\mathcal{I}$ as structure sheaf is $\mathrm{Specan}(\mathcal{Q}) \subset Q^*$. We denote the natural projections by $\pi_{Q^*} : Q^* \to M$ and $\pi : \mathrm{Specan}(\mathcal{Q}) \to M$. A part of the following theorem is already clear from the discussion. A complete proof and thorough discussion can be found in [**Hou**, ch. 3].

Theorem 2.3 *The set* $\bigcup_{p\in M} \mathrm{Hom}_{\mathbb{C}-alg}(Q(p), \mathbb{C})$ *is the support of the analytic spectrum* $\mathrm{Specan}(\mathcal{Q}) =: L$. *The composition of maps*

$$\mathbf{a} : \mathcal{Q} \hookrightarrow (\pi_{Q^*})_* \mathcal{O}_{Q^*} \to \pi_* \mathcal{O}_L \tag{2.2}$$

is an isomorphism of $\mathcal{O}_M$*-algebras and of free* $\mathcal{O}_M$*-modules of rank n, here n is the fibre dimension of* $Q \to M$. *The projection* $\pi : L \to M$ *is finite and flat of degree n.*

Consider a point $p \in M$ *with eigenspace decomposition* $Q(p) = \bigoplus_{k=1}^{l(p)} Q_k(p)$ *and* $L \cap \pi^{-1}(p) = \{\lambda_1, \ldots, \lambda_{l(p)}\}$. *The restriction of the isomorphism*

$$\mathcal{Q}_p \xrightarrow{\mathbf{a}} (\pi_* \mathcal{O}_L)_p \cong \bigoplus_{k=1}^{l(p)} \mathcal{O}_{L,\lambda_k} \tag{2.3}$$

to the fibre over p yields isomorphisms

$$Q_k(p) \cong \mathcal{O}_{L,\lambda_k} \otimes_{\mathcal{O}_{M,p}} \mathbb{C}. \tag{2.4}$$

Corollary 2.4 *In a sufficiently small neighbourhood U of a point* $p \in M$, *the eigenspace decomposition* $Q(p) = \bigoplus_{k=1}^{l(p)} Q_k(p)$ *of the fibre* $Q(p)$ *extends uniquely to a decomposition of the bundle* $Q|_U \to U$ *into multiplication invariant holomorphic subbundles.*

Proof of Corollary 2.4: The $\mathcal{O}_{M,p}$-free submodules $\mathcal{O}_{L,\lambda_k}$ in the decomposition in (2.3) of $(\pi_* \mathcal{O}_L)_p$ are obviously multiplication invariant. Via the isomorphism $\mathbf{a}$ one obtains locally a decomposition of the sheaf $\mathcal{Q}$ of sections of $Q \to M$ into multiplication invariant free $\mathcal{O}_M$-submodules. □

Of course, the induced decomposition of $Q(q)$ for a point q in the neighbourhood of p may be coarser than the eigenspace decomposition of $Q(q)$.

The base is naturally stratified with respect to the numbers and dimensions of the components of the eigenspace decompositions of the fibres of $Q \to M$. To make this precise we introduce a partial ordering $\succ$ on the set $\mathcal{P}$ of partitions of n:

$$\mathcal{P} := \left\{ \beta = (\beta_1, \dots, \beta_{l(\beta)}) \mid \beta_i \in \mathbb{N},\ \beta_i \geq \beta_{i+1},\ \sum_{i=1}^{l(\beta)} \beta_i = n \right\};$$

for $\beta, \gamma \in \mathcal{P}$ define

$$\beta \succ \gamma : \iff \exists\, \sigma : \{1, \dots, l(\gamma)\} \to \{1, \dots, l(\beta)\} \text{ s.t. } \beta_j = \sum_{i \in \sigma^{-1}(j)} \gamma_i .$$

The partition $P(p)$ of a fibre $Q(p)$ is the partition of $n = \dim Q(p)$ by the dimensions of the subspaces of the eigenspace decomposition.

Proposition 2.5 *Fix a partition $\beta \in \mathcal{P}$. The subset $\{p \in M \mid P(p) \succ \beta\}$ is empty or an analytic subset of M.*

Proof: The partition $P(f)$ of a polynomial of degree n is the partition of n by the multiplicity of the zeros of f.

Fact: *The space $\{a \in \mathbb{C}^n \mid P(z^n + \sum_{i=1}^n a_i z^{n-i}) \succ \beta\}$ is an algebraic subvariety of $\mathbb{C}^n$ with normalization isomorphic to $\mathbb{C}^{l(\beta)}$.*

For the proof one only has to regard the finite map $\mathbb{C}^n \to \mathbb{C}^n$, $u \mapsto ((-1)^i \sigma_i(u))_{i=1,\dots,n}$ ($\sigma_i(u)$ is the i-th symmetric polynomial).

A section $X \in \mathcal{Q}(U)$, $U \subset M$ open, induces via the coefficients of the characteristic polynomial $p_{ch,X\circ}$ of multiplication by X a holomorphic map $U \to \mathbb{C}^n$. Hence the set $\{q \in U \mid P(p_{ch,X\circ}) \succ \beta\}$ is analytic. The intersection of such analytic sets for a basis of sections in U is $\{q \in U \mid P(q) \succ \beta\}$. □

Suppose that M is connected. Then there is a unique partition β_0 such that $\{p \in M \mid P(p) = \beta_0\}$ is open in M. The complement $\mathcal{K} := \{p \in M \mid P(p) \neq \beta_0\}$ will be called the *caustic* in M; this name is motivated by the Lagrange maps (sections 3.1, 3.3, 3.4) and the hypersurface singularities (section 5.1). The multiplication is generically semisimple if and only if $\beta_0 = (1, \dots, 1)$.

Proposition 2.6 *The caustic $\mathcal{K}$ is a hypersurface or empty.*

Proof: Locally in $M - \mathcal{K}$ there is a holomorphically varying decomposition $Q(p) = \bigoplus_{k=1}^l Q_k(p)$ with partial unit fields $e_1, \dots, e_l$.

Suppose $\dim(\mathcal{K}, p) \leq \dim M - 2$ for some point $p \in \mathcal{K}$. Then in a neighbourhood U the complement $U - \mathcal{K}$ is simply connected. There is no monodromy for the locally defined vector fields $e_1, \ldots, e_l$ in $U - \mathcal{K}$. They extend to vector fields in U. For $p \in U - \mathcal{K}$ the map $e_k \circ : Q(p) \to Q(p)$ is the projection to $Q_k(p)$. Because of $e_1 + \cdots + e_l = e$ these projections extend to all of U and yield a decomposition of $Q(p)$ as above for all $p \in U$. Hence $\mathcal{K} \cap U = \emptyset$. □

2.3 Definition of F-manifolds

An F-manifold is a manifold M with a multiplication on the tangent bundle TM which harmonizes with the Lie bracket in the most natural way. A neat formulation of this property requires the Lie derivative of tensors.

Remark 2.7 Here the sheaf of (k, l)-tensors ($k, l \in \mathbb{N}_0$) on a manifold M is the sheaf of $\mathcal{O}_M$-module homomorphisms $\mathrm{Hom}_{\mathcal{O}_M}(\bigotimes_{i=1}^k \mathcal{T}_M, \bigotimes_{i=1}^l \mathcal{T}_M)$. A $(0, l)$-tensor $T : \mathcal{O}_M \to \bigotimes_{i=1}^l \mathcal{T}_M$ can be identified with $T(1)$. Vector fields are $(0, 1)$-tensors, a (commutative) multiplication on $\mathcal{T}_M$ is a (symmetric) $(2, 1)$-tensor.

The Lie derivative Lie_X with respect to a vector field X is a derivation on the sheaf of (k, l)-tensors. It is $\mathrm{Lie}_X(f) = X(f)$ for functions f, $\mathrm{Lie}_X(Y) = [X, Y]$ for vector fields Y, $\mathrm{Lie}_X(Y_1 \otimes \ldots \otimes Y_l) = \sum_i Y_1 \otimes ..[X, Y_i].. \otimes Y_l$ for $(0, l)$-tensors, and $(\mathrm{Lie}_X T)(Y) = \mathrm{Lie}_X(T(Y)) - T(\mathrm{Lie}_X(Y))$ for (k, l)-tensors T. One can always write it explicitly with Lie brackets. Because of the Jacobi identity the Lie derivative satisfies $\mathrm{Lie}_{[X,Y]} = [\mathrm{Lie}_X, \mathrm{Lie}_Y]$.

Definition 2.8 (a) An *F-manifold* is a triple $(M, \circ, e)$ where M is a complex connected manifold of dimension ≥ 1, $\circ$ is a commutative and associative $\mathcal{O}_M$-bilinear multiplication $\mathcal{T}_M \times \mathcal{T}_M \to \mathcal{T}_M$, e is a global unit field, and the multiplication satisfies for any two local vector fields X, Y

$$\mathrm{Lie}_{X \circ Y}(\circ) = X \circ \mathrm{Lie}_Y(\circ) + Y \circ \mathrm{Lie}_X(\circ). \tag{2.5}$$

(b) Let $(M, \circ, e)$ be an F-manifold. An *Euler field E of weight $d \in \mathbb{C}$* is a global vector field E which satisfies

$$\mathrm{Lie}_E(\circ) = d \cdot \circ. \tag{2.6}$$

(If no weight is mentioned, an Euler field will usually mean an Euler field of weight 1.)

Remarks 2.9 (i) We do not require that the algebras $(T_pM, \circ, e(p))$ are Frobenius algebras (cf. section 2.1). Nevertheless, this is a distinguished class. Frobenius manifolds are F-manifolds [**HM**][**Man1**, I§5]. This will be discussed in section 2.5.

(ii) Definition 2.8 differs slightly from the definition of F-manifolds in [**HM**] by the addition of a global unit field e. This unit field is important, for example, for the definition of Specan($\mathcal{T}_M$). Also, the Euler fields were called weak Euler fields in [**HM**] in order to separate them from the Euler fields with stronger properties of Frobenius manifolds. This is not necessary here.

(iii) Formula (2.5) is equivalent to

$$\begin{aligned}[X \circ Y, Z \circ W] &- [X \circ Y, Z] \circ W - [X \circ Y, W] \circ Z \\ &- X \circ [Y, Z \circ W] + X \circ [Y, Z] \circ W + X \circ [Y, W] \circ Z \\ &- Y \circ [X, Z \circ W] + Y \circ [X, Z] \circ W + Y \circ [X, W] \circ Z = 0 \end{aligned} \tag{2.7}$$

for any four (local) vector fields X, Y, Z, W. Formula (2.6) is equivalent to

$$[E, X \circ Y] - [E, X] \circ Y - X \circ [E, Y] - d \cdot X \circ Y = 0 \tag{2.8}$$

for any two (local) vector fields X, Y. The left hand side of (2.8) is $\mathcal{O}_M$-polylinear with respect to X and Y, because $\text{Lie}_E(\circ)$ is a (2, 1)–tensor. The left hand side of (2.7) is $\mathcal{O}_M$–polylinear with respect to X, Y, Z, W. Hence it defines a (4, 1)–tensor. In order to check (2.5) and (2.6) for a manifold with multiplication, it suffices to check (2.7) and (2.8) for a basis of vector fields.

(iv) The unit field e in an F-manifold $(M, \circ, e)$ plays a distinguished role. It is automatically nowhere vanishing. It is an Euler field of weight 0,

$$\text{Lie}_e(\circ) = 0 \cdot \circ, \tag{2.9}$$

because of (2.5) for $X = Y = e$. So, the multiplication of the F-manifold is constant along the unit field.

(v) An Euler field E of weight $d \neq 0$ is not constant along the unit field. But one has for any $d \in \mathbb{C}$

$$[e, E] = d \cdot e, \tag{2.10}$$

because of (2.8) for $X = Y = e$. More generally, in [**HM**][**Man1**, I§5] the identity

$$[E^{\circ n}, E^{\circ m}] = d(m - n) \cdot E^{\circ(m+n-1)} \tag{2.11}$$

is proved. Section 3.1 will show how intrinsic the notion of an Euler field is for an F-manifold.

(vi) The sheaf of Euler fields of an F-manifold $(M, \circ, e)$ is a Lie subalgebra of $\mathcal{T}_M$. If E_1 and E_2 are Euler fields of weight d_1 and d_2, then $c \cdot E_1$ ($c \in \mathbb{C}$) is an Euler field of weight $c \cdot d_1$, $E_1 + E_2$ is an Euler field of weight $d_1 + d_2$, and $[E_1, E_2]$ is an Euler field of weight 0. The last holds because of $\text{Lie}_{[E_1, E_2]} = [\text{Lie}_{E_1}, \text{Lie}_{E_2}]$ (cf. Remark 2.7 and [**HM**][**Man1**, I§5]).

(vii) The caustic $\mathcal{K}$ of an F-manifold is the subvariety of points $p \in M$ such that the eigenspace decomposition of T_pM has fewer components than for generic points (cf. section 2.2). The caustic is invariant with respect to e because of (2.9).

2.4 Decomposition of F-manifolds and examples

Proposition 2.10 *The product of two F-manifolds* $(M_1, \circ_1, e_1)$ *and* $(M_2, \circ_2, e_2)$ *is an F-manifold* $(M, \circ, e) = (M_1 \times M_2, \circ_1 \oplus \circ_2, e_1 + e_2)$.

If E_1 *and* E_2 *are Euler fields on* M_1 *and* M_2 *of the same weight* d *then the sum* $E_1 + E_2$ *(of the lifts to* M*) is an Euler field of weight* d *on* M.

Proof: The tangent sheaf decomposes,

$$\mathcal{T}_M = \mathcal{O}_M \cdot pr_1^{-1}\mathcal{T}_{M_1} \oplus \mathcal{O}_M \cdot pr_2^{-1}\mathcal{T}_{M_2}.$$

Any vector fields $X_i, Y_i \in pr_i^{-1}\mathcal{T}_{M_i}, i = 1, 2$, satisfy

$$X_i \circ Y_i \in pr_i^{-1}\mathcal{T}_{M_i},\ [X_i, Y_i] \in pr_i^{-1}\mathcal{T}_{M_i},$$
$$X_1 \circ Y_2 = 0,\ [X_1, Y_2] = 0.$$

This together with (2.7) for vector fields in $\mathcal{T}_{M_1}$ and for vector fields in $\mathcal{T}_{M_2}$ gives (2.7) for vector fields in $pr_1^{-1}\mathcal{T}_{M_1} \cup pr_2^{-1}\mathcal{T}_{M_2}$. Because of the $\mathcal{O}_M$–polylinearity then (2.7) holds for all vector fields. For the same reasons, $E_1 + E_2$ satisfies (2.8). □

Theorem 2.11 *Let* (M, p) *be the germ in* $p \in M$ *of an F-manifold* $(M, \circ, e)$.

Then the eigenspace decomposition $T_pM = \bigoplus_{k=1}^{l}(T_pM)_k$ *of the algebra* T_pM *extends to a unique decomposition*

$$(M, p) = \prod_{k=1}^{l}(M_k, p)$$

of the germ (M, p) *into a product of germs of F-manifolds. These germs* (M_k, p) *are irreducible germs of F-manifolds, as the algebras* $T_pM_k \cong (T_pM)_k$ *are already irreducible.*

An Euler field E *on* (M, p) *decomposes into a sum of Euler fields of the same weights on the germs* (M_k, p) *of F-manifolds.*

Proof: By Corollary 2.4 the eigenspace decomposition of T_pM extends in some neighbourhood of p to a decomposition of the tangent bundle into a sum of multiplication invariant subbundles. First we have to show that these subbundles and any sum of them are integrable.

Accordingly, let $\mathcal{T}_{M,p} = \bigoplus_{k=1}^{l}(\mathcal{T}_{M,p})_k$ be the decomposition of $\mathcal{T}_{M,p}$ into multiplication invariant free $\mathcal{O}_{M,p}$-submodules, and $e = \sum_k e_k$ with $e_k \in (\mathcal{T}_{M,p})_k$. Then $e_k \circ : \mathcal{T}_{M,p} \to (\mathcal{T}_{M,p})_k$ is the projection; $e_j \circ e_k = \delta_{jk} e_k$.

Claim:

(i) With respect to e_k the multiplication is invariant, $\text{Lie}_{e_k}(\circ) = 0 \cdot \circ$;
(ii) the vectorfields e_j and e_k commute, $[e_j, e_k] = 0$;
(iii) they leave the subsheaves invariant, $[e_j, (\mathcal{T}_{M,p})_k] \subset (\mathcal{T}_{M,p})_k$;
(iv) the subsheaves satisfy $[(\mathcal{T}_{M,p})_j, (\mathcal{T}_{M,p})_k] \subset (\mathcal{T}_{M,p})_j + (\mathcal{T}_{M,p})_k$.

Proof of the claim: (i) The equality

$$\delta_{jk} \cdot \text{Lie}_{e_k}(\circ) = \text{Lie}_{e_j \circ e_k}(\circ) = e_j \circ \text{Lie}_{e_k}(\circ) + e_k \circ \text{Lie}_{e_j}(\circ).$$

implies for $j \neq k$ as well as for $j = k$ that $e_j \circ \text{Lie}_{e_k}(\circ) = 0 \cdot \circ$. Thus $\text{Lie}_{e_k}(\circ) = 0 \cdot \circ$.

(ii) The equality

$$0 = \text{Lie}_{e_j}(\circ)(e_k, e_k) = [e_j, e_k \circ e_k] - 2e_k \circ [e_j, e_k]$$

shows that $[e_j, e_k] \in (\mathcal{T}_{M,p})_k$, so for $j \neq k$ we have $[e_j, e_k] = 0$, for $j = k$ this holds anyway.

(iii) Suppose $X = e_k \circ X \in (\mathcal{T}_{M,p})_k$; then

$$0 = \text{Lie}_{e_j}(\circ)(e_k, X) = [e_j, X] - e_k \circ [e_j, X].$$

(iv) Suppose $X \in (\mathcal{T}_{M,p})_j,\ Y \in (\mathcal{T}_{M,p})_k,\ k \neq i \neq j$; then $e_i \circ X = 0$ and

$$\begin{aligned} 0 &= \text{Lie}_{e_i \circ X}(\circ)(e_i, Y) = e_i \circ \text{Lie}_X(\circ)(e_i, Y) \\ &= e_i \circ [X, e_i \circ Y] - e_i \circ [X, e_i] \circ Y - e_i \circ [X, Y] \circ e_i \\ &= -e_i \circ [X, Y]. \end{aligned}$$ ◊

Claim (iv) shows that for any k the subbundle with germs of sections $\bigoplus_{j \neq k}(\mathcal{T}_{M,p})_j$ is integrable. According to the Frobenius theorem (cf. for example [**War**, p. 41]) there is a (germ of a) submersion $f_k : (M, p) \to (\mathbb{C}^{\dim(T_pM)_k}, 0)$ such that the fibres are the integral manifolds of this subbundle. Then $\bigoplus_k f_k : (M, p) \to (\mathbb{C}^{\dim M}, 0)$ is an isomorphism.

The submanifolds $(M_k, p) := ((\bigoplus_{j \neq k} f_j)^{-1}(0), p)$ yield the decomposition $(M, p) \cong \prod_{k=1}^{l}(M_k, p)$ with

$$\mathcal{T}_{M,p} = \bigoplus_k \mathcal{O}_{M,p} \cdot pr_k^{-1}\mathcal{T}_{M_k,p} = \bigoplus_k (\mathcal{T}_{M,p})_k.$$

Claim:
(v) If $X, Y \in pr_k^{-1}\mathcal{T}_{M_k,p}$ then $X \circ Y \in pr_k^{-1}\mathcal{T}_{M_k,p}$.
(vi) If E is an Euler field then $e_k \circ E \in pr_k^{-1}\mathcal{T}_{M_k,p}$.

Proof of the claim: (v) The product $X \circ Y$ is contained in $\mathcal{O}_{M,p} \cdot pr_k^{-1}\mathcal{T}_{M_k,p}$ because this sheaf is multiplication invariant. Now $X \circ Y \in pr_k^{-1}\mathcal{T}_{M_k,p}$ is true if and only if $[Z, X \circ Y] \in (\mathcal{T}_{M,p})_j$ for any j and any $Z \in (\mathcal{T}_{M,p})_j$; but

$$
\begin{aligned}
[Z, X \circ Y] &= \mathrm{Lie}_Z(X \circ Y) = \mathrm{Lie}_{e_j \circ Z}(X \circ Y) \\
&= e_j \circ \mathrm{Lie}_Z(X \circ Y) \in (\mathcal{T}_{M,p})_j.
\end{aligned}
$$

(vi) Analogously, for any $Z \in (\mathcal{T}_{M,p})_j$

$$
\begin{aligned}
-[Z, e_k \circ E] &= \mathrm{Lie}_{e_k \circ E}(e_j \circ Z) \\
&= \mathrm{Lie}_{e_k \circ E}(\circ)(e_j, Z) + \mathrm{Lie}_{e_k \circ E}(e_j) \circ Z + \mathrm{Lie}_{e_k \circ E}(Z) \circ e_j \\
&= e_k \circ e_j \circ Z + \mathrm{Lie}_{e_k \circ E}(e_j) \circ Z + \mathrm{Lie}_{e_k \circ E}(Z) \circ e_j \in (\mathcal{T}_{M,p})_j.
\end{aligned}
$$

◊

Claim (v) and (vi) show that the multiplication on (M, p) and an Euler field E come from multiplication and vector fields on the submanifolds (M_k, p) via the decomposition. These satisfy (2.7) and (2.8): this is just the restriction of (2.7) and (2.8) to $pr_k^{-1}\mathcal{T}_{M_k,p}$. □

Examples 2.12 (i) The manifold $M = \mathbb{C}$ with coordinate u and unit field $e = \frac{\partial}{\partial u}$ with multiplication $e \circ e = e$ is an F-manifold. The field $E = u \cdot e = u\frac{\partial}{\partial u}$ is an Euler field of weight 1. The space of all Euler fields of weight d is $d \cdot E + \mathbb{C} \cdot e$. One has only to check (2.7) and (2.8) for $X = Y = Z = W = e$ and compare (2.10).

Any 1-dimensional F-manifold is locally isomorphic to an open subset of this F-manifold $(\mathbb{C}, \circ, e)$. It will be called A_1.

(ii) From (i) and Proposition 2.10 one obtains the F-manifold $A_1^n = (\mathbb{C}^n, \circ, e)$ with coordinates $u_1, \ldots, u_n$, idempotent vector fields $e_i = \frac{\partial}{\partial u_i}$, semisimple multiplication $e_i \circ e_j = \delta_{ij} e_i$, unit field $e = \sum_i e_i$ and an Euler field $E = \sum_i u_i \cdot e_i$ of weight 1. Because of Theorem 2.11, the space of Euler fields of weight d is $d \cdot E + \sum_i \mathbb{C} \cdot e_i$.

Also because of Theorem 2.11, any F-manifold M with semisimple multiplication is locally isomorphic to an open subset of the F-manifold A_1^n. The induced local coordinates $u_1, \ldots, u_n$ on M are unique up to renumbering and shift. They are called *canonical coordinates*, following Dubrovin. They are the eigenvalues of a locally defined Euler field of weight 1.

(iii) Any Frobenius manifold is an F-manifold [**HM**][**Man1**, I§5], see section 2.5.

(iv) Especially, the complex orbit space of a finite Coxeter group carries the structure of a Frobenius manifold [**Du2**][**Du3**, Lecture 4]. The F-manifold structure will be discussed in section 5.3, the Frobenius manifold structure in section 5.4. Here we only give the multiplication and the Euler fields for the 2-dimensional F-manifolds $I_2(m)$, $m \geq 2$, with $I_2(2) = A_1^2$, $I_2(3) = A_2$, $I_2(4) = B_2 = C_2$, $I_2(5) =: H_2$, $I_2(6) = G_2$.

The manifold is $M = \mathbb{C}^2$ with coordinates t_1, t_2; we denote $\delta_i := \frac{\partial}{\partial t_i}$. Unit field e and multiplication $\circ$ are given by $e = \delta_1$ and $\delta_2 \circ \delta_2 = t_2^{m-2} \cdot \delta_1$. An Euler field E of weight 1 is $E = t_1\delta_1 + \frac{2}{m}t_2\delta_2$. The space of global Euler fields of weight d is $d \cdot E + \mathbb{C} \cdot e$ for $m \geq 3$. The caustic is $\mathcal{K} = \{t \in M \mid t_2 = 0\}$ for $m \geq 3$ and $\mathcal{K} = \emptyset$ for $m = 2$. The multiplication is semisimple outside of $\mathcal{K}$; the germ (M, t) is an irreducible germ of an F-manifold if and only if $t \in \mathcal{K}$. One can check all of this by hand. We will come back to it in Theorem 4.7, when more general results allow more insight.

(v) Another 2-dimensional F-manifold is $\mathbb{C}^2$ with coordinates t_1, t_2, unit field $e = \delta_1$ and multiplication $\circ$ given by $\delta_2 \circ \delta_2 = 0$. Here all germs (M, t) are irreducible and isomorphic. $E_1 := t_1\delta_1$ is an Euler field of weight 1. Contrary to the above cases with generically semisimple multiplication, here the space of Euler fields of weight 0 is infinite dimensional, by (2.8):

$$
\begin{aligned}
&\{E \mid \mathrm{Lie}_E(\circ) = 0 \cdot \circ\} \\
&= \{E \mid [\delta_1, E] = 0,\ \delta_2 \circ [\delta_2, E] = 0\} \\
&= \big\{\varepsilon_1\delta_1 + \varepsilon_2(t_2)\delta_2 \mid \varepsilon_1 \in \mathbb{C},\ \varepsilon_2 \in \mathcal{O}_{\mathbb{C}^2}(\mathbb{C}^2),\ \delta_1(\varepsilon_2) = 0\big\}. \qquad (2.12)
\end{aligned}
$$

(vi) The base space of a semiuniversal unfolding of an isolated hypersurface singularity is (a germ of) an F-manifold. The multiplication was defined first by K. Saito [**SK6**, (1.5)] [**SK9**, (1.3)]. A good part of the geometry of F-manifolds that will be developed in the next sections is classical in the case of hypersurface singularities, from different points of view. We will discuss this in section 5.1.

(vii) Also the base of a semiuniversal unfolding of a boundary singularity is (a germ of) an F-manifold, compare section 5.2. There are certainly more classes of semiuniversal unfoldings which carry the structure of F-manifolds.

2.5 F-manifolds and potentiality

The integrability condition (2.5) for the multiplication in F-manifolds and the potentiality condition in Frobenius manifolds are closely related. For semisimple multiplication this has been known previously (with Theorem 3.2

(i) $\iff$ (ii) and e.g. [**Hi**, Theorem 3.1]). Here we give a general version, requiring neither semisimple multiplication nor flatness of the metric.

Remarks 2.13 (a) In this section we need some basic notions from differential geometry: connections, covariant derivative of vector fields, torsion freeness, metric, Levi–Civita connection. For the real C^∞-case one finds these in any textbook on differential geometry. The translation to the complex and holomorphic case here is straightforward.

(b) Let M be a manifold with a connection ∇. The covariant derivative $\nabla_X T$ of a (k,l)-tensor with respect to a vector field is defined exactly as the Lie derivative $\mathrm{Lie}_X T$ in Remark 2.7, starting with the covariant derivatives of vector fields. The operator ∇_X is a derivation on the sheaf of (k,l)-tensors just as Lie_X. But ∇_X is also $\mathcal{O}_M$-linear in X, opposite to Lie_X. Therefore ∇T is a $(k+1,l)$-tensor.

Theorem 2.14 *Let $(M, \circ, \nabla)$ be a manifold M with a commutative and associative multiplication $\circ$ on TM and with a torsion free connection ∇. By definition, $\nabla \circ (X, Y, Z)$ is symmetric in Y and Z.*

If the $(3,1)$*-tensor $\nabla\circ$ is symmetric in all three arguments, then the multiplication satisfies for any local vector fields X and Y*

$$\mathrm{Lie}_{X\circ Y}(\circ) = X \circ \mathrm{Lie}_Y(\circ) + Y \circ \mathrm{Lie}_X(\circ). \tag{2.13}$$

Proof: The term $\nabla \circ (X, Y, Z) = \nabla_X(Y \circ Z) - \nabla_X(Y) \circ Z - Y \circ \nabla_X(Z)$ is symmetric in Y and Z. The $(4,1)$-tensor

$$\begin{aligned}(X, Y, Z, W) \mapsto\ & \nabla \circ (X, Y \circ Z, W) + W \circ \nabla \circ (X, Y, Z) \\ =\ & \nabla_X(Y \circ Z \circ W) - \nabla_X(Y) \circ Z \circ W \\ & - Y \circ \nabla_X(Z) \circ W - Y \circ Z \circ \nabla_X(W)\end{aligned} \tag{2.14}$$

is symmetric in Y, Z, W. A simple calculation using the torsion freeness of ∇ shows

$$\begin{aligned}&(\mathrm{Lie}_{X\circ Y}(\circ) - X \circ \mathrm{Lie}_Y(\circ) - Y \circ \mathrm{Lie}_X(\circ))(Z, W) \\ &\quad = \nabla \circ (X \circ Y, Z, W) - X \circ \nabla \circ (Y, Z, W) - Y \circ \nabla \circ (X, Z, W) \\ &\qquad - \nabla \circ (Z \circ W, X, Y) + Z \circ \nabla \circ (W, X, Y) + W \circ \nabla \circ (Z, X, Y).\end{aligned} \tag{2.15}$$

If $\nabla\circ$ is symmetric in all three arguments then

$$\nabla \circ (X \circ Y, Z, W) + Z \circ \nabla \circ (W, X, Y) + W \circ \nabla \circ (Z, X, Y) \tag{2.16}$$

is symmetric in X, Y, Z, W because of the symmetry of the tensor in (2.14). Then the right hand side of (2.15) vanishes. □

Theorem 2.15 *Let* $(M, \circ, e, g)$ *be a manifold with a commutative and associative multiplication* $\circ$ *on* TM*, a unit field* e*, and a metric* g *(a symmetric nondegenerate bilinear form) on* TM *which is multiplication invariant, i.e. the* $(3, 0)$*-tensor* A*,*

$$A(X, Y, Z) := g(X, Y \circ Z), \tag{2.17}$$

is symmetric in all three arguments. ∇ *denotes the Levi–Civita connection of the metric. The coidentity* ε *is the 1–form which is defined by* $\varepsilon(X) = g(X, e)$*. The following conditions are equivalent:*

(i) The manifold with multiplication and unit $(M, \circ, e)$ *is an F-manifold and* ε *is closed.*
(ii) The $(4, 0)$*-tensor* ∇A *is symmetric in all four arguments.*
(iii) The $(3, 1)$*-tensor* $\nabla\circ$ *is symmetric in all three arguments.*

Proof: The Levi–Civita connection satisfies $\nabla g = 0$. Therefore

$$\begin{aligned} \nabla A(X, Y, Z, W) &= Xg(Y, Z \circ W) - g(\nabla_X Y, Z \circ W) \\ &\quad - g(Y, W \circ \nabla_X Z) - g(Y, Z \circ \nabla_X W) \\ &= g(Y, \nabla_X(Z \circ W) - W \circ \nabla_X Z - Z \circ \nabla_X W) \\ &= g(Y, \nabla \circ (X, Z, W)). \end{aligned} \tag{2.18}$$

The metric g is nondegenerate and $\nabla A(X, Y, Z, W)$ is always symmetric in Y, Z, W. Equation (2.18) shows (ii) $\iff$ (iii). Because of $\nabla g = 0$ and the torsion freeness $\nabla_X Y - \nabla_Y X = [X, Y]$, the 1–form ε satisfies

$$\begin{aligned} d\varepsilon(X, Y) &= X(\varepsilon(Y)) - Y(\varepsilon(X)) - \varepsilon([X, Y]) \\ &= g(Y, \nabla_X e) - g(X, \nabla_Y e) \\ &= -\nabla A(X, Y, e, e) + \nabla A(Y, X, e, e). \end{aligned} \tag{2.19}$$

Hence (ii) $\Rightarrow$ $d\varepsilon = 0$; with Theorem 2.14 this gives (ii) $\Rightarrow$ (i). It remains to show (i) $\Rightarrow$ (ii).

The equations (2.20) and (2.21) follow from the definition of $\nabla\circ$ and from (2.18),

$$\nabla \circ (X, Y, e) = Y \circ \nabla \circ (X, e, e), \tag{2.20}$$

$$\nabla A(X, U, Y, e) = \nabla A(X, U \circ Y, e, e). \tag{2.21}$$

One calculates with (2.15) and (2.18)

$$g(e, (\mathrm{Lie}_{X\circ Y}(\circ) - X \circ \mathrm{Lie}_Y(\circ) - Y \circ \mathrm{Lie}_X(\circ))(Z, W))$$
$$= \nabla A(X \circ Y, e, Z, W) - \nabla A(Y, X, Z, W) - \nabla A(X, Y, Z, W)$$
$$- \nabla A(Z \circ W, e, X, Y) + \nabla A(W, Z, X, Y) + \nabla A(Z, W, X, Y). \quad (2.22)$$

If (i) holds then (2.19), (2.21), and (2.22) imply

$$\nabla A(Y, X, Z, W) - \nabla A(W, Z, X, Y)$$
$$= -\nabla A(X, Y, Z, W) + \nabla A(Z, W, X, Y). \quad (2.23)$$

The left hand side is symmetric in X and Z, the right hand side is skewsymmetric in X and Z, so both sides vanish. ∇A is symmetric in all four arguments. □

Lemma 2.16 *Let (M, g, ∇) be a manifold with metric g and Levi–Civita connection ∇. Then a vector field Z is flat, i.e. $\nabla Z = 0$, if and only if* $\mathrm{Lie}_Z(g) = 0$ *and the 1–form $\varepsilon_Z := g(Z, .)$ is closed.*

Proof: The connection ∇ is torsion free and satisfies $\nabla g = 0$. Therefore (cf. (2.19))

$$d\varepsilon_Z(X, Y) = g(Y, \nabla_X Z) - g(X, \nabla_Y Z), \quad (2.24)$$

$$\mathrm{Lie}_Z(g)(X, Y) = g(Y, \nabla_X Z) + g(X, \nabla_Y Z). \quad (2.25)$$

□

Remarks 2.17 (a) Let $(M, \circ, e, g)$ satisfy the hypotheses of Theorem 2.15 and let g be flat. Then condition (ii) in Theorem 2.15 is equivalent to the existence of a local potential $\Phi \in \mathcal{O}_{M,p}$ (for any $p \in M$) with $(XYZ)\Phi = A(X, Y, Z)$ for any flat local vector fields X, Y, Z.

(b) In view of this the conditions (ii) and (iii) in Theorem 2.15 are called *potentiality* conditions.

(c) The manifold $(M, \circ, e, E, g)$ is a Frobenius manifold if it satisfies the hypotheses and conditions in Theorem 2.15, if g is flat, if $\mathrm{Lie}_e(g) = 0$ (respectively e is flat), and if E is an Euler field (of weight 1, with respect to M as F-manifold), with $\mathrm{Lie}_E(g) = D \cdot g$ for some $D \in \mathbb{C}$ (cf. Definition 9.1).

Chapter 3
Massive F-manifolds and Lagrange maps

In this section the relation between F-manifolds and symplectic geometry is discussed. The most crucial fact is shown in section 3.1: the analytic spectrum of a massive (i.e. with generically semisimple multiplication) F-manifold M is a Lagrange variety $L \subset T^*M$; and a Lagrange variety $L \subset T^*M$ in the cotangent bundle of a manifold M supplies the manifold M with the structure of an F-manifold if and only if the map $\mathbf{a} : \mathcal{T}_M \to \pi_* \mathcal{O}_L$ from (3.1) is an isomorphism.

The condition that this map $\mathbf{a} : \mathcal{T}_M \to \pi_* \mathcal{O}_L$ is an isomorphism is close to Givental's notion of a miniversal Lagrange map [**Gi2**, ch. 13]. In section 3.4 the correspondence between massive F-manifolds and Lagrange maps is rewritten using this notion.

If E is an Euler field in a massive F-manifold M then the holomorphic function $F := \mathbf{a}^{-1}(E) : L \to \mathbb{C}$ satisfies $\mathrm{d}F|L_{reg} = \alpha|L_{reg}$ (here α is the canonical 1-form on T^*M). But as L may have singularities, the global existence of E and of such a holomorphic function is not clear. This is discussed in section 3.2.

Much weaker than the existence of E is the existence of a continuous function $F : L \to \mathbb{C}$ which is holomorphic on L_{reg} with $\mathrm{d}F|L_{reg} = \alpha|L_{reg}$. This is called a generating function for the massive F-manifold. It gives rise to the three notions bifurcation diagram, Lyashko–Looijenga map, and discriminant. They turn out to be holomorphic even if F is only continuous along L_{sing}. This is discussed in section 3.5 for F-manifolds and in section 3.3 more generally for Lagrange maps.

A good reference for the basic notions from symplectic geometry which are used in this chapter (Lagrange variety, Lagrange fibration, Lagrange map, generating function) is [**AGV1**, ch. 18].

3.1 Lagrange property of massive F-manifolds

Consider an n-dimensional manifold $(M, \circ, e)$ with commutative and associative multiplication on the tangent bundle and with unit field e. Its analytic spectrum $L := \text{Specan}(\mathcal{T}_M)$ is a subvariety of the cotangent bundle T^*M. The cotangent bundle carries a natural symplectic structure, given by the 2–form $d\alpha$. Here α is the canonical 1–form, which is written as $\alpha = \sum_i y_i dt_i$ in local coordinates $t_1, \ldots, t_n$ for the base and dual coordinates $y_1, \ldots, y_n$ for the fibres $(\mathcal{T}_M \to (\pi_{T^*M})_*\mathcal{O}_{T^*M}, \frac{\partial}{\partial t_i} \mapsto y_i)$.

The isomorphism $\mathbf{a} : \mathcal{T}_M \to \pi_*\mathcal{O}_L$ from (2.2) can be expressed with α by

$$\mathbf{a}(X) = \alpha(\widetilde{X})|_L, \tag{3.1}$$

where $X \in \mathcal{T}_M$ and $\widetilde{X}$ is any lift of X to a neighbourhood of L in T^*M.

The values of the function $\mathbf{a}(X)$ on $\pi^{-1}(p)$ are the eigenvalues of $X\circ$ on T_pM; this follows from Theorem and Lemma 2.1.

Definition 3.1. A manifold $(M, \circ, e)$ with commutative and associative multiplication on the tangent bundle and with unit field e is *massive* if the multiplication is generically semisimple.

Then the set of points where the multiplication is not semisimple is empty or a hypersurface, which is the caustic $\mathcal{K}$ (Proposition 2.6). In the rest of the paper we will study the local structure of massive F-manifolds at points where the multiplication is not semisimple.

It is known that the analytic spectrum of a massive Frobenius manifold is Lagrange (compare [**Au**] and the references cited there). Theorem 3.2 together with Theorem 2.15 make the relations between the different conditions transparent.

Theorem 3.2 *Let $(M, \circ, e)$ be a massive n-dimensional manifold M.*

The analytic spectrum $L = \text{Specan}(\mathcal{T}_M) \subset T^*M$ is an everywhere reduced subvariety. The map $\pi : L \to M$ is finite and flat. It is a branched covering of degree n, branched above the caustic $\mathcal{K}$. The following conditions are equivalent.

(i) The manifold $(M, \circ, e)$ is a massive F-manifold;
(ii) At any semisimple point $p \in M - \mathcal{K}$, the idempotent vector fields $e_1, \ldots, e_n \in \mathcal{T}_{M,p}$ commute.
(iii) The subvariety $L \subset T^*M$ is a Lagrange variety, i.e. $\alpha|L_{reg}$ is closed.

Proof: The variety $L - \pi^{-1}(\mathcal{K})$ is smooth, and $\pi : L - \pi^{-1}(\mathcal{K}) \to M - \mathcal{K}$ is a covering. The sheaf $\pi_*\mathcal{O}_L$ ($\cong \mathcal{T}_M$) is a free $\mathcal{O}_M$–module, so a Cohen–Macaulay $\mathcal{O}_M$–module and a Cohen–Macaulay ring. Then L is Cohen–Macaulay and everywhere reduced, as it is reduced at generic points (cf. [**Lo2**, pp. 49–51] for the notion Cohen–Macaulay and details of these arguments).

It remains to show the equivalences (i) $\iff$ (ii) $\iff$ (iii).

(i) $\Rightarrow$ (ii) follows from Theorem 2.11 and has been discussed in Example 2.12 (ii).

(ii) $\Rightarrow$ (i) is clear because (2.8) holds everywhere if it holds at generic points (in fact, one point suffices).

(ii) $\Rightarrow$ (iii) We fix canonical coordinates u_i with $\frac{\partial}{\partial u_i} = e_i$ on (M, p) for a point $p \in M - \mathcal{K}$ and the dual coordinates x_i on the fibres of the cotangent bundle $(\mathcal{T}_M \to (\pi_{T^*M})_*\mathcal{O}_{T^*M},\ e_i \mapsto x_i)$. Then locally above (M, p) the analytic spectrum L is in these coordinates

$$\begin{aligned} L &\cong \{(x_j, u_j) \mid x_1 + \cdots + x_n = 1,\ x_i x_j = \delta_{ij} x_j\} \\ &= \bigcup_{i=1}^{n} \{(x_j, u_j) \mid x_j = \delta_{ij}\}. \end{aligned} \tag{3.2}$$

The 1–form $\alpha = \sum x_i du_i$ is closed in $L - \pi^{-1}(\mathcal{K})$. This set is open and dense in L_{reg}, hence L is a Lagrange variety.

(iii) $\Rightarrow$ (ii) Above a small neighbourhood U of $p \in M - \mathcal{K}$ the analytic spectrum consists of n smooth components L_k, $k = 1, \ldots, n$, with $\pi : L_k \xrightarrow{\cong} U$. An idempotent vector field e_i can be lifted to vector fields $\widetilde{e_i}$ in neighbourhoods U_k of L_k in T^*M such that they are tangent to all L_k. The commutator $[\widetilde{e_i}, \widetilde{e_j}]$ is a lift of the commutator $[e_i, e_j]$ in these neighbourhoods U_k.

$$\begin{aligned} \mathbf{a}([e_i, e_j]|_U) &= \bigcup_k (\alpha([\widetilde{e_i}, \widetilde{e_j}]))|_{L_k} \\ &= \bigcup_k (\widetilde{e_i}(\alpha(\widetilde{e_j})) - \widetilde{e_j}(\alpha(\widetilde{e_i})) - d\alpha(\widetilde{e_i}, \widetilde{e_j}))|_{L_k} \\ &= \bigcup_k (\widetilde{e_i}|_{L_k}(\delta_{jk}) - \widetilde{e_j}|_{L_k}(\delta_{ik}) - d\alpha|_{L_k}(\widetilde{e_i}|_{L_k}, \widetilde{e_j}|_{L_k})) = 0. \end{aligned}$$

But $\mathbf{a} : \mathcal{T}_M \to \pi_*\mathcal{O}_L$ is an isomorphism, so $[e_i, e_j] = 0$ □

Theorem 3.3 *(a) Let $(M, \circ, e)$ be a massive F-manifold.*

A vector field E is an Euler field of weight $c \in \mathbb{C}$ if and only if

$$d(\mathbf{a}(E))|L_{reg} = c \cdot \alpha|L_{reg}. \tag{3.3}$$

(b) Let $(M, p) = \prod_{k=1}^{l}(M_k, p)$ be the decomposition of the germ of a massive F-manifold into irreducible germs of F-manifolds $(M_k, \circ, e_k)$.

(i) The space of (germs of) Euler fields of weight 0 for (M, p) is the abelian Lie algebra $\sum_{k=1}^{l} \mathbb{C} \cdot e_k$.

(ii) There is a unique continuous function $F : (L, \pi^{-1}(p)) \to (\mathbb{C}, 0)$ *on the multigerm* $(L, \pi^{-1}(p))$ *which has value 0 on* $\pi^{-1}(p)$*, is holomorphic on* L_{reg} *and satisfies* $(\mathrm{d}F)|L_{reg} = \alpha|L_{reg}$.

(iii) An Euler field of weight $c \neq 0$ *for* (M, p) *exists if and only if this function* F *is holomorphic. In that case,* $c \cdot \mathbf{a}^{-1}(F)$ *is an Euler field of weight c and* $\mathbb{C} \cdot \mathbf{a}^{-1}(F) + \sum_{k=1}^{l} \mathbb{C} \cdot e_k$ *is the Lie algebra of all Euler fields on the germ* (M, p).

Proof: (a) It is sufficient to prove this locally for a germ (M, p) with $p \in M - \mathcal{K}$. This germ is isomorphic to A_1^n. A vector field $E = \sum_{i=1}^{n} \varepsilon_i e_i$, $\varepsilon_i \in \mathcal{O}_{M,p}$, is an Euler field of weight c if and only if $\mathrm{d}\varepsilon_i = c \cdot \mathrm{d}u_i$ (Theorem 2.11 and Example 2.12 (ii)).

Going into the proof of 3.2 (ii) $\Rightarrow$ (iii), one sees that this is equivalent to (3.3).

(b) The multigerm $(L, \pi^{-1}(p))$ has l components, and the space of locally constant functions on it has dimension l. The function (multigerm) F exists because $\alpha|L_{reg}$ is closed. This will be explained in the next section (Lemma 3.4). All statements follow now with (a). □

3.2 Existence of Euler fields

By Theorem 3.2, the analytic spectrum (L, λ) of an irreducible germ (M, p) of a massive F-manifold is a germ of an (often singular) Lagrange variety, and $(L, \lambda) \hookrightarrow (T^*M, \lambda) \to (M, p)$ is a germ of a Lagrange map. The paper **[Gi2]** of Givental is devoted to such objects. It contains implicitly many results on massive F-manifolds. It will be extremely useful and often cited in the following.

The question when does a germ of a massive F-manifold have an Euler field of weight 1 is reduced by Theorem 3.3 (b)(iii) to the question when is the function germ F holomorphic. Partial answers are given in Corollary 3.8 and Lemma 3.9. We start with a more general situation, as in [**Gi2**, chapter 1.1].

Let $(L, 0) \subset (\mathbb{C}^N, 0)$ be a reduced complex space germ. Statements on germs will often be formulated using representatives, but they are welldefined for the germs, e.g. '$\alpha|_{L_{reg}}$ is closed' for $\alpha \in \Omega^k_{\mathbb{C}^N,0}$.

Lemma 3.4 *Let* $\alpha \in \Omega^1_{\mathbb{C}^N,0}$ *be closed on* L_{reg}*. Then there exists a unique function germ* $F : (L, 0) \to (\mathbb{C}, 0)$ *which is holomorphic on* L_{reg}*, continuous on L and satisfies* $\mathrm{d}F|_{L_{reg}} = \alpha|_{L_{reg}}$.

Proof: The germ $(L, 0)$ is homeomorphic to a cone as it admits a Whitney stratification. One can integrate α along paths corresponding to such a cone structure, starting from 0. One obtains a continuous function F on L, which is holomorphic on L_{reg} because of $d\alpha|_{L_{reg}} = 0$ and which satisfies $\mathrm{d}F|_{L_{reg}} = \alpha|_{L_{reg}}$. The unicity of F with value $F(0) = 0$ is clear. □

Which germs $(L, 0)$ have the property that all such function germs are holomorphic on $(L, 0)$? This property has not been studied much. It can be seen to be in line with the normality and maximality of complex structures and is weaker than maximality.

It can be rephrased as $H^1_{Giv}((L, 0)) = 0$. Here $H^*_{Giv}((L, 0))$ is the cohomology of the de Rham complex

$$(\Omega^*_{\mathbb{C}^N,0}/\{\omega \in \Omega^*_{\mathbb{C}^N,0} \mid \omega|_{L_{reg}} = 0\}, d), \tag{3.4}$$

which is considered in [**Gi2**, chapter 1.1]. We state some known results on this cohomology.

Theorem 3.5 *(a)* (Poincaré-Lemma, [**Gi2**, chapter 1.1]) *If* $(L, 0)$ *is weighted homogeneous with positive weights then* $H^*_{Giv}((L, 0)) = 0$.

(b) ([**Va5**]) *Suppose that* $(L, 0)$ *is a germ of a hypersurface with an isolated singularity,* $(L, 0) = (f^{-1}(0), 0) \subset (\mathbb{C}^{n+1}, 0)$ *and* $f : (\mathbb{C}^{n+1}, 0) \to (\mathbb{C}, 0)$ *is a holomorphic function with an isolated singularity. Then*

$$\begin{aligned}\dim H^n_{Giv}((L, 0)) &= \mu - \tau \\ &= \dim \mathcal{O}_{\mathbb{C}^{n+1},0} \Big/ \left(\frac{\partial f}{\partial x_i}\right) - \dim \mathcal{O}_{\mathbb{C}^{n+1},0} \Big/ \left(f, \frac{\partial f}{\partial x_i}\right).\end{aligned} \tag{3.5}$$

(c) (essentially Varchenko and Givental, [**Gi2**, chapter 1.2]) *Let* $(L, 0)$ *be as in (b) with* $\mu - \tau \neq 0$. *The class* $[\eta] \in H^n_{Giv}((L, 0))$ *of* $\eta \in \Omega^n_{\mathbb{C}^{n+1},0}$ *is not vanishing if* $\mathrm{d}\eta$ *is a volume form, i.e.* $\mathrm{d}\eta = h\mathrm{d}x_0 \dots \mathrm{d}x_n$ *with* $h(0) \neq 0$.

Remarks 3.6 (i) The proofs of (b) and (c) use the Gauß–Manin connection for isolated hypersurface singularities.

(ii) Theorem 3.5 (c) was formulated in [**Gi2**, chapter 1.2] only for $n = 1$. The missing piece of the proof for all n was the following fact, which at that time was only known for $n = 1$: *The exponent of a form* $h\mathrm{d}x_0 \dots \mathrm{d}x_n$ *is the minimal exponent if and only if* $h(0) \neq 0$.

This fact has been established by M. Saito [**SM4**, (3.11)] for all n.

(iii) By a result of K. Saito [**SK1**], an isolated hypersurface singularity $(L, 0) = (f^{-1}(0), 0) \subset (\mathbb{C}^{n+1}, 0)$ is weighted homogeneous (with positive weights) if and only if $\mu - \tau = 0$.

(iv) For us only the case $n = 1$ in Theorem 3.5 (b) and (c) is relevant. Proposition 3.7, which is also due to Givental, implies the following:

Of all isolated hypersurface singularities $(L, 0) = (f^{-1}(0), 0) \subset (\mathbb{C}^{n+1}, 0)$ only the curve singularities $(n = 1)$ turn up as germs of Lagrange varieties. These are, of course, germs of Lagrange varieties with respect to any volume form on $(\mathbb{C}^2, 0)$.

(v) If $(L, 0) \subset ((S, 0), \omega)$ is the germ of a Lagrange variety in a symplectic space S with symplectic form ω, then the class $[\alpha] \in H^1_{Giv}((L, 0))$ of some α with $d\alpha = \omega$ is independent of the choice of α. It is called the characteristic class of $(L, 0) \subset ((S, 0), \omega)$.

(vi) Givental made the conjecture [**Gi2**, chapter 1.2]: *Let $(L, 0)$ be an n-dimensional Lagrange germ. If $H^n_{Giv}((L, 0)) \neq 0$ then $H^1_{Giv}((L, 0)) \neq 0$ and the characteristic class $[\alpha] \in H^1_{Giv}((L, 0))$ is nonzero.*

It is true for $n = 1$ because of Theorem 3.5 and Remark 3.6 (iii). Givental sees the conjecture to be analogous to a conjecture of Arnold's which was proved by Gromov 1985 (cf. [**Gi2**, chapter 1.2]): *any real closed Lagrange manifold $L \subset T^*\mathbb{R}^n$ has a nonvanishing characteristic class $[\alpha] \in H^1(L, \mathbb{R})$.*

Proposition 3.7 ([**Gi2**, chapter 1.1]) *An n-dimensional germ $(L, 0)$ of a Lagrange variety with embedding dimension* $\operatorname{embdim}(L, 0) = n + k$ *is a product of a k-dimensional Lagrange germ $(L', 0)$ with* $\operatorname{embdim}(L', 0) = 2k$ *and a smooth $(n - k)$-dimensional Lagrange germ $(L'', 0)$; here the decomposition of $(L, 0)$ corresponds to a decomposition*

$$((S, 0), \omega) \cong ((S', 0), \omega') \times ((S'', 0), \omega'') \tag{3.6}$$

of the symplectic space germ $(S, 0) \supset (L, 0)$.

Proof: If $k < n$ then a holomorphic function f on S exists with smooth fibre $f^{-1}(0) \supset L$. The Hamilton flow of this function f respects L and the fibres of f. The spaces of orbits in $f^{-1}(0)$ and L give a symplectic space germ of dimension $2n - 2$ and in it a Lagrange germ (e.g. [**AGV1**, 18.2]).

To obtain a decomposition as in (3.6) one chooses a germ $(\Sigma, 0) \subset (S, 0)$ of a $2n - 1$-dimensional submanifold Σ in S which is transversal to the Hamilton field H_f of f. There is a unique section v in $(TS)|_\Sigma$ with $\omega(H_f, v) = 1$ and $\omega(T_p\Sigma, v) = 0$ for $p \in \Sigma$. The shift $\widetilde{v}$ of v with the Hamilton flow of f forms together with H_f a 2-dimensional integrable distribution on S, because of $0 = \mathrm{Lie}_{H_f}(\widetilde{v}) = [H_f, \widetilde{v}]$.

This distribution is everywhere complementary and orthogonal to the integrable distribution whose integral manifolds are the intersections of the fibres of f with the shifts of Σ by the Hamilton flow of f. This yields a decomposition $(S, 0) \cong (\mathbb{C}^2, 0) \times (\Sigma \cap f^{-1}(0), 0)$. One can check that the symplectic form decomposes as required. If $k < n - 1$ one repeats this process. □

Corollary 3.8 *(a) Let $(L, 0)$ be an n-dimensional Lagrange germ isomorphic to $(L', 0) \times (\mathbb{C}^{n-1}, 0)$ as complex space germ. Then $(L', 0)$ is a plane curve singularity. The characteristic class $[\alpha] \in H^1_{Giv}((L, 0))$ is vanishing if and only if $(L', 0)$ is weighted homogeneous.*

(b) Let $(L, \lambda) \subset (T^*M, \lambda)$ be the analytic spectrum of an irreducible germ (M, p) of a massive F-manifold. Suppose $(L, \lambda) \cong (L', 0) \times (\mathbb{C}^{n-1}, 0)$. Then there exists an Euler field of weight 1 on (M, p) if and only if $(L', 0)$ is weighted homogeneous.

Proof: (a) Proposition 3.7, Theorem 3.5, and Remark 3.6 (iii).

(b) Part (a) and Theorem 3.3 (b)(iii). □

In Proposition 5.27 for any plane curve singularity $(L', 0)$ irreducible germs of F-manifolds with analytic spectrum $(L, \lambda) \cong (L', 0) \times (\mathbb{C}^{n-1}, 0)$ for some n will be constructed. So, often there exists no Euler field of weight 1 on a germ of a massive F-manifold. On the other hand, the Poincaré-Lemma 3.5 (a) and Proposition 3.7 say that an Euler field of weight 1 exists on a germ of a massive F-manifold (M, p) if the multigerm $(L, \pi^{-1}(p))$ of the analytic spectrum is at all points of $\pi^{-1}(p)$ a product of a smooth germ and a germ which is weighted homogeneous with positive weights. Also, we have the following.

Lemma 3.9 *Let M be a massive F-manifold and $F : L \to \mathbb{C}$ a continuous function with $\mathrm{d}F|_{L_{reg}} = \alpha|_{L_{reg}}$. Then $\mathbf{a}^{-1}(F|(L - \pi^{-1}(\mathcal{K})))$ is an Euler field of weight 1 on $M - \mathcal{K}$. It extends to an Euler field on M if (L, λ) is at all points $\lambda \in L$ outside of a subset of codimension ≥ 2 a product of a smooth germ and a germ which is weighted homogeneous with positive weights.*

Proof: Suppose $K \subset L$ is a subset of codimension ≥ 2 with this property. Then F is holomorphic in $L - K$ because of the Poincaré-Lemma 3.5 (a) and Proposition 3.7. The Euler field extends to $M - \pi(K)$. But $\pi(K)$ also has codimension ≥ 2. So the Euler field extends to M (and F is holomorphic on L). □

3.3 Lyashko–Looijenga maps and graphs of Lagrange maps

In this section classical facts on Lagrange maps are presented, close to [**Gi2**, chapter 1.3], but slightly more general. They will be used in sections 3.4–4.1.

Let $L \subset T^*M$ be a Lagrange variety (not necessarily smooth) in the cotangent bundle of an m-dimensional connected manifold M. We assume:

(a) The projection $\pi : L \to M$ is a branched covering of degree n, that is, there exists a subvariety $K \subset M$ such that $\pi : L - \pi^{-1}(K) \to M - K$ is a covering of degree n ($\pi : L \to M$ is not necessarily flat).
(b) There exists a *generating function* $F : L \to \mathbb{C}$, that is, a continuous function which is holomorphic on L_{reg} with $\mathrm{d}F|_{L_{reg}} = \alpha|_{L_{reg}}$ (locally such a function exists by Lemma 3.4).

Such a function F will be fixed. It can be considered as a multivalued function on $M - K$; the 1–graph of this multivalued function is $L - \pi^{-1}(K)$.

The *Lyashko–Looijenga map* $\Lambda = (\Lambda_1, \ldots, \Lambda_n) : M \to \mathbb{C}^n$ of $L \subset T^*M$ and F is defined as follows: for $q \in M - K$, the roots of the unitary polynomial $z^n + \sum_{i=1}^n \Lambda_i(q) z^{n-i}$ are the values of F on $\pi^{-1}(q)$. It extends to a holomorphic map on M because F is holomorphic on L_{reg} and continuous on L.

The *reduced Lyashko–Looijenga map* $\Lambda^{(red)} = (\Lambda_2^{(red)}, \ldots, \Lambda_n^{(red)}) : M \to \mathbb{C}^{n-1}$ of $L \subset T^*M$ and F is defined as follows: for $q \in M - K$, the roots of the unitary polynomial $z^n + \sum_{i=2}^n \Lambda_i^{(red)}(q) z^{n-i}$ are the values of F on $\pi^{-1}(q)$, shifted by their centre $-\frac{1}{n}\Lambda_1(q) = \frac{1}{n}\sum_{\lambda \in \pi^{-1}(q)} F(\lambda)$. It also extends to a holomorphic map on M. Its significance will be discussed after Remarks 3.11.

The *front* Φ_L of $L \subset T^*M$ and F is the image $\mathrm{Im}(F, pr) \subset \mathbb{C} \times M$ of $(F, pr) : L \to \mathbb{C} \times M$. It is the zero set of the polynomial $z^n + \sum_{i=1}^n \Lambda_i \cdot z^{n-i}$. So, it is an analytic hypersurface even if F is not holomorphic on all of L.

Following Teissier ([**Te2**, 2.4, 5.5], [**Lo2**, 4.C]), the *development* $\widetilde{\Phi_L} \subset \mathbb{P}T^*(\mathbb{C} \times M)$ of this hypersurface Φ_L in $\mathbb{C} \times M$ is defined as the closure in $\mathbb{P}T^*(\mathbb{C} \times M)$ of the set of tangent hyperplanes at the smooth points of Φ_L. It is an analytic subvariety and a Legendre variety with respect to the canonical contact structure on $\mathbb{P}T^*(\mathbb{C} \times M)$.

The map

$$\mathbb{C} \times T^*M \to \mathbb{P}T^*(\mathbb{C} \times M), \quad (c, \lambda) \mapsto ((\mathrm{d}z - \lambda)^{-1}(0), (c, p)) \tag{3.7}$$

($\lambda \in T_p^*M$ and $\mathrm{d}z - \lambda \in T_{(c,p)}^*(\mathbb{C} \times M) \cong T_c^*\mathbb{C} \times T_p^*M$) identifies $\mathbb{C} \times T^*M$ with the open subset in $\mathbb{P}T^*(\mathbb{C} \times M)$ of hyperplanes in the tangent spaces which do not contain $\mathbb{C} \cdot \frac{\partial}{\partial z}$. The induced contact structure on $\mathbb{C} \times T^*M$ is given by the 1–form $\mathrm{d}z - \alpha$.

The following fact is well known. It is one way in which the relation between Lagrange and Legendre maps can be made explicit (e.g. [**AGV1**, ch. 18–20]). To check it, one has to consider F as a multivalued function on $M - K$ and Φ_L as its graph.

Proposition 3.10 *The embedding $\mathbb{C} \times T^*M \hookrightarrow \mathbb{P}T^*(\mathbb{C} \times M)$ identifies the graph* $\mathrm{Im}(F, id) \subset \mathbb{C} \times T^*M$ *of $F : L \to \mathbb{C}$ with the development $\widetilde{\Phi_L}$ of the front Φ_L.*

Remarks 3.11 It has some nontrivial consequences.

(i) The development $\widetilde{\Phi_L}$ is contained in the open subset of $\mathbb{P}T^*(\mathbb{C} \times M)$ of hyperplanes in the tangent spaces which do not contain $\frac{\partial}{\partial z}$. Therefore the vector field $\frac{\partial}{\partial z}$ is everywhere transversal to the front $\widetilde{\Phi_L}$.

(ii) The polynomial $z^n + \sum_{i=1}^n \Lambda_i z^{n-i}$ has no multiple factors and the branched covering $\Phi_L \to M$ has degree n: over any point $p \in M - K$ the varieties L and $\widetilde{\Phi_L}$ have n points and the points of Φ_L have n tangent planes; so, also Φ_L has n points over a generic point $p \in M - K$.

(iii) The graph $\mathrm{Im}(F, id) \cong \widetilde{\Phi_L}$ is an analytic variety even if F is not holomorphic on all of L.

(iv) The composition of maps $\widetilde{\Phi_L} \xrightarrow{\cong} \mathrm{Im}(F, id) \xrightarrow{pr} L$ is a bijective morphism. It is an isomorphism if and only if F is holomorphic on L. Also, the continuous map $L \to \Phi_L$ is a morphism if and only if F is holomorphic on L.

(v) The Lagrange variety $L \subset T^*M$ together with the values of F at one point of each connectivity component of L and any of the following data determine each other uniquely: the front Φ_L, the development $\widetilde{\Phi_L}$, the Lyashko–Looijenga map Λ, the generating function F as a multivalued function on the base M.

To motivate the reduced Lyashko–Looijenga map, we have to talk about Lagrange maps and their isomorphisms ([**AGV1**, ch. 18], [**Gi2**, 3.1]).

A Lagrange map is a diagram $L \hookrightarrow (S, \omega) \to M$ where L is a Lagrange variety in a symplectic manifold (S, ω) and $S \to M$ is a Lagrange fibration. An isomorphism between two Lagrange maps is given by an isomorphism of the Lagrange fibrations which maps one Lagrange variety to the other.

An automorphism of $T^*M \to M$ as a Lagrange fibration which fixes the base is given by a shift in the fibres,

$$T^*M \to T^*M, \quad \lambda \mapsto \lambda + dS,$$

where $S : M \to \mathbb{C}$ is holomorphic ([**AGV1**, 18.5]). So, regarding $T^*M \to M$ as a Lagrange fibration means to forget the 0-section and the 1–form α, but to keep the Lagrange fibration and the class $\alpha + \{\mathrm{d}S \mid S : M \to \mathbb{C} \text{ holomorphic}\}$ of 1–forms.

Corollary 3.12 *Let $L \hookrightarrow T^*M \to M$ be as above (satisfying the assumptions (a) and (b)) with l connectivity components and points $\lambda_1, \ldots, \lambda_l$, one in each connectivity component. The data in (i)–(iii) are equivalent.*

(i) The diagram $L \hookrightarrow T^*M \to M$ as a Lagrange map and the differences $F(\lambda_i) - F(\lambda_j) \in \mathbb{C}$ of values of F,
(ii) the generating function F modulo addition of a function on the base,
(iii) the reduced Lyashko–Looijenga map $\Lambda^{(red)} : M \to \mathbb{C}^{n-1}$.

Proof: (i)$\Rightarrow$(ii): Integrating the 1–forms in $\alpha + \{dS \mid S : M \to \mathbb{C}$ holomorphic$\}$ gives (ii).

(ii) $\Rightarrow$ (iii) Definition of $\Lambda^{(red)}$.

(iii) $\Rightarrow$ (i): The map $(0, \Lambda^{(red)}) = (0, \Lambda_2^{(red)}, \ldots, \Lambda_n^{(red)}) : M \to \mathbb{C}^n$ is the Lyashko–Looijenga map of a Lagrange variety in T^*M which differs from the original Lagrange variety only by the shift of $d(\frac{-1}{n}\Lambda_1)$ in the fibres. The map $(0, \Lambda^{(red)})$ determines this Lagrange variety and a generating function for it because of Remark 3.11 (v). Then (i) is obtained. □

3.4 Miniversal Lagrange maps and F-manifolds

The notion of a miniversal germ of a Lagrange map is central in Givental's paper [**Gi2**]. We need a slight generalization to multigerms, taking a semilocal viewpoint.

Let $L \subset T^*M$ be a Lagrange variety with finite projection $\pi : L \to M$. The *germ at the base point* $p \in M$ of $L \hookrightarrow T^*M \to M$ is the diagram

$$(L, \pi^{-1}(p)) \hookrightarrow (T^*M, T_p^*M) \to (M, p).$$

Here $(L, \pi^{-1}(p))$ is a multigerm. The map of germs $(T^*M, T_p^*M) \to (M, p)$ is the cotangent bundle of the germ (M, p); it is a germ in the base, but not in the fibre.

For this diagram the morphisms

$$\mathcal{O}_{M,p} \oplus \mathcal{T}_{M,p} \to \pi_*(\mathcal{O}_L)_p, \ (c, X) \mapsto (c + \alpha(\widetilde{X}))|_L \tag{3.8}$$

and

$$\begin{aligned} \mathbb{C} \oplus T_pM &\to \pi_*(\mathcal{O}_L)_p/\mathbf{m}_p \cdot \pi_*(\mathcal{O}_L)_p, \\ (c, X) &\mapsto (c + \alpha(\widetilde{X}))|\pi_*(\mathcal{O}_L)_p/\mathbf{m}_p \cdot \pi_*(\mathcal{O}_L)_p \end{aligned} \tag{3.9}$$

are welldefined. Here $\widetilde{X}$ is in both cases a lift of X to T^*M. These morphisms are not invariants of the diagram as a *germ at the base point $p \in M$ of a Lagrange map* because the identification of the Lagrange fibration with the cotangent bundle of (M, p) is unique only up to shifts in the fibres and only the class of 1–forms $\alpha + \{dS \mid S : M \to \mathbb{C}$ holomorphic$\}$ is uniquely determined (cf. section 3.3).

But being an isomorphism or epimorphism in (3.8) and (3.9) is clearly a property of the germ at p of the Lagrange map.

Definition 3.13 The germ at $p \in M$ of $L \hookrightarrow T^*M \to M$ as a Lagrange map is called *miniversal* (*versal*) if the morphism in (3.9) is an isomorphism (epimorphism) (cf. [**Gi2**, chapter 1.3]).

We are only interested in the case of a flat projection $\pi : L \to M$. Well known criteria of flatness for finite maps (cf. [**Fi**, 3.13]) give the next lemma.

Lemma 3.14 *The following conditions are equivalent.*

(i) The germ at $p \in M$ of $L \hookrightarrow T^*M \to M$ as a Lagrange map is miniversal with flat projection $\pi : (L, \pi^{-1}(p)) \to (M, p)$,
(ii) it is miniversal with $\deg \pi = 1 + \dim M$,
(iii) the morphism in (3.8) *is an isomorphism,*
(iv) the Lagrange map is miniversal at all points in a neighbourhood of $p \in M$.

Example 3.15 A miniversal germ at a base point of a Lagrange map with a projection $\pi : L \to M$ which is not flat is given by the germ at $0 \in \mathbb{C}^2$ of the Lagrange fibration $\mathbb{C}^4 \to \mathbb{C}^2$, $(y_2, y_3, t_2, t_3) \mapsto (t_2, t_3)$ with $\omega = \mathrm{d}y_2\mathrm{d}t_2 + \mathrm{d}y_3\mathrm{d}t_3$ and by the Lagrange variety L which is the union of two appropriate planes and which is defined by the ideal

$$(y_2, y_3) \cap (y_2 - t_2, y_3 - t_3) = (y_2, y_3) \cdot (y_2 - t_2, y_3 - t_3). \tag{3.10}$$

Now let M be a massive n-dimensional F-manifold with analytic spectrum $L \subset T^*M$. Then $(L, \pi^{-1}(p)) \hookrightarrow (T^*M, T_p^*M) \to (M, p)$ is for any $p \in M$ a versal, but not a miniversal germ at the base point $p \in M$ of a Lagrange map. But there is a miniversal one.

The germ of the fibration at p whose fibres are the orbits of the unit field e is denoted by $pr_e : (M, p) \to (M^{(r)}, p^{(r)})$. The fibrewise linear function on T^*M which corresponds to e is called y_1. Its Hamilton field $\widetilde{e} := H_{y_1}$ is a lift of e to T^*M. It leaves the hypersurface $y_1^{-1}(1) \subset T^*M$ and the Lagrange variety $L \subset y_1^{-1}(1)$ invariant. The orbits of $\widetilde{e}$ in $y_1^{-1}(1)$ form a germ of a $2n - 2$-dimensional symplectic manifold with a Lagrange fibration, which can be identified with the cotangent bundle

$$\left(T^*M^{(r)}, T^*_{p^{(r)}}M^{(r)}\right) \to \left(M^{(r)}, p^{(r)}\right).$$

But this identification is only unique up to shifts in the fibres. The orbits of $\widetilde{e}$ in L form a Lagrange variety $L^{(r)} \subset T^*M^{(r)}$.

The germ at $p^{(r)} \in M^{(r)}$ of the diagram $L^{(r)} \hookrightarrow T^*M^{(r)} \to M^{(r)}$ is unique up to isomorphism of *germs in the base of Lagrange maps*. It will be called the *restricted Lagrange map* of the germ (M, p) of the F-manifold M. An explicit description will be given in the proof of the next result.

Theorem 3.16 *(a) The restricted Lagrange map of the germ (M, p) of a massive F-manifold is miniversal with flat projection $\pi^{(r)} : L^{(r)} \to M^{(r)}$.*

(b) It determines the germ (M, p) of the F-manifold uniquely.

(c) Any miniversal germ at a base point of a Lagrange map $L' \hookrightarrow T^*M' \to M'$ with flat projection $L' \to M'$ is the restricted Lagrange map of a germ of a massive F-manifold.

Proof: (a) In order to be as explicit as possible we choose coordinates $t = (t_1, t') = (t_1, \ldots, t_n) : (M, p) \to (\mathbb{C}^n, 0)$ with $e(t_1) = 1$. The dual coordinates on T^*M are $(y_1, \ldots, y_n) = (y_1, y') = y$. The multiplication is given by $\frac{\partial}{\partial t_i} \circ \frac{\partial}{\partial t_j} = \sum_k a_{ij}^k(t') \frac{\partial}{\partial t_k}$ and the analytic spectrum L is

$$L \cong \left\{(y, t) \in \mathbb{C}^n \times (\mathbb{C}^n, 0) \mid y_1 = 1,\ y_i y_j = \sum_k a_{ij}^k(t') y_k\right\}. \tag{3.11}$$

The restricted Lagrange map is represented by the Lagrange fibration

$$\mathbb{C}^{n-1} \times (\mathbb{C}^{n-1}, 0) \to (\mathbb{C}^{n-1}, 0), \quad (y', t') \mapsto t' \tag{3.12}$$

with canonical 1–form $\alpha' := \sum_{i \geq 2} y_i dt_i$ and by the Lagrange variety

$$\left\{(y', t') \in \mathbb{C}^{n-1} \times (\mathbb{C}^{n-1}, 0) \mid y_i y_j = a_{ij}^1(t') + \sum_{k \geq 2} a_{ij}^k(t') y_k \text{ for } i, j \geq 2\right\} \cong L^{(r)}. \tag{3.13}$$

The equations for the Lagrange variety in (3.13) show that the morphism in (3.8) for this Lagrange map with fixed canonical 1–form α' is an isomorphism. This implies that the restricted Lagrange map for (M, p) is miniversal with flat projection.

(b) and (c) Any miniversal germ at a base point $p' \in M'$ of a Lagrange map $L' \hookrightarrow T^*M' \to M'$ with flat projection $\pi' : L' \to M'$ can be represented by a Lagrange fibration as in (3.12) and a Lagrange variety as in (3.13).

Defining L by (3.11) and $M := \mathbb{C} \times (M', p')$ and $e := \frac{\partial}{\partial t_1}$, one obtains an F-manifold with unit field e and analytic spectrum L.

It remains to show that this does not depend on the way in which the Lagrange fibration is identified with the cotangent bundle of $(\mathbb{C}^{n-1}, 0)$ in (3.12). But one sees easily that a shift in the fibres of (3.12) of the type $y_i \mapsto y_i + \frac{\partial S}{\partial t_i}$ for some holomorphic function $S : (\mathbb{C}^{n-1}, 0) \to \mathbb{C}$ on the base corresponds only to a change of the coordinate fields $\frac{\partial}{\partial t_2}, \ldots, \frac{\partial}{\partial t_n}$ and the coordinate t_1 in M and thus to a shift of the section $\{0\} \times (M', p')$ in $M \to (M', p')$. It does not affect L and the multiplication on (M, p). □

Let (M, p) be a germ of a massive F-manifold. The germ

$$\mathcal{G}_{p^{(r)}} := \{X \in \mathcal{T}_{M,p} \mid [e, X] = 0\} \tag{3.14}$$

is a free $\mathcal{O}_{M^{(r)},p^{(r)}}$-module of rank n. It is an $\mathcal{O}_{M^{(r)},p^{(r)}}$-algebra because of $\text{Lie}_e(\circ) = 0 \cdot \circ$. The functions $\mathbf{a}(X)$ for $X \in \mathcal{G}_{p^{(r)}}$ are invariant with respect to $\widetilde{e}$ and induce holomorphic functions on $L^{(r)}$. One obtains a map

$$\mathbf{a}^{(r)} : \mathcal{G}_{p^{(r)}} \to \left(\pi_*^{(r)} \mathcal{O}_{L^{(r)}}\right)_{p^{(r)}}. \tag{3.15}$$

Lemma 3.17 *The map $\mathbf{a}^{(r)}$ is an isomorphism of $\mathcal{O}_{M^{(r)},p^{(r)}}$-algebras.*

Proof: The isomorphism $\mathbf{a} : \mathcal{T}_{M,p} \to (\pi_* \mathcal{O}_L)_p$ maps the e-invariant vector fields in (M, p) to the $\widetilde{e}$-invariant functions in $(\pi_* \mathcal{O}_L)_p$. □

This isomorphism $\mathbf{a}^{(r)}$ is closely related to (3.8) for the restricted Lagrange map of (M, p):

An isomorphism as in (3.8) requires the choice of a 1–form for its Lagrange fibration. The choice of a function $t_1 : (M, p) \to (\mathbb{C}, 0)$ with $e(t_1) = 1$ yields such a 1–form: the 1–form which is induced by $\alpha - \mathrm{d}t_1$ ($\alpha - \mathrm{d}t_1$ on T^*M is $\widetilde{e}$-invariant and vanishes on $\widetilde{e}$ and induces a 1–form on the space of $\widetilde{e}$-orbits of $y_1^{-1}(1)$).

The choice of t_1 also yields an isomorphism

$$\mathcal{O}_{M^{(r)},p^{(r)}} \oplus \mathcal{T}_{M^{(r)},p^{(r)}} \to \mathcal{O}_{M^{(r)},p^{(r)}} \cdot e \oplus \{X \in \mathcal{G}_{p^{(r)}} \mid X(t_1) = 0\} = \mathcal{G}_{p^{(r)}}. \tag{3.16}$$

One sees with the proof of Theorem 3.16 (a) that the composition of this isomorphism with $\mathbf{a}^{(r)}$ gives the isomorphism in (3.8) for the restricted Lagrange map of (M, p) (the germ (M, p) in (3.8) in this case is $(M^{(r)}, p^{(r)})$).

3.5 Lyashko–Looijenga map of an F-manifold

Definition 3.18. Let $(M, \circ, e)$ be a massive n-dimensional F-manifold with analytic spectrum $L \subset T^*M$.

(a) A *generating function* F for $(M, \circ, e)$ is a generating function for L, that is, a continuous function $F : L \to \mathbb{C}$ which is holomorphic on L_{reg} with $\mathrm{d}F|_{L_{reg}} = \alpha|_{L_{reg}}$.

(b) Let F be a generating function for $(M, \circ, e)$.

(i) The *bifurcation diagram* $\mathcal{B} \subset M$ of $(M, \circ, e, F)$ is the set of points $p \in M$ such that F has less than n different values on $\pi^{-1}(p)$.
(ii) The *Lyashko–Looijenga map* $\Lambda = (\Lambda_1, \ldots, \Lambda_n) : M \to \mathbb{C}^n$ of $(M, \circ, e, F)$ is the Lyashko–Looijenga map of F as the generating function for $L \subset T^*M$ (cf. section 3.3).
(iii) The *discriminant* $\mathcal{D} \subset M$ of $(M, \circ, e, F)$ is $\mathcal{D} := \Lambda_n^{-1}(0)$.

The discriminant will be discussed in section 4.1. A generating function for an F-manifold exists locally (Lemma 3.4), but not necessarily globally.

A holomorphic generating function F corresponds to an Euler field $E := \mathbf{a}^{-1}(F)$ of weight 1 (Theorem 3.3); then the values of F on $\pi^{-1}(p)$, $p \in M$, are the eigenvalues of $E\circ : T_pM \to T_pM$. The objects $\mathcal{B},\ \Lambda,\ \mathcal{D}$ of Definition 3.18 (b) are welldefined for $(M, \circ, e, E)$ if E is such an Euler field.

The restriction of $\pi : L \to M$ to the complement of the caustic $\mathcal{K}$ is a covering $\pi : \pi^{-1}(M - \mathcal{K}) \to M - \mathcal{K}$ of degree n, and $\pi^{-1}(M - \mathcal{K})$ is smooth. Hence a generating function F is holomorphic on $\pi^{-1}(M - \mathcal{K})$ and corresponds to an Euler field E on $M - \mathcal{K}$. Results and examples about the extendability of E to M are given in Lemma 3.9 and Theorem 5.30.

The bifurcation diagram $\mathcal{B}$ of $(M, \circ, e, F)$ contains the caustic $\mathcal{K}$. The caustic is a hypersurface or empty (Proposition 2.6) and invariant with respect to the unit field e (Remark 2.9 (vii)). The bifurcation diagram has the same properties: the restriction of F to an open set $U \subset M - \mathcal{K}$ with canonical coordinates $(u_1, \ldots, u_n)$ corresponds to an Euler field $E = \sum (u_i + r_i)e_i$ for some $r_i \in \mathbb{C}$, and the bifurcation diagram is the hypersurface

$$\mathcal{B} \cap U = U \cap \{u \mid u_i + r_i = u_j + r_j \text{ for some } i \neq j\}.$$

It is invariant with respect to e because of $e(u_i - u_j) = 0$.

The Lyashko–Looijenga map for the F-manifold $A_1^n = (\mathbb{C}^n, \circ, e)$ (Example 2.12 (ii)) with Euler field $E = \sum u_i e_i$ and Euler field-function $F := \mathbf{a}(E)$ is

$$\Lambda^{(n)} : \mathbb{C}^n \to \mathbb{C}^n,\ u \mapsto ((-1)^i \sigma_i(u))\ ; \tag{3.17}$$

here $\sigma_1(u), \ldots, \sigma_n(u)$ are the symmetric polynomials. The group of automorphisms of the F-manifold A_1^n which respect the Euler field E is the symmetric group S_n which permutes the coordinates $u_1, \ldots, u_n$. The map $\Lambda^{(n)}$

is the quotient map for this group. It is branched along the bifurcation diagram $\mathcal{B} = \{u \mid u_i = u_j \text{ for some } i \neq j\}$. The image $\Lambda^{(n)}(\mathcal{B})$ is the hypersurface

$$\mathcal{D}^{(n)} := \left\{ a \in \mathbb{C}^n \mid z^n + \sum a_i z^{n-i} \text{ has multiple roots} \right\} \subset \mathbb{C}^n. \quad (3.18)$$

The restriction $\Lambda^{(n)} : \mathbb{C}^n - \mathcal{B} \to \mathbb{C}^n - \mathcal{D}^{(n)}$ induces an F-manifold structure on $\mathbb{C}^n - \mathcal{D}^{(n)}$, with unit field

$$e^{(n)} := \mathrm{d}\Lambda^{(n)}(e) = -n\frac{\partial}{\partial a_1} - \sum_{i \geq 2}(n - i + 1)a_{i-1}\frac{\partial}{\partial a_i} \quad (3.19)$$

and Euler field

$$E^{(n)} := \mathrm{d}\Lambda^{(n)}(E) = \sum_i i a_i \frac{\partial}{\partial a_i}. \quad (3.20)$$

This F-manifold $(\mathbb{C}^n - \mathcal{D}^{(n)}, \circ, e^{(n)})$ will be denoted by A_1^n/S_n.

Theorem 3.19 *Let $(M, \circ, e)$ be a massive F-manifold with generating function $F : L \to \mathbb{C}$ and Lyashko–Looijenga map $\Lambda : M \to \mathbb{C}^n$.*

Then $\Lambda^{-1}(\mathcal{D}^{(n)}) = \mathcal{B}$ and $\mathrm{d}\Lambda(e) = e^{(n)}$. The restriction $\Lambda : M - \mathcal{B} \to \mathbb{C}^n - \mathcal{D}^{(n)}$ is an immersion and locally an isomorphism of F-manifolds. It maps the Euler field $\mathbf{a}^{-1}(F|_{M-\mathcal{B}})$ on $M - \mathcal{B}$ to the Euler field $E^{(n)}$.

Proof: In $M - \mathcal{K}$ the multiplication is semisimple and locally the values of the generating function are canonical coordinates. The map Λ factors on $M - \mathcal{B}$ locally into an isomorphism to A_1^n and into the map $\Lambda^{(n)}$. □

The most important part of Theorem 3.19 is that $\Lambda : M - \mathcal{B} \to \mathbb{C}^n - \mathcal{D}^{(n)}$ is locally biholomorphic. The following statements for germs will also be useful.

Lemma 3.20 *Let $(M, p) = \prod_{k=1}^{l}(M_k, p)$ be a germ of a massive F-manifold with analytic spectrum L and with decomposition into irreducible germs (M_k, p) of dimension n_k, $\sum n_k = n$.*

(a) There exists precisely one generating function on the multigerm $(L, \pi^{-1}(p))$ for any choice of its values on $\pi^{-1}(p) = \{\lambda_1, \ldots, \lambda_l\}$.

(b) Choose a function $t_1 : (M, p) \to \mathbb{C}$ with $e(t_1) = 1$.

The values of a generating function for the points in L above an orbit of e are of the form t_1 + a constant.

The entry Λ_i of a Lyashko–Looijenga map $\Lambda : (M, p) \to \mathbb{C}^n$ is a polynomial of degree i in t_1 with coefficients in $\{g \in \mathcal{O}_{M,p} \mid e(g) = 0\}$ and leading coefficient $(-1)^i \binom{m}{i}$.

(c) Choose representatives M_k for the germs (M_k, p) and Lyashko–Looijenga maps $\Lambda^{[k]} : M_k \to \mathbb{C}^{n_k}$. Then the function $\Lambda = (\Lambda_1, \ldots, \Lambda_n) : \prod_k M_k \to \mathbb{C}^n$ which is defined by

$$z^n + \sum_{i=1}^{n} \Lambda_i z^{n-i} = \prod_{k=1}^{l} \left(z^{n_k} + \sum_{i=1}^{n_k} \Lambda_i^{[k]} z^{n-i} \right) \tag{3.21}$$

is a Lyashko–Looijenga map for the representative $\prod M_k$ of the germ (M, p). Any Lyashko–Looijenga map for (M, p) is of this type.

Proof: (a) Lemma 3.4.

(b) It suffices to prove the first part for an orbit of e in $M - \mathcal{K}$. There the generating function comes from an Euler field. The formulas (2.9) and (2.10) imply $\mathrm{Lie}_e(E\circ) = id$. The values of F are the eigenvalues of $E\circ$.

(c) The map $\Lambda^{[k]}$ corresponds to an Euler field $E^{[k]}$ (at least) on $M_k - \mathcal{B}_{M_k}$. The sum $\sum_k E^{[k]}$ is an Euler field on $\prod(M_k - \mathcal{B}_{M_k})$ by Proposition 2.10. The corresponding generating function extends to $\prod M_k$ and has the given Λ as its Lyashko–Looijenga map. The last statement follows with (a). □

Consider the projection $pr_e : (M, p) \to (M^{(r)}, p^{(r)})$ whose fibres are the orbits of e (section 3.4). The e-invariant hypersurfaces $\mathcal{B}$ and $\mathcal{K}$ project to hypersurfaces in $M^{(r)}$, which are called the *restricted bifurcation diagram* $\mathcal{B}^{(r)}$ and the *restricted caustic* $\mathcal{K}^{(r)}$. In section 3.3 the restricted Lagrange map was defined as the germ at $p^{(r)} \in M^{(r)}$ of a Lagrange map $L^{(r)} \hookrightarrow T^*M^{(r)} \to M^{(r)}$. Because of Corollary 3.12 the notion of a reduced Lyashko–Looijenga map is welldefined for the restricted Lagrange map (independently of the identification of the Lagrange fibration with the cotangent bundle of $M^{(r)}$).

The space of orbits of the field $e^{(n)}$ (formula (3.19)) in $\mathbb{C}^n$ can be identified with $\{a \in \mathbb{C}^n \mid a_1 = 0\} = \{0\} \times \mathbb{C}^{n-1} \cong \mathbb{C}^{n-1}$ and is equipped with the coordinate system $(a_2, \ldots, a_n) = a'$. The projection to this orbit space is denoted by $pr^{(n)} : \mathbb{C}^n \to \mathbb{C}^{n-1}$, and the image of $\mathcal{D}^{(n)}$ is

$$\mathcal{D}(A_{n-1}) := \left\{ a' \in \mathbb{C}^{n-1} \mid z^n + \sum_{i=2}^{n} a_i z^{n-i} \text{ has multiple roots} \right\} \tag{3.22}$$

(it is isomorphic to the discriminant of the singularity or F-manifold A_{n-1}, cf. section 5.1).

Corollary 3.21 *Let (M, p) be the germ of a massive F-manifold, F a generating function, Λ $(\Lambda^{(red)})$ the (reduced) Lyashko–Looijenga map of $(M, \circ, e, F)$.*

The map $\Lambda^{(red)} : (M, p) \to \mathbb{C}^{n-1}$ is constant along the orbits of e. The induced map $\Lambda^{(red)(r)} : (M^{(r)}, p^{(r)}) \to \mathbb{C}^{n-1}$ is a reduced Lyashko–Looijenga *map for the restricted Lagrange map. The following diagram commutes, the diagonal morphism is* $\Lambda^{(red)} = pr^{(n)} \circ \Lambda = \Lambda^{(red)(r)} \circ pr_e$*,*

$$\begin{array}{ccc} (M, p) & \xrightarrow{\Lambda} & \mathbb{C}^n \\ \big\downarrow{\scriptstyle pr_e} & & \big\downarrow{\scriptstyle pr^{(n)}} \\ \big(M^{(r)}, p^{(r)}\big) & \xrightarrow{\Lambda^{(red)(r)}} & \mathbb{C}^{n-1} \end{array} \tag{3.23}$$

The restriction

$$\Lambda^{(red)(r)} : M^{(r)} - \mathcal{B}^{(r)} \to \mathbb{C}^{n-1} - \mathcal{D}(A_{n-1}) \tag{3.24}$$

is locally biholomorphic.

Proof: The map $\Lambda^{(red)}$ is constant along the orbits of e because of Lemma 3.20 (b). The formulas (3.11), (3.12), (3.13) show that $\Lambda^{(red)(r)}$ is a reduced Lyashko–Looijenga map for the restricted Lagrange map. The rest follows from Theorem 3.19. □

Chapter 4

Discriminants and modality of F-manifolds

Discriminants play a central role in singularity theory. Usually they have a rich geometry and say a lot about the mappings or other objects from which they are derived. The discriminant $\mathcal{D}$ of a massive F-manifold M with a generating function (cf. Definition 3.18) is an excellent model case of such discriminants, having many typical properties.

Together with the unit field it determines the whole F-manifold in a nice geometric way. This is discussed in section 4.1 (cf. Corollary 4.6). In section 4.3 results from singularity theory are adapted to show that the discriminant and also the bifurcation diagram are free divisors under certain hypotheses.

The classification of germs of 2-dimensional massive F-manifolds is nice and is carried out in section 4.2. Already for 3-dimensional germs it is vast (cf. section 5.5). In section 4.4 the Lyashko–Looijenga map is used to prove that the automorphism group of a germ of a massive F-manifold is finite. There also the notions modality and μ-constant stratum from singularity theory are adapted to F-manifolds. In section 4.5 the relation between analytic spectrum and multiplication is generalized. This allows F-manifolds to be found in natural geometric situations (e.g. hypersurface and boundary singularities) and to be written down in an economic way (e.g. in 5.22, 5.27, 5.30, 5.32).

4.1 Discriminant of an F-manifold

Let $(M, \circ, e, F)$ be a massive n-dimensional F-manifold with a generating function $F : L \to \mathbb{C}$ and Lyashko–Looijenga map $\Lambda = (\Lambda_1, \ldots, \Lambda_n) : M \to \mathbb{C}^n$; the discriminant of $(M, \circ, e, F)$ is the hypersurface $\mathcal{D} = \Lambda_n^{-1}(0) \subset M$ (Definition 3.18).

If F is holomorphic and $E = \mathbf{a}^{-1}(F)$ is its Euler field then $\Lambda_n = (-1)^n \cdot \det(E\circ)$, and the discriminant is the set of points where the multiplication with E is not invertible.

By the definition of Λ_n, the discriminant is $\mathcal{D} = \pi(F^{-1}(0))$. Theorem 4.1 will give an isomorphism between $F^{-1}(0)$ and the development $\widetilde{\mathcal{D}} \subset \mathbb{P}T^*M$ of $\mathcal{D}$.

We need an identification of subsets of T^*M and $\mathbb{P}T^*M$. The fibrewise linear function on T^*M which corresponds to e is called y_1. The canonical map

$$y_1^{-1}(1) \longrightarrow \mathbb{P}T^*M \tag{4.1}$$

identifies $y_1^{-1}(1) \subset T^*M$ with the open subset in $\mathbb{P}T^*M$ of hyperplanes in the tangent spaces of M which do not contain $\mathbb{C} \cdot e$.

The restriction to $y_1^{-1}(1)$ of the canonical 1–form α on T^*M gives the contact structure on $y_1^{-1}(1)$ which is induced by the canonical contact structure on $\mathbb{P}T^*M$.

Theorem 4.1 *Let $(M, \circ, e, F)$ and $F^{-1}(0) \subset L \subset y_1^{-1}(1) \subset T^*M$ be as above.*

The canonical map $y_1^{-1}(1) \hookrightarrow \mathbb{P}T^*M$ identifies $F^{-1}(0)$ with the development $\widetilde{\mathcal{D}} \subset \mathbb{P}T^*M$ of the discriminant $\mathcal{D}$.

Proof: We want to make use of the discussion of fronts and graphs (section 3.3) and of the restricted Lagrange map (section 3.4). It is sufficient to consider the germ (M, p) for some $p \in \mathcal{D}$. We choose coordinates $(t_1, \ldots, t_n)$ and $(y_1, \ldots, y_n)$ as in the proof of Theorem 3.16. The generating function F for $L \hookrightarrow T^*M \to M$ takes the form

$$F(y', t) = t_1 + F^{(r)}(y', t'), \tag{4.2}$$

where $F^{(r)}$ is a generating function of $L^{(r)}$ with respect to (3.13), (3.12), and $\alpha' = \sum_{i \geq 2} y_i dt_i$. The isomorphism with a sign

$$(-t_1, pr_e) : (M, p) \to \big(\mathbb{C} \times M^{(r)}, 0 \times p^{(r)}\big) \tag{4.3}$$

maps the discriminant to the front $\mathrm{Im}(F^{(r)}, \pi^{(r)})$ of $L^{(r)}$ and $F^{(r)}$.

The development of this front is identified with the graph $\mathrm{Im}(F^{(r)}, id) \subset \mathbb{C} \times T^*M^{(r)}$ of $F^{(r)} : L^{(r)} \to \mathbb{C}$, by Proposition 3.10 and the embedding (cf. formula (3.7))

$$\begin{aligned} \mathbb{C} \times T^*M^{(r)} &\hookrightarrow \mathbb{P}T^*(\mathbb{C} \times M^{(r)}), \\ (-t_1, \lambda) &\mapsto \big((dt_1 + \lambda)^{-1}(0), \big(-t_1, \pi^{(r)}(\lambda)\big)\big). \end{aligned} \tag{4.4}$$

But (4.1), (4.3), and (4.4) together also yield an isomorphism $y_1^{-1}(1) \to \mathbb{C} \times T^*M^{(r)}$, which maps $F^{-1}(0)$ to this graph. □

Remarks 4.2 (i) The most important consequence is that $\widetilde{\mathcal{D}} \subset \mathbb{P}T^*M$ does not contain a hyperplane which contains $\mathbb{C} \cdot e$ (cf. Remark 3.11 (i)). Therefore the unit field e is everywhere transversal to the discriminant $\mathcal{D}$.

(ii) In the proof only the choice of t_1 is essential. It is equivalent to various other choices: the choice of a section of the projection $(M, p) \to (M^{(r)}, p^{(r)})$, the choice of a 1–form α' for the Lagrange fibration in the restricted Lagrange map.

(iii) Fixing such a choice of t_1, one obtains together with α' and $F^{(r)}$ a Lyashko–Looijenga map $\Lambda^{(r)} : (M^{(r)}, p^{(r)}) \to \mathbb{C}^{n-1}$ for $F^{(r)}$. Then (4.3) identifies the entry Λ_n of the Lyashko–Looijenga map $\Lambda : (M, p) \to \mathbb{C}^n$ with the polynomial $(-t_1)^n + \sum_{i=2}^n \Lambda_i^{(r)}(-t_1)^{n-i}$.

(iv) The set $F^{-1}(0) \subset L$ is not an analytic hypersurface of L at points of L_{sing} where F is not holomorphic. But Theorem 4.1 shows that it is everywhere a subvariety of L of pure codimension 1. Examples where F is not holomorphic will be given in section 5.5.

(v) Let $(M, p) = \prod_{k=1}^l (M_k, p)$ be the decomposition of a germ (M, p) into irreducible germs of F-manifolds. A Lyashko–Looijenga map $\Lambda : (M, p) \to \mathbb{C}^n$ corresponds to Lyashko–Looijenga maps $\Lambda^{[k]} : (M_k, p) \to \mathbb{C}^{n_k}$ for the irreducible germs, in a way which was described in Lemma 3.20 (c). One obtains especially $\Lambda_n = \prod_k \Lambda_{n_k}^{[k]}$, and the germ $(\mathcal{D}, p)$ of the discriminant is

$$(\mathcal{D}, p) = \bigcup_k \left(\prod_{j<k} M_j \times (\Lambda_{n_k}^{[k]})^{-1}(0) \times \prod_{j>k} M_j, p \right), \tag{4.5}$$

the union of products of smooth germs with the discriminants for the irreducible germs (of course, it is possible that $((\Lambda_{n_k}^{[k]})^{-1}(0), p) = \emptyset$ for some or all k).

(vi) The development $\widetilde{\mathcal{D}}$ of the discriminant $\mathcal{D}$ gives the tangent hyperplanes to $\mathcal{D}$. Let $(M, p) = \prod_{k=1}^l (M_k, p)$ be as in (iv), and $\pi^{-1}(p) = \{\lambda_1, \ldots, \lambda_l\}$.

Theorem 4.1 says that the tangent hyperplanes to $(\mathcal{D}, p)$ are those hyperplanes $\lambda_k^{-1}(0) \subset T_pM$ for which $F(\lambda_k) = 0$.

Especially, if $l = 1$ and $F(\lambda_1) = 0$, then $\lambda_1^{-1}(0) \subset T_pM$ is the nilpotent subalgebra of T_pM and the unique tangent hyperplane to $(\mathcal{D}, p)$. The general case fits with Lemma 2.1 (iv) and (4.5).

(vii) The equality $\alpha|_{L_{reg}} = \mathrm{d}F|_{L_{reg}}$ shows immediately that $F^{-1}(0) \subset y_1^{-1}(1)$ is a Legendre subvariety.

(viii) If $\lambda \in F^{-1}(0) \cap T_p^*M$, there is a canonical projection from the Legendre germ $(F^{-1}(0), \lambda)$ to one component of the Lagrange multigerm $(L^{(r)}, {\pi^{(r)}}^{-1}(p^{(r)}))$. It is a bijective morphism. It is an isomorphism if and only if F is holomorphic at (L, λ) (see Remark 3.11 (iv)).

One can recover the multiplication on a massive n-dimensional F-manifold M from the unit field e and a discriminant $\mathcal{D}$ if the orbits of e are sufficiently large. To make this precise, we introduce the following notion.

Definition 4.3 A massive F-manifold $(M, \circ, e, F)$ with generating function F is in *standard form* if there exists globally a projection $pr_e : M \to M^{(r)}$ to a manifold $M^{(r)}$ such that

(α) the fibres are the orbits of e (and thus connected),
(β) they are with their affine linear structure isomorphic to an open (connected) subset of $\mathbb{C}$,
(γ) the projection $pr_e : \mathcal{D} \to M^{(r)}$ is a branched covering of degree n.

Remarks 4.4 (i) If $(M, \circ, e, F)$ is a massive F-manifold with generating function F and properties (α) and (β) then

$$\left(-\frac{1}{n}\Lambda_1, pr_e\right) : M \to \mathbb{C} \times M^{(r)} \tag{4.6}$$

is an embedding because of $e(-\frac{1}{n}\Lambda_1) = 1$ (Lemma 3.20 (b)).

The F-manifold M can be extended uniquely to an F-manifold isomorphic to $\mathbb{C} \times M^{(r)}$. Also the generating function F can be extended. The discriminant of this extended F-manifold satisfies (γ) because of $\mathcal{D} = \Lambda_n^{-1}(0)$ and Lemma 3.20 (b).

(ii) For $(M, \circ, e, F)$ as in (i) the coordinate $t_1 := -\frac{1}{n}\Lambda_1$ is distinguished and, up to the addition of a constant, even independent of the choice of F. Nevertheless it does not seem to have good properties: In the case of the simplest 3-dimensional irreducible germs of F-manifolds, A_3, B_3, H_3, it is not part of the coordinate system of a nice normal form (section 5.3). Using the data in **[Du2]** one can also check that $-\frac{1}{n}\Lambda_1$ is not a flat coordinate of the Frobenius manifolds A_3, B_3, H_3.

Corollary 4.5 *Let $(M, \circ, e, F)$ be a massive F-manifold with generating function F and in standard form.*

(a) The branch locus of the branched covering $pr_e : \mathcal{D} \to M^{(r)}$ is $\mathcal{D}_{sing}$, the set $pr_e(\mathcal{D}_{sing})$ of critical values is the restricted bifurcation diagram $\mathcal{B}^{(r)} = pr_e(\mathcal{B})$.

(b) The union of the shifts of $F^{-1}(0)$ with the Hamilton field $\widetilde{e} = H_{y_1}$ is the analytic spectrum $L \subset T^*M$.

(c) The data $(M, \circ, e, F)$ and $(M, e, \mathcal{D})$ are equivalent.

Proof: (a) Theorem 4.1 implies that all tangent hyperplanes to $\mathcal{D}$ are transversal to the unit field. Therefore the branch locus is only $\mathcal{D}_{sing}$.

(b) and (c) The discriminant $\mathcal{D}$ and the unit field e determine $F^{-1}(0) \subset y_1^{-1}(1)$ because of Theorem 4.1. The union of shifts of $F^{-1}(0)$ with the Hamilton field $\widetilde{e} = H_{y_1}$ is finite of degree n over M because of (γ) and it is contained in L, so it is L. The function $F : L \to \mathbb{C}$ is determined by $F^{-1}(0) \subset L$ and by the linearity of F along the orbits of $\widetilde{e}$. □

Theorem 4.1 and Corollary 4.5 (b) give one way to recover the multiplication of a massive F-manifold $(M, \circ, e, F)$ in standard form from the discriminant $\mathcal{D}$ and the unit field e. The following is a more elementary way.

Corollary 4.6 *Let $(M, \circ, e, F)$ be a massive n-dimensional F-manifold with generating function F and in standard form. The multiplication can be recovered from the discriminant $\mathcal{D}$ and the unit field e in the following way. The multiplication is semisimple outside of the bifurcation diagram $\mathcal{B} = pr_e^{-1}(pr_e(\mathcal{D}_{sing}))$. For a point $p \in M - \mathcal{B}$, the idempotent vectors $e_i(p) \in T_pM$ with $e_i(p) \circ e_j(p) = \delta_{ij} e_i(p)$ are uniquely determined by (i) and (ii):*

(i) the unit vector is $e(p) = \sum_{i=1}^{n} e_i(p)$,

(ii) the multigerm $(\mathcal{D}, \mathcal{D} \cap pr_e^{-1}(pr_e(p)))$ has exactly n tangent hyperplanes; their shifts to T_pM with e are the hyperplanes $\bigoplus_{i \neq k} \mathbb{C} \cdot e_i \subset T_pM$, $k = 1, \ldots, n$.

Proof: Remark 4.2 (vi). □

4.2 2-dimensional F-manifolds

The only 1-dimensional germ of an F-manifold is A_1 (Example 2.12 (i)). The class of 3-dimensional germs of massive F-manifolds is already vast. Examples and a partial classification will be given in section 5.5. But the classification of 2-dimensional germs of F-manifolds is nice.

Theorem 4.7 *(a) The only germs of 2-dimensional massive F-manifolds are, up to isomorphism, the germs $I_2(m)$, $m \in \mathbb{N}_{\geq 2}$, with $I_2(2) = A_1^2$, $I_2(3) = A_2$, $I_2(4) = B_2$, $I_2(5) =: H_2$, $I_2(6) = G_2$, from Example 2.12 (iv):*

The multiplication on $(M, p) = (\mathbb{C}^2, 0)$ with coordinates t_1, t_2 and $\delta_i := \frac{\partial}{\partial t_i}$ is given by $e := \delta_1$ and $\delta_2 \circ \delta_2 = t_2^{m-2} \cdot \delta_1$. An Euler field of weight 1 is $E = t_1\delta_1 + \frac{2}{m} t_2 \delta_2$. Its discriminant is $\mathcal{D} = \{t \mid t_1^2 - \frac{4}{m^2} t_2^m = 0\}$. The germ $I_2(m)$ of an F-manifold is irreducible for $m \geq 3$ with caustic and bifurcation diagram $\mathcal{K} = \mathcal{B} = \{t \mid t_2 = 0\}$. The space of Euler fields of weight d is $d \cdot E + \mathbb{C} \cdot e$ for $m \geq 3$.

(b) The only germ of a 2-dimensional not massive F-manifold is the germ $(\mathbb{C}^2, 0)$ from Example 2.12 (v) with multiplication given by $e := \delta_1$ and $\delta_2 \circ \delta_2 = 0$. The caustic is empty. An Euler field of weight 1 is $E = t_1 \delta_1$. The space of all Euler fields of weight 0 is $\{\varepsilon_1 \delta_1 + \varepsilon_2(t_2)\delta_2 \mid \varepsilon_1 \in \mathbb{C},\ \varepsilon_2(t_2) \in \mathbb{C}\{t_2\}\}$.

Proof: (a) Givental [**Gi2**, 1.3, p. 3253] classified the 1-dimensional miniversal germs of Lagrange maps with flat projection. Together with Theorem 3.16 this yields implicitly the classification of the 2-dimensional irreducible germs of massive F-manifolds. But we can recover this in a simple way and we need to be more explicit.

Let (M, p) be a 2-dimensional germ of a massive F-manifold with projection $pr_e : (M, p) \to (M^{(r)}, p^{(r)})$ to the space of orbits of e. There is a unique generating function $F : (L, \pi^{-1}(p)) \to (\mathbb{C}, 0)$. Its bifurcation diagram $\mathcal{B} \subset M$ and restricted bifurcation diagram $\mathcal{B}^{(r)} \subset M^{(r)}$ are the hypersurfaces $\mathcal{B} = pr_e^{-1}(p^{(r)})$ and $\mathcal{B}^{(r)} = \{p^{(r)}\}$.

By Corollary 3.21, the reduced Lyashko–Looijenga map $\Lambda^{(red)(r)}$: $(M^{(r)}, p^{(r)}) \to (\mathbb{C}, 0)$ of the restricted Lagrange map is a cyclic branched covering of some order $\widetilde{m}$, and the Lyashko–Looijenga map $\Lambda : M \to \mathbb{C}^2$ is also a cyclic branched covering of order $\widetilde{m}$, branched along $\mathcal{B}$.

Because of Corollary 3.12 and Theorem 3.16, this branching order $\widetilde{m}$ of $\Lambda^{(red)(r)}$ determines the germ (M, p) of the F-manifold up to isomorphism. It remains to determine the allowed $\widetilde{m}$ and explicit formulas for the F-manifolds.

Now consider the manifolds $\mathbb{C}^2$ with multiplication on $T\mathbb{C}^2$ given by $e = \delta_1$ and $\delta_2 \circ \delta_2 = t_2^{m-2} \cdot \delta_1$. The analytic spectrum $L \subset T^*\mathbb{C}^2$ is

$$L = \left\{(y_1, y_2, t_1, t_2) \mid y_1 = 1,\ y_2 \cdot y_2 = t_2^{m-2}\right\}. \tag{4.7}$$

It is an exercise to see that L is a Lagrange variety with generating function $F = t_1 + \frac{2}{m} y_2 t_2$ with respect to $\alpha = y_1 \mathrm{d}t_1 + y_2 \mathrm{d}t_2$, i.e. one has $(\alpha - \mathrm{d}F)|_{L_{reg}} = 0$. Then this gives an F-manifold and $E = \mathbf{a}^{-1}(F) = t_1\delta_1 + \frac{2}{m} t_2\delta_2$ is an Euler field of weight 1. The Lyashko–Looijenga map Λ is

$$\Lambda : \mathbb{C}^2 \to \mathbb{C}^2,\ (t_1, t_2) \mapsto \left(-2t_1, t_1^2 - \frac{4}{m^2} t_2^m \right). \tag{4.8}$$

It is branched along $\mathcal{B} = \{t \mid t_2 = 0\}$ of degree m.

So $I_2(m)$ is the desired F-manifold for any branching order $m = \widetilde{m} \geq 2$. The same calculation yields for $m = \widetilde{m} = 1$ the F-manifold A_1^2/S_2 on $\mathbb{C}^2 - \{t \mid t_2 = 0\}$ (section 3.5) with meromorphic multiplication along $\{t \mid t_2 = 0\}$.

(b) Let $(M, p) = (\mathbb{C}^2, 0)$ be the germ of a 2-dimensional not massive F-manifold with $e = \delta_1$. Then $\mathrm{Lie}_e(\circ) = 0$ and $[e, \delta_2] = 0$ imply $\mathrm{Lie}_e(\delta_2 \circ \delta_2) = 0$. Hence $\delta_2 \circ \delta_2 = \beta(t_2)\delta_1 + \gamma(t_2)\delta_2$ for some $\beta(t_2), \gamma(t_2) \in \mathbb{C}\{t_2\}$.

The field $\widetilde{\delta_2} := \delta_2 - \frac{1}{2}\gamma(t_2)\delta_1$ satisfies $\widetilde{\delta_2} \circ \widetilde{\delta_2} = (\beta(t_2) + \frac{1}{4}\gamma(t_2)^2)\delta_1$ and $[\delta_1, \widetilde{\delta_2}] = 0$. Changing coordinates we may suppose $\delta_2 = \widetilde{\delta_2}$, $\gamma(t_2) = 0$.

The analytic spectrum is $L = \{(y_1, y_2, t_1, t_2) \mid y_1 = 1,\ y_2 \cdot y_2 = \beta(t_2)\}$. The F-manifold is not massive, hence $\beta(t_2) = 0$. One checks with (2.7) and (2.8) easily that this multiplication gives an F-manifold and that the space of Euler fields is as claimed. □

Let us discuss the role which 2-dimensional germs of F-manifolds can play for higher dimensional massive F-manifolds. The set

$$\mathcal{D}^{(n,3)} := \left\{ a \in \mathbb{C}^n \mid z^n + \sum_{i=1}^{n} a_i z^{n-i} \text{ has a root of multiplicity} \geq 3 \right\} \subset \mathcal{D}^{(n)} \subset \mathbb{C}^n \tag{4.9}$$

is an algebraic subvariety of $\mathbb{C}^n$ of codimension 2 (see the proof of Proposition 2.5). Given a massive F-manifold M, the space

$$\mathcal{K}^{(3)} := \{p \in M \mid P(T_pM) \succ (3, 1, \dots, 1)\} \subset \mathcal{K} \subset M \tag{4.10}$$

of points p such that (M, p) does not decompose into 1- and 2-dimensional germs of F-manifolds is empty or an analytic subvariety (Propostion 2.5).

Theorem 4.8 *Let $(M, \circ, e)$ be a massive F-manifold with generating function F.*

(a) The function F is holomorphic on $\pi^{-1}(M - \mathcal{K}^{(3)})$ and gives rise to an Euler field of weight 1 on $M - \mathcal{K}^{(3)}$.

(b) If codim$\mathcal{K}^{(3)} \geq 2$ *then F is holomorphic on L and $E = \mathbf{a}^{-1}(F)$ is an Euler field of weight 1 on M.*

(c) One has $\mathcal{K}^{(3)} \subset \Lambda^{-1}(\mathcal{D}^{(n,3)})$, and $\overline{\Lambda^{-1}(\mathcal{D}^{(n,3)}) - \mathcal{K}^{(3)}}$ is analytic of pure codimension 2. Thus codim$\mathcal{K}^{(3)} \geq 2 \iff$ codim$\Lambda^{-1}(\mathcal{D}^{(n,3)}) \geq 2$.

(d) The restriction of the Lyashko–Looijenga map

$$\Lambda : M - \Lambda^{-1}\left(\mathcal{D}^{(n,3)}\right) \to \mathbb{C}^n - \mathcal{D}^{(n,3)}$$

is locally a branched covering, branched along $\mathcal{B} - \Lambda^{-1}(\mathcal{D}^{(n,3)})$.

If $p \in \mathcal{B} - \Lambda^{-1}(\mathcal{D}^{(n,3)})$ and $\Lambda(p) \in \mathcal{D}^{(n)} - \mathcal{D}^{(n,3)}$ are smooth points of the hypersurfaces $\mathcal{B}$ and $\mathcal{D}^{(n)}$ and if there the branching order is m, then (M, p) is the germ of an F-manifold of type $I_2(m) \times A_1^{n-2}$.

Proof: (a) and (b) Each germ (L, λ_k) of the analytic spectrum $(L, \pi^{-1}(p)) = (L, \{\lambda_1, \dots, \lambda_l\})$ of a reducible germ $(M, p) = \prod_k (M_k, p)$ is the product of a smooth germ with the analytic spectrum of (M_k, p). The analytic spectrum of $I_2(m)$ $(m \geq 2)$ is isomorphic to $(\mathbb{C}, 0) \times (\{y_2, t_2) \mid y_2^2 = t_2^{m-2}\}, 0)$. One applies Lemma 3.9.

(c) A Lyashko–Looijenga map of $I_2(m)$ is a cyclic branched covering of order m, branched along the bifurcation diagram. This together with Lemma 3.20 (c) implies that locally around a point $p \in M - \mathcal{K}^{(3)}$ the fibres of the Lyashko–Looijenga map $\Lambda : M \to \mathbb{C}^n$ are finite. Therefore $\mathrm{codim}_M(\Lambda^{-1}(\mathcal{D}^{(n,3)}), p) = \mathrm{codim}_{\mathbb{C}^n}(\mathcal{D}^{(n,3)}) = 2$.

(d) The map Λ determines the multiplication of the F-manifold M (Theorem 3.19). One uses this, Lemma 3.20 (c) and properties of $I_2(m)$. □

Many interesting F-manifolds, e.g. those for hypersurface singularities, boundary singularities, finite Coxeter groups (sections 5.1, 5.2, 5.3), satisfy the property $\mathrm{codim}\mathcal{K}^{(3)} \geq 2$ and have an Euler field of weight 1.

4.3 Logarithmic vector fields

K. Saito [**SK4**] introduced the notions of logarithmic vector fields and free divisors. Let $H \subset M$ be a reduced hypersurface in an n-dimensional manifold M. The sheaf $\mathrm{Der}_M(\log H) \subset \mathcal{T}_M$ of *logarithmic vector fields* consists of those holomorphic vector fields which are tangent to H_{reg}. This sheaf is discussed in detail in section 8.1. There it is shown that it is a coherent and reflexive $\mathcal{O}_M$-module. The hypersurface H is a *free divisor* if $\mathrm{Der}_M(\log H)$ is a free $\mathcal{O}_M$-module of rank n.

The results in this section are not really new. They had been established in various generality by Bruce [**Bru**], Givental [**Gi2**, chapter 1.4]), Lyashko [**Ly1**][**Ly3**], K. Saito [**SK6**][**SK9**], Terao [**Ter**], and Zakalyukin [**Za**] as results for hypersurface singularities, boundary singularities or miniversal Lagrange maps. But the formulation using the multiplication of F-manifolds is especially nice.

Theorem 4.9 *Let $(M, \circ, e)$ be a massive F-manifold with Euler field E of weight 1, generating function $F = \mathbf{a}(E)$ and discriminant $\mathcal{D} = (\det(E\circ))^{-1}(0) = \pi(F^{-1}(0))$.*

(a) The discriminant is a free divisor with $\mathrm{Der}_M(\log \mathcal{D}) = E \circ \mathcal{T}_M$.

(b) The kernel of the map

$$\mathbf{a}_{\mathcal{D}} : \mathcal{T}_M \to \pi_* \mathcal{O}_{F^{-1}(0)}, \quad X \mapsto \mathbf{a}(X)|_{F^{-1}(0)} \tag{4.11}$$

is

$$\ker \mathbf{a}_{\mathcal{D}} = E \circ \mathcal{T}_M = \mathrm{Der}_M(\log \mathcal{D}). \tag{4.12}$$

Proof: (a) The sheaf $E \circ \mathcal{T}_M$ is a free $\mathcal{O}_M$-module of rank n. Therefore (a) follows from (4.12).

(b) The $\mathcal{O}_M$-module $\pi_*\mathcal{O}_{F^{-1}(0)}$ has support $\mathcal{D}$. Equation (4.12) holds in $M-\mathcal{D}$. The set $\mathcal{D}\cap\mathcal{B}=\mathcal{D}_{sing}$ (cf. Corollary 4.5 (a)) has codimension 2 in M. The following shows that it is sufficient to prove (4.12) in $\mathcal{D}-\mathcal{D}_{sing}$.

Let $M-\mathcal{D}_{sing}\hookrightarrow M$ be the inclusion. The Riemann extension theorem says $\mathcal{O}_M=i_*(\mathcal{O}_{M-\mathcal{D}_{sing}})$ (cf. for example [**Fi**, 2.23]). Now $E\circ\mathcal{T}_M$ satisfies

$$E\circ\mathcal{T}_M=i_*(E\circ\mathcal{T}_M|_{M-\mathcal{D}_{sing}}) \tag{4.13}$$

because it is a free $\mathcal{O}_M$-module. The sheaves $\ker\mathbf{a}_{\mathcal{D}}$ and $\mathrm{Der}_M(\log\mathcal{D})$ satisfy the analogous equations because of their definition. Hence (4.12) holds in M if it holds in $\mathcal{D}-\mathcal{D}_{sing}$.

Let $p\in\mathcal{D}-\mathcal{D}_{sing}$. We choose a small neighbourhood U of p with canonical coordinates $u_1,\dots,u_n$ centred at p, with $\mathcal{D}\cap U=\{u\mid u_1=0\}$ and with Euler field $E=u_1e_1+\sum_{i\geq 2}(u_i+r_i)e_i$ for some $r_i\in\mathbb{C}-\{0\}$. With the notation of the proof of Theorem 3.2 (ii) $\Rightarrow$ (iii) we have $\alpha=\sum x_i du_i$,

$$F^{-1}(0)\cap\pi^{-1}(U)=\{(x,u)\mid x_j=\delta_{1j},u_1=0\},$$

and for any vector field $X=\sum\xi_ie_i\in\mathcal{T}_M(U)$

$$\mathbf{a}(X)|F^{-1}(0)\cap\pi^{-1}(U)=\xi_1(0,u_2,\dots,u_n).$$

Therefore

$$\begin{aligned}(\ker\mathbf{a}_{\mathcal{D}})_p&=\mathcal{O}_{M,p}\cdot u_1e_1\oplus\bigoplus_{i=2}^n\mathcal{O}_{M,p}\cdot e_i\\&=E\circ\mathcal{T}_{M,p}=\mathrm{Der}_{M,p}(\log\mathcal{D}).\end{aligned} \tag{4.14}$$

□

Remark 4.10 One can see Theorem 4.9 (a) in a different way: there is a criterion of K. Saito [**SK4**, Lemma (1.9)]. To apply it, one has to show

$$[E\circ\mathcal{T}_M,E\circ\mathcal{T}_M]\subset E\circ\mathcal{T}_M. \tag{4.15}$$

With (2.5) and (2.6) one calculates for any two (local) vector fields X,Y

$$[E\circ X,E\circ Y]=E\circ([X,E\circ Y]-[Y,E\circ X]-E\circ[X,Y]). \tag{4.16}$$

In the rest of this section $(M,\circ,e,E)$ will be a massive F-manifold which is equipped with an Euler field E of weight 1 and which is in standard form (Definition 4.3). The map $pr_e:M\to M^{(r)}$ is the projection to the space of orbits of e. The sheaf of e-invariant vector fields

$$\mathcal{G}:=\{X\in(pr_e)_*\mathcal{T}_M\mid[e,X]=0\} \tag{4.17}$$

is a free $\mathcal{O}_{M^{(r)}}$-module of rank n. Because of $\mathrm{Lie}_e(\circ) = 0$ it is also an $\mathcal{O}_{M^{(r)}}$-algebra.

Theorem 4.11 *Let $(M, \circ, e, E)$ be a massive F-manifold with Euler field E of weight 1 and in standard form.*

(a)

$$(pr_e)_*\mathcal{T}_M = \mathcal{G} \oplus (pr_e)_*(E \circ \mathcal{T}_M). \tag{4.18}$$

(b) The kernel of the map

$$(pr_e)_*\mathbf{a}_\mathcal{D} : (pr_e)_*\mathcal{T}_M \to (pr_e \circ \pi)_*\mathcal{O}_{F^{-1}(0)}, \quad X \mapsto \mathbf{a}(X)|_{F^{-1}(0)} \tag{4.19}$$

is $(pr_e)_*(E \circ \mathcal{T}_M)$. The restriction

$$(pr_e)_*\mathbf{a}_\mathcal{D} : \mathcal{G} \to (pr_e \circ \pi)_*\mathcal{O}_{F^{-1}(0)} \tag{4.20}$$

is an isomorphism of $\mathcal{O}_{M^{(r)}}$-algebras.

Proof: (a) It follows from (b).

(b) The kernel of $(pr_e)_*\mathbf{a}_\mathcal{D}$ is $(pr_e)_*(E \circ \mathcal{T}_M)$ because of Theorem 4.9 (b). The F-manifold in standard form has a global restricted Lagrange map $L^{(r)} \hookrightarrow T^*M^{(r)} \to M^{(r)}$ (the identification of its Lagrange fibration with $T^*M^{(r)} \to M^{(r)}$ is unique only up to shifts in the fibres).

The canonical projection $F^{-1}(0) \to L^{(r)}$ is bijective (Corollary 4.5 (b)), and then an isomorphism because F is holomorphic. It induces an isomorphism $(pr_e \circ \pi)_*\mathcal{O}_{F^{-1}(0)} \cong (\pi^{(r)})_*\mathcal{O}_{L^{(r)}}$. The composition with (4.20) is the isomorphism

$$\mathbf{a}^{(r)} : \mathcal{G} \to \left(\pi^{(r)}\right)_* \mathcal{O}_{L^{(r)}} \tag{4.21}$$

from Lemma 3.17. □

Theorem 4.9 and Theorem 4.11 are translations to F-manifolds of statements in [**SK6**, (1.6)][**SK9**, (1.7)] for hypersurface singularities. In fact, K. Saito essentially used (4.20) to define the multiplication on $\mathcal{G}$ for hypersurface singularities.

The arguments in Lemma 4.12 and Theorem 4.13 are due to Lyashko [**Ly1**] [**Ly3**] and Terao [**Ter**], see also Bruce [**Bru**].

Again $(M, \circ, e, E)$ is a massive F-manifold with Euler field of weight 1 and in standard form. We choose a function $t_1 : M \to \mathbb{C}$ with $e(t_1) = 1$ (e.g. $t_1 = -\frac{1}{n}\Lambda_1$, cf. Lemma 3.20 (b)). This choice simplifies the formulation of the results in Lemma 4.12. The vector fields in $\mathcal{T}_{M^{(r)}}$ will be identified with their (unique) lifts in $\{X \in \mathcal{G} \mid X(t_1) = 0\} \subset \mathcal{G} \subset (pr_e)_*\mathcal{T}_M$. The projection to $\mathcal{T}_{M^{(r)}}$ of all possible lifts to M of vector fields in $M^{(r)}$ is

$$\mathrm{d}(pr_e) : \mathcal{T}_{M^{(r)}} \oplus (pr_e)_*\mathcal{O}_M \cdot e \to \mathcal{T}_{M^{(r)}}. \tag{4.22}$$

Lemma 4.12 *Let $(M, \circ, e, E, t_1)$ be as above.*

(a)

$$(pr_e)_*(E \circ \mathcal{T}_M) \cap \left(\mathcal{T}_{M^{(r)}} \oplus \bigoplus_{k=0}^{n-1} \mathcal{O}_{M^{(r)}} \cdot t_1^k \cdot e \right)$$
$$= \bigoplus_{k=1}^{n-1} \mathcal{O}_{M^{(r)}} \cdot \left(t_1^k \cdot e - (t_1 e - E)^{\circ k}\right). \tag{4.23}$$

(b) Each vector field in $M^{(r)}$ which lifts to a vector field in $(pr_e)_*\mathrm{Der}_M(\log \mathcal{D})$ lifts to a unique vector field in (4.23).

(c) The vector fields in $M^{(r)}$ which lift to vector fields in $(pr_e)_*\mathrm{Der}_M(\log \mathcal{D})$ are tangent to the restricted bifurcation diagram $\mathcal{B}^{(r)} = pr_e(\mathcal{B}) \subset M^{(r)}$ and form the free $\mathcal{O}_{M^{(r)}}$-module of rank $n-1$

$$\bigoplus_{k=1}^{n-1} \mathcal{O}_{M^{(r)}} \cdot \mathrm{d}(pr_e)((t_1 e - E)^{\circ k}) \subset Der_{M^{(r)}}\left(\log \mathcal{B}^{(r)}\right). \tag{4.24}$$

Proof: (a) One sees inductively by multiplication with $t_1 e - E$ that for any $k \geq 1$

$$t_1^k e - (t_1 e - E)^{\circ k} \in (pr_e)_*(E \circ \mathcal{T}_M) \tag{4.25}$$

holds. The inclusion $t_1 e - E \in \mathcal{G}$ and $\mathrm{Lie}_e(\circ) = 0$ imply $(t_1 e - E)^{\circ k} \in \mathcal{G}$, therefore

$$\mathcal{T}_{M^{(r)}} \oplus \bigoplus_{k=0}^{n-1} \mathcal{O}_{M^{(r)}} t_1^k e = \mathcal{G} \oplus \bigoplus_{k=1}^{n-1} \mathcal{O}_{M^{(r)}} \left(t_1^k e - (t_1 e - E)^{\circ k}\right). \tag{4.26}$$

Now the decomposition (4.18) yields (4.23).

(b) The map $pr_e : \mathcal{D} \to M^{(r)}$ is a branched covering of degree n (Definition 4.3), so

$$\bigoplus_{k=0}^{n-1} \mathcal{O}_{M^{(r)}} \cdot t_1^k \to (pr_e)_*(\mathcal{O}_\mathcal{D}) \tag{4.27}$$

is an isomorphism. Therefore any lift $h \cdot e + X, h \in (pr_e)_*\mathcal{O}_M$, of $X \in \mathcal{T}_{M^{(r)}}$ can be replaced by a unique lift $\widetilde{h} \cdot e + X$ with $\widetilde{h} \in \bigoplus_{k=0}^{n-1} \mathcal{O}_{M^{(r)}} t_1^k$ and $(h - \widetilde{h})|_\mathcal{D} = 0$. If $h \cdot e + X$ is tangent to $\mathcal{D}$, then $(h - \widetilde{h})|_\mathcal{D} = 0$ is necessary and sufficient for $\widetilde{h} \cdot e + X$ to be tangent to $\mathcal{D}$.

(c) A generic point $p^{(r)} \in (\mathcal{B}^{(r)})_{reg}$ has a preimage $p \in (\mathcal{D}_{sing})_{reg}$ such that the projection of germs $pr_e : (\mathcal{D}_{sing}, p) \to (\mathcal{B}^{(r)}, p^{(r)})$ is an isomorphism (Corollary 4.5 (a)). A vector field $h \cdot e + X$, $X \in \mathcal{T}_{M^{(r)}}$, which is tangent to $\mathcal{D}_{reg}$ is also tangent to $(\mathcal{D}_{sing})_{reg}$. Then X is tangent to $(\mathcal{B}^{(r)})_{reg}$. One obtains the generators in (4.24) by projection to $\mathcal{T}_{M^{(r)}}$ of the generators in (4.23). □

The set $\mathcal{K}^{(3)} \subset \mathcal{K} \subset M$ is the set of points $p \in M$ such that (M, p) does not decompose into 1- and 2-dimensional germs of F-manifolds (section 4.2).

Theorem 4.13 *Let $(M, \circ, e, E)$ be a massive F-manifold with Euler field E of weight 1 and in standard form. Suppose that* $\mathrm{codim}\mathcal{K}^{(3)} \geq 2$.

Then the restricted bifurcation diagram $\mathcal{B}^{(r)}$ is a free divisor and (4.24) *is an equality.*

Proof: In view of Lemma 4.12 (c) it is sufficient to show that any vector field tangent to $\mathcal{B}^{(r)}$ lifts to a vector field tangent to $\mathcal{D}$.

The projection $pr_e : \mathcal{D} - \mathcal{B} \to M^{(r)} - \mathcal{B}^{(r)}$ is a covering of degree n. For any vector field $X \in \mathcal{T}_{M^{(r)}}$ there exists a unique function $h_X \in (pr_e)_* \mathcal{O}_{\mathcal{D}-\mathcal{B}}$ such that $\widetilde{h} \cdot e + X$ is tangent to $\mathcal{D} - \mathcal{B}$ if and only if $\widetilde{h}|_{\mathcal{D}-\mathcal{B}} = h_X$.

One has to show that h_X extends to a function in $(pr_e)_* \mathcal{O}_{\mathcal{D}}$ if $X \in \mathrm{Der}_{M^{(r)}}(\log \mathcal{B}^{(r)})$. Then the unique lift $\widetilde{h} \cdot e + X$ with $\widetilde{h} \in \bigoplus_{k=0}^{n-1} \mathcal{O}_{M^{(r)}} t_1^k$ and $\widetilde{h}|_{\mathcal{D}} = h_X$ is tangent to $\mathcal{D}$.

Let p be a point in the set

$$\left\{p \in (\mathcal{D}_{sing})_{reg} \mid p^{(r)} \in \left(\mathcal{B}^{(r)}\right)_{reg}, pr_e : (\mathcal{D}, p) \to \left(M^{(r)}, p^{(r)}\right) \text{ has degree } 2\right\}. \tag{4.28}$$

Then the germ $(\mathcal{D}, p)$ is the product of $(\mathbb{C}^{n-2}, 0)$ and the discriminant of the germ of an F-manifold of type $I_2(m)$ $(m \geq 2)$ (Remark 4.2 (v)). One can find coordinates $(t_1, t') = (t_1, \ldots, t_n)$ around $p \in M$ such that $(\mathcal{D}, p) \subset (M, p) \to (M^{(r)}, p^{(r)})$ corresponds to

$$\left(\left\{(t_1, t') \mid t_1^2 - t_2^m = 0\right\}, 0\right) \subset (\mathbb{C}^n, 0) \to (\mathbb{C}^{n-1}, 0), \; t \mapsto t'. \tag{4.29}$$

Then $(\mathcal{B}^{(r)}, p^{(r)}) \cong (\{t' \mid t_2 = 0\}, 0)$. Obviously the vector fields tangent to $(\mathcal{B}^{(r)}, p^{(r)})$ locally have lifts to vector fields tangent to $(\mathcal{D}, p)$. The function h_X of a field $X \in \mathrm{Der}_{M^{(r)}}(\log \mathcal{B}^{(r)})$ extends holomorphically to the set in (4.28). The complement in $\mathcal{D}$ of $\mathcal{D}_{reg} = \mathcal{D} - \mathcal{B}$ and of the set in (4.28) has codimension ≥ 2 because of $\mathrm{codim}\mathcal{K}^{(3)} \geq 2$. Therefore $h_X \in (pr_e)_* \mathcal{O}_{\mathcal{D}}$. □

4.4 Isomorphisms and modality of germs of F-manifolds

The following three results are applications of Theorem 3.19 for the Lyashko–Looijenga map. They will be proved together. The tuple $((M, p), \circ, e, \Lambda)$ denotes the germ of an F-manifold with the function germ $\Lambda : (M, p) \to \mathbb{C}^n$ as additional structure. A map germ $\varphi : (M, p) \to (M, p)$ respects Λ if $\Lambda \circ \varphi = \Lambda$.

Theorem 4.14 *The automorphism group of a germ (M, p) of a massive F-manifold is finite.*

Theorem 4.15 *Let $(M, \circ, e, F)$ be a massive F-manifold with generating function F and Lyashko–Looijenga map $\Lambda : M \to \mathbb{C}^n$. For any $p_1 \in M$ the set*

$$\{q \in M \mid ((M, p_1), \circ, e, \Lambda) \cong ((M, q), \circ, e, \Lambda)\}$$

is discrete and closed in M.

Corollary 4.16 *Let $(M, \circ, e, E)$ be a massive F-manifold with Euler field E of weight 1. For any $p_1 \in M$ the set*

$$\{q \in M \mid ((M, p_1), \circ, e, E) \cong ((M, q), \circ, e, E)\}$$

is discrete and closed in M.

Proof: Corollary 4.16 follows from Theorem 4.15. For Theorem 4.14, it suffices to regard an irreducible germ of a massive F-manifold. The automorphisms of an irreducible germ respect a given Lyashko–Looijenga map Λ because of Lemma 3.20 (a). So we may fix for Theorem 4.14 and Theorem 4.15 a massive F-manifold $(M, \circ, e)$ and a Lyashko–Looijenga map $\Lambda : M \to \mathbb{C}^n$.

The set

$$\Delta := \{(p, p') \in M \times M \mid \Lambda(p) = \Lambda(p')\}$$

has a reduced complex structure. It is a subset of $(M - \mathcal{B}) \times (M - \mathcal{B}) \cup \mathcal{B} \times \mathcal{B}$ and the intersection $\Delta \cap (M - \mathcal{B}) \times (M - \mathcal{B})$ is smooth of dimension n. This follows from Theorem 3.19.

Now consider an isomorphism $\varphi : ((M, p), \circ, e, \Lambda) \to ((M, p'), \circ, e, \Lambda)$. The graph germ

$$(\mathcal{G}(\varphi), (p, p')) := (\{(q, \varphi(q)) \in M \times M \mid q \text{ near } p\}, (p, p'))$$

is a smooth analytic germ of dimension n and is contained in the germ $(\Delta, (p, p'))$. It meets $\Delta \cap (M - \mathcal{B}) \times (M - \mathcal{B})$. Because of the purity of the dimension of an irreducible analytic germ, it is an irreducible component of the analytic germ $(\Delta, (p, p'))$. One can recover the map germ φ from the graph germ $(\mathcal{G}(\varphi), (p, p'))$. The germ $(\Delta, (p, p'))$ consists of finitely many irreducible components. The case $p = p'$ together with the remarks at the beginning of the proof give Theorem 4.14.

For Theorem 4.15, we assume that there is an infinite sequence $(p_i, \varphi_i)_{i\in\mathbb{N}}$ of different points $p_i \in M$ and map germs

$$\varphi_i : ((M, p_1), \circ, e, \Lambda) \stackrel{\cong}{\to} ((M, p_i), \circ, e, \Lambda)$$

and one accumulation point $p_\infty \in M$. The set $\overline{\Delta - \mathcal{B} \times \mathcal{B}}$ is analytic of pure dimension n. It contains the germs $(\mathcal{G}(\varphi_i), (p_1, p_i))$ and the point (p_1, p_∞).

We can choose a suitable open neighbourhood U of (p_1, p_∞) in $M \times M$ and a stratification

$$\bigcup_\alpha S_\alpha = U \cap \overline{\Delta - \mathcal{B} \times \mathcal{B}}$$

of $U \cap \overline{\Delta - \mathcal{B} \times \mathcal{B}}$ which consists of finitely many disjoint smooth connected constructible sets S_α and satisfies the boundary condition: *The boundary $\overline{S_\alpha} - S_\alpha$ of a stratum S_α is a union of other strata.*

The germ $(\mathcal{G}(\varphi_i), (p_1, p_i))$ is an n-dimensional irreducible component of the n-dimensional germ $(\overline{\Delta - \mathcal{B} \times \mathcal{B}}, (p_1, p_i))$. There is a unique n-dimensional stratum whose closure contains $(\mathcal{G}(\varphi_i), (p_1, p_i))$. If $(p_1, p_i) \in S_\alpha$ then this together with the boundary condition implies $(S_\alpha, (p_1, p_i)) \subset (\mathcal{G}(\varphi_i), (p_1, p_i))$. The germ $(\mathcal{G}(\varphi_i), (p_1, p_i))$ is the graph of the isomorphism φ_i. Therefore it intersects the germ $(\{p_1\} \times M, (p_1, p_i))$ only in (p_1, p_i); the same holds for $(S_\alpha, (p_1, p_i))$.

Now there exists at least one stratum S_{α_0} which contains infinitely many of the points (p_1, p_i). The intersection of the analytic sets $\overline{S_{\alpha_0}}$ and $U \cap (\{p_1\} \times M)$ contains these points as isolated points. This is impossible. The above assumption was wrong. □

In singularity theory there are the notions of μ-constant stratum and (proper) modality of an isolated hypersurface singularity. One can define versions of them for the germ (M, p) of an F-manifold $(M, \circ, e)$ (massive or not massive): The *μ-constant stratum* (S_μ, p) is the analytic germ of points $q \in M$ such that the eigenspace decompositions of T_qM and T_pM have the same partition (cf. Proposition 2.5).

The idempotent fields $e_1, \dots, e_l$ of the decomposition $(M, p) = \prod_{k=1}^l (M_k, p)$ into irreducible germs of F-manifolds commute and satisfy $\mathrm{Lie}_{e_i}(\circ) = 0 \cdot \circ$. So the germs (M, q) of points q in one integral manifold of $e_1, \dots, e_l$ are isomorphic. This motivates the definition of the *modality*:

$$\mathrm{mod}_\mu(M, p) := \dim(S_\mu, p) - l. \tag{4.30}$$

Let $(S_\mu^{[k]}, p)$ denote the μ-constant stratum of (M_k, p); Then Theorem 2.11 implies

$$(S_\mu, p) = \prod_k \left(S_\mu^{[k]}, p\right) \text{ and } \tag{4.31}$$

$$\operatorname{mod}(M, p) = \sum_k \operatorname{mod}_\mu(M_k, p). \tag{4.32}$$

For massive F-manifolds, Theorem 4.15 and Lemma 3.20 give more information:

Corollary 4.17 *Let $(M, p) = \prod_{k=1}^{l}(M_k, p)$ be the germ of a massive F-manifold and $\Lambda : (M, p) \to \mathbb{C}^n$ a Lyashko–Looijenga map.*

(a) There exist a representative S_μ of the μ-constant stratum (S_μ, p), a neighbourhood $U \subset \mathbb{C}^l$ of 0 and an isomorphism

$$\psi : S_\mu \to (S_\mu \cap \Lambda^{-1}(\Lambda(p))) \times U \tag{4.33}$$

such that $\psi^{-1}(\{q\} \times U)$ is the integral manifold of $e_1, \ldots, e_l$ which contains q. Any subset of points in $S_\mu \cap \Lambda^{-1}(\Lambda(p))$ with isomorphic germs of F-manifolds is discrete and closed.

(b)

$$\operatorname{mod}_\mu(M, p) = \dim(S_\mu \cap \Lambda^{-1}(\Lambda(p)), p), \tag{4.34}$$

$$\sup(\operatorname{mod}_\mu(M, q) \mid q \text{ near } p) = \dim(\Lambda^{-1}(\Lambda(p)), p). \tag{4.35}$$

Proof: (a) For $l = 1$, the existence of ψ follows from the e-invariance of S_μ and from $e(-\frac{1}{n}\Lambda_1) = 1$ (Lemma 3.20 (b)).

For arbitrary l, one uses (4.31) and Lemma 3.20 (c): the maps Λ and $(\Lambda^{[1]}, \ldots, \Lambda^{[l]})$ have the same germs of fibres, especially

$$S_\mu \cap \Lambda^{-1}(\Lambda(p)) = \prod_k S_\mu^{[k]} \cap \Lambda^{[k]-1}(\Lambda^{[k]}(p)). \tag{4.36}$$

A germ (M, q) has only a finite number of Lyashko–Looijenga maps with fixed value at q (Lemma 3.20 (a)). The finiteness statement in Corollary 4.17 (a) follows from this and Theorem 4.15.

(b) Equation (4.34) follows from (a). A representative of the germ $(\Lambda^{-1}(\Lambda(p)), p)$ is stratified into constructible subsets which consist of the points q with the same partition for the eigenspace decomposition of $T_q M$ (Proposition 2.5). A point $q \in \Lambda^{-1}(\Lambda(p))$ in an open stratum with maximal dimension satisfies

$$\begin{aligned}\operatorname{mod}_\mu(M, q) &= \dim(S_\mu(q) \cap \Lambda^{-1}(\Lambda(p)), q) \\ &= \dim(\Lambda^{-1}(\Lambda(p)), q) = \dim(\Lambda^{-1}(\Lambda(p)), p).\end{aligned} \tag{4.37}$$

This shows

$$\sup(\text{mod}_\mu(M, q) \mid q \in \Lambda^{-1}(\Lambda(p)) \text{ near } p) = \dim(\Lambda^{-1}(\Lambda(p)), p). \quad (4.38)$$

The upper semicontinuity of the fibre dimension of Λ gives (4.35). □

Remark 4.18 Gabrielov [**Ga**] proved in the case of isolated hypersurface singularities the upper semicontinuity of the modality,

$$\text{mod}_\mu(M, q) \leq \text{mod}_\mu(M, p) \quad \text{for } q \text{ near } p \quad (4.39)$$

(and the equality with another version of modality which was defined by Arnold). He used (4.34), (4.35), and a result of himself, Lazzeri, and Lê, which, translated to the F-manifold of a singularity (section 5.1), says:

$$(S_\mu \cap \Lambda^{-1}(\Lambda(p)), p) = (\Lambda^{-1}(\Lambda(p)), p). \quad (4.40)$$

The inequality (4.39) is an immediate consequence of (4.34), (4.35) and (4.40). But for other F-manifolds (4.40) and (4.39) are not clear.

In the case of the simple hypersurface singularities, the base of the semiuniversal unfolding is an F-manifold $M \cong \mathbb{C}^n$ and the map $\Lambda : M - \mathcal{B} \to \mathbb{C}^n - \mathcal{D}^{(n)}$ is a finite covering. Therefore the complement $M - \mathcal{B}$ is a $K(\pi, 1)$ space and the fundamental group is a subgroup of finite index of the braid group $Br(n)$. This is the application of Looijenga [**Lo1**] and Lyashko [**Ar1**] of the map Λ, which led to the name Lyashko–Looijenga map.

It can be generalized to F-manifolds. We call a massive F-manifold M *simple* if $\text{mod}_\mu(M, p) = 0$ for all $p \in M$. This fits with the notions of simple hypersurface singularities, simple boundary singularities, and simple Lagrange maps ([**Gi2**, 1.3, p. 3251]).

A distinguished class of simple F-manifolds are the F-manifolds of the finite Coxeter groups (section 5.3 and [**Lo1**][**Ar1**][**Ly1**][**Ly3**][**Gi2**]). There are other examples (Proposition 5.32 and Remark 5.33).

A Lyashko–Looijenga map of a massive F-manifold is locally a branched covering if and only if M is simple ((4.35) and Theorem 3.19). A detailed proof of the following result had been given by Looijenga [**Lo1**, Theorem 2.1] (cf. also [**Gi2**, 1.4, Theorem 5]).

Theorem 4.19 *Let $(M, p) = (\mathbb{C}^n, p)$ be the germ of a simple F-manifold with fixed coordinates. Then, if $\varepsilon < \varepsilon_0$ for some ε_0, the space $\{z \in \mathbb{C}^n \mid |z| < \varepsilon\} - \mathcal{B}$ is a $K(\pi, 1)$ space. Its fundamental group is a subgroup of finite index of the braid group $Br(n)$.*

4.5 Analytic spectrum embedded differently

The analytic spectrum $L \subset T^*M$ of an F-manifold determines the multiplication on $\mathcal{T}_M$ via the isomorphism ((3.1) and (2.2))

$$\mathbf{a} : \mathcal{T}_M \to \pi_* \mathcal{O}_L, \quad X \mapsto \alpha(\widetilde{X})|_L. \tag{4.41}$$

One can generalize this and replace L, T^*M, and α by other spaces and other 1–forms. This allows F-manifolds to be found in natural geometric situations and to be encoded in an economic way. Corollary 4.21 and Definition 4.23 are the two most interesting special cases of Theorem 4.20.

Theorem 4.20 *Let the following data be given:*
manifolds Z and M, where M is connected and n-dimensional;
a surjective map $\pi_Z : Z \to M$ which is everywhere a submersion;
an everywhere n-dimensional reduced subvariety $C \subset Z$ such that the restriction $\pi_C : C \to M$ is finite;
a 1–form α_Z on Z with the property:

$$\text{any local lift } \widetilde{X} \in \mathcal{T}_Z \text{ of the zero vector field } 0 \in \mathcal{T}_M \text{ satisfies } \alpha_Z(\widetilde{X})|_C = 0. \tag{4.42}$$

Then

(a) The map

$$\mathbf{a}_C : \mathcal{T}_M \to (\pi_C)_* \mathcal{O}_C, \quad X \mapsto \alpha_Z(\widetilde{X})|_C \tag{4.43}$$

is welldefined; here $\widetilde{X} \in \mathcal{T}_Z$ is any lift of X to a neighbourhood of C in Z.

(b) The image $L \subset T^*M$ of the map

$$\mathbf{q} : C \to T^*M, \quad z \mapsto \mathbf{q}(z) = (X \mapsto \mathbf{a}_C(X)(z)) \in T^*_{\pi_C(z)}M \tag{4.44}$$

is a (reduced) variety. The map $\mathbf{q} : C \to L$ is a finite map, the projections $\pi : L \to M$ and $\pi_C = \pi \circ \mathbf{q}$ are branched coverings. The composition of the maps $\hat{\mathbf{q}} : \pi_* \mathcal{O}_L \to (\pi_C)_* \mathcal{O}_C$ and

$$\mathbf{a} : \mathcal{T}_M \to \pi_* \mathcal{O}_L, \quad X \mapsto \alpha(\widetilde{X})|_L \tag{4.45}$$

is $\mathbf{a}_C = \hat{\mathbf{q}} \circ \mathbf{a}$. All three are $\mathcal{O}_M$-module homomorphisms.

(c) The 1–forms α and α_Z satisfy $(\mathbf{q}^*\alpha)|_{C_{reg}} = \alpha_Z|_{C_{reg}}$. Therefore L is a Lagrange variety if and only if $\alpha_Z|_{C_{reg}}$ is exact.

(d) The map $\mathbf{a} : \mathcal{T}_M \to \pi_* \mathcal{O}_L$ is an isomorphism if and only if

(i) the map $\mathbf{a}_C$ is injective,

(ii) its image $\mathbf{a}_C(\mathcal{T}_M) \subset (\pi_C)_*\mathcal{O}_C$ *is multiplication invariant,*
(iii) the image $\mathbf{a}_C(\mathcal{T}_M)$ *contains the unit* $1_C \in (\pi_C)_*\mathcal{O}_C$.

In this case $\mathbf{a}_C : \mathcal{T}_M \to (\pi_C)_*\mathcal{O}_C$ *induces a (commutative and associative and) generically semisimple multiplication on* $\mathcal{T}_M$ *with global unit field and with analytic spectrum* L.

(e) The map $\mathbf{a}_C : \mathcal{T}_M \to (\pi_C)_*\mathcal{O}_C$ *provides* M *with the structure of a massive F-manifold if and only if* $\alpha_Z|_{C_{reg}}$ *is exact and the conditions (i)–(iii) in (d) are satisfied.*

Proof: (a) This follows from (4.42).

(b) The equality $\dim C = n = \dim M$ and π_C finite imply that π_C is open. M is connected, thus π_C is a branched covering. Using local coordinates for M and T^*M one sees that $\mathbf{q} : C \to T^*M$ is an analytic map. The equality $\pi_C = \pi \circ \mathbf{q}$ is clear and shows that $\mathbf{q}$ is finite. Then $L = \mathbf{q}(C)$ is a variety and π is a branched covering. The equality $\mathbf{a}_C = \hat{\mathbf{q}} \circ \mathbf{a}$ follows from the definition of $\mathbf{q}$.

(c) There is an open subset $M^{(0)} \subset M$ with analytic complement $M - M^{(0)}$ such that $\pi_C^{-1}(M^{(0)}) \subset C$ and $\pi^{-1}(M^{(0)}) \subset L$ are smooth, $\pi_C : \pi_C^{-1}(M^{(0)}) \to M^{(0)}$ and $\pi : \pi^{-1}(M^{(0)}) \to M^{(0)}$ are coverings and $\mathbf{q} : \pi_C^{-1}(M^{(0)}) \to \pi^{-1}(M^{(0)})$ is a covering on each component of $\pi^{-1}(M^{(0)})$. Now $\mathbf{a}_C = \hat{\mathbf{q}} \circ \mathbf{a}$ implies $\mathbf{q}^*\alpha|\pi_C^{-1}(M^{(0)}) = \alpha_Z|\pi_C^{-1}(M^{(0)})$.

(d) The map $\hat{\mathbf{q}} : \pi_*\mathcal{O}_L \to (\pi_C)_*\mathcal{O}_C$ is an injective homomorphism of $\mathcal{O}_M$-algebras. If $\mathbf{a} : \mathcal{T}_M \to \pi_*\mathcal{O}_L$ is an isomorphism then (i)–(iii) are obviously satisfied.

Suppose that (i)–(iii) are satisfied. Then $\mathbf{a} : \mathcal{T}_M \to \pi_*\mathcal{O}_L$ is injective with multiplication invariant image $\mathbf{a}(\mathcal{T}_M) \subset \pi_*\mathcal{O}_L$ and with $1_L \in \mathbf{a}(\mathcal{T}_M)$. The maps $\mathbf{a}$ and $\mathbf{a}_C$ induce the same (commutative and associative) multiplication with global unit field on TM.

We have to show that this multiplication is generically semisimple with analytic spectrum L. Then $\mathbf{a} : \mathcal{T}_M \to \pi_*\mathcal{O}_L$ is an isomorphism and the proof of (d) is complete.

If for each $p \in M$ the linear forms in $\pi^{-1}(p) \subset T_p^*M$ would generate a subspace of T_p^*M of dimension $< n$ then $\mathbf{a}$ would not be injective. So, for a generic point $p \in M$ there exist n elements in $\pi^{-1}(p) \subset T_p^*M$ which form a basis of T_p^*M. We claim that $\pi^{-1}(p)$ contains no elements other than these: $\pi^{-1}(p)$ does not contain $0 \in T_p^*M$ because of $1_L \in \mathbf{a}(\mathcal{T}_M)$. From the multiplication invariance of $\mathbf{a}(\mathcal{T}_M)$ one derives easily that $\pi^{-1}(p)$ does not contain any further elements.

This extends to a small neighbourhood U of the generic point $p \in M$: $\pi^{-1}(U)$ consists of n sheets which form a basis of sections of T^*M; the map

$\mathbf{a}|_U : \mathcal{T}_U \to \pi_*(\pi^{-1}(U))$ is an isomorphism and induces a semisimple multiplication on TM with analytic spectrum $\pi^{-1}(U)$.

Then L is the analytic spectrum of the multiplication on $\mathcal{T}_M$ because M is connected.

(e) By (c) and (d) and Theorem 3.2. □

In Theorem 4.20 the map $\pi : L \to M$ has degree n, but $\pi_C : C \to M$ can have degree $> n$; and even if $\pi_C : C \to M$ has degree n the map $\mathbf{q} : C \to L$ does not need to be an isomorphism. Examples will be discussed below (Examples 4.24, Lemma 5.17). But the most important special case is the following.

Corollary 4.21 *Let $Z, M, \pi_Z, C \subset Z, \alpha_Z, \mathbf{a}_C, L$, and $\mathbf{q}$ be as in Theorem 4.20. Suppose that $\alpha_Z|_{C_{reg}}$ is exact and $\mathbf{a}_C : \mathcal{T}_M \to (\pi_C)_*\mathcal{O}_C$ is an isomorphism.*

Then $\mathbf{q} : C \to L$ is an isomorphism and $\mathbf{a}_C = \hat{\mathbf{q}} \circ \mathbf{a}$ provides M with the structure of a massive F-manifold with analytic spectrum L.

Proof: Theorem 4.20 (e) gives all of the corollary except for the isomorphism $\mathbf{q} : C \to L$. This follows from the isomorphism $\hat{\mathbf{q}} : \pi_*\mathcal{O}_L \to (\pi_C)_*\mathcal{O}_C$ and a universal property of the analytic spectrum. □

One can encode an irreducible germ of a massive F-manifold with data as in Corollary 4.21 such that the dimension of Z is minimal.

Lemma 4.22 *Let (M, p) be an irreducible germ of a massive n-dimensional F-manifold. Let $\mathbf{m} \subset T_pM$ denote the maximal ideal in T_pM.*

(a) Then $\dim Z \geq n + \dim \mathbf{m}/\mathbf{m}^2$ *for any data as in Corollary 4.21.*

(b) There exist data as in Corollary 4.21 for (M, p) with $\dim Z = n + \dim \mathbf{m}/\mathbf{m}^2$ *(the construction will be given in the proof).*

Proof: (a) $\pi_C^{-1}(p) = \pi_Z^{-1}(p) \cap C$ consists of one fat point with structure ring T_pM. Its embedding dimension $\dim \mathbf{m}/\mathbf{m}^2$ is bounded by the dimension $\dim \pi_Z^{-1}(p) = \dim Z - n$ of the smooth fibre $\pi_Z^{-1}(p)$.

(b) One can choose coordinates $(t_1, \ldots, t_n) = (t_1, t') = t$ for (M, p) with $e = \frac{\partial}{\partial t_1}$ as usual and with

$$\bigoplus_{i=2}^{n} \mathbb{C} \cdot \left[\frac{\partial}{\partial t_i}\right] = \mathbf{m} \subset T_pM \quad \text{and} \tag{4.46}$$

$$\bigoplus_{i=m+1}^{n} \mathbb{C} \cdot \left[\frac{\partial}{\partial t_i}\right] = \mathbf{m}^2 \subset T_pM \tag{4.47}$$

for $m = 1 + \dim \mathbf{m}/\mathbf{m}^2$. The dual coordinates on (T^*M, T_p^*M) are $y_1, \ldots, y_n$, the analytic spectrum is (cf. (2.1))

$$L = \left\{ (y, t) \mid y_1 = 1,\ y_i y_j = \sum a_{ij}^k(t') y_k \right\}. \tag{4.48}$$

Because of (4.47) there exist functions $b_i \in \mathbb{C}\{t'\}[y_2, \ldots, y_m]$ with

$$y_i|_L = b_i(y_2, \ldots, y_m, t')|_L \quad \text{for } i = m+1, \ldots, n. \tag{4.49}$$

We identify (M, p) and $(\mathbb{C}^n, 0)$ using $(t_1, \ldots, t_n)$ and define $(Z, 0) = (\mathbb{C}^{m-1} \times \mathbb{C}^n, 0)$. The embedding

$$\iota : (Z, 0) = (\mathbb{C}^{m-1} \times \mathbb{C}^n, 0) \hookrightarrow T^*M, \tag{4.50}$$

$$(x_1, \ldots, x_{m-1}, t) \mapsto (y, t) = (1, x_1, \ldots, x_{m-1}, b_{m+1}(x, t'), \ldots, b_n(x, t'), t)$$

provides canonical choices for the other data,

$$\pi_Z : (Z, 0) \to (M, p), (x, t) \mapsto t, \tag{4.51}$$

$$C = \iota^{-1}(L), \tag{4.52}$$

$$\alpha_Z = \iota^* \alpha = \mathrm{d}t_1 + \sum_{i=2}^{m} x_{i-1} \mathrm{d}t_i + \sum_{i=m+1}^{n} b_i(x, t') \mathrm{d}t_i. \tag{4.53}$$

The conditions in Corollary 4.21 are obviously satisfied. □

The notion of a generating family for a Lagrange map ([**AGV1**, ch. 19], [**Gi2**, 1.4]) motivates us to single out another special case of Theorem 4.20.

Definition 4.23 Let Z, M, π_Z, C, α_Z, and $\mathbf{a}_C$ be as in Theorem 4.20 with $\alpha_Z|_{C_{reg}}$ exact and $\mathbf{a}_C : \mathcal{T}_M \to (\pi_C)_* \mathcal{O}_C$ injective with multiplication invariant image $\mathbf{a}_C(\mathcal{T}_M) \supset \{1_C\}$. These data yield a massive F-manifold $(M, \circ, e)$.

A function $F : Z \to \mathbb{C}$ is a *generating family* for this F-manifold if $\alpha_Z = \mathrm{d}F$ and if C is the critical set of the map $(F, \pi_Z) : Z \to \mathbb{C} \times M$.

There are two reasons for the name *generating family*:

(1) The function F is considered as a family of functions on the fibres $\pi_Z^{-1}(p)$, $p \in M$.

(2) The restriction of F to C is the lift of a generating function $\widetilde{F} : L \to \mathbb{C}$, i.e. $F = \widetilde{F} \circ \mathbf{q}$; so the 1-graph of F as a multivalued function on M is L.

In the case of a generating family the conditions (4.42) and α_Z exact are obvious. The most difficult condition is the multiplication invariance of $\mathbf{a}(\mathcal{T}_M)$.

It is not clear whether for any massive F-manifold M data (Z, π_Z, F) as in Definition 4.23 exist. But even many nonisomorphic data often exist. We illustrate this for the 2-dimensional germs $I_2(m)$ of F-manifolds (section 4.1).

Examples 4.24 Always $(Z,0)=(\mathbb{C}\times\mathbb{C}^2,0)$ and $(M,p)=(\mathbb{C}^2,0)$ with projection $\pi_Z:(Z,0)\to(M,0)$, $(x,t_1,t_2)\mapsto(t_1,t_2)$ and $e:=\delta_1:=\frac{\partial}{\partial t_1}$, $\delta_2:=\frac{\partial}{\partial t_2}$.

(a) $C=\{(x,t)\mid x^{m-2}-t_2^2=0\}$, $\alpha_Z=\mathrm{d}t_1+x\mathrm{d}t_2$.
These are data as in Corollary 4.21 for $I_2(m)$.

(b) Generating family $F=t_1+\int_0^x(t_2-u^2)^k\mathrm{d}u$ $(k\geq 1)$, $C=\{(x,t)\mid t_2-x^2=0\}$, $\alpha_Z|_C=\mathrm{d}F|_C=(\mathrm{d}t_1+c\cdot x^{2k-1}\mathrm{d}t_2)|_C$ for some $c\in\mathbb{C}-\{0\}$, $\mathbf{a}_C(\delta_2)\cdot\mathbf{a}_C(\delta_2)=c^2\cdot t_2^{2k-1}\cdot 1_C$.
These are data as in Definition 4.23 for $I_2(2k+1)$, the map $\pi_C:C\to M$ has degree 2, the map $\mathbf{q}:C\to L$ is the normalization and the maximalization of L (cf. [**Fi**, 2.26 and 2.29] for these notions).

(c) Generating family $F=t_1+x^{k+1}t_2-\frac{k+1}{k+2}x^{k+2}$ $(k\geq 1)$, $C=\{(x,t)\mid(t_2-x)x=0\}$, $\alpha_Z|_{C_{reg}}=\mathrm{d}F|_{C_{reg}}=(\mathrm{d}t_1+x^{k+1}\mathrm{d}t_2)|_{C_{reg}}$, $\mathbf{a}_C(\delta_2-\frac{1}{2}t_2^{k+1}\cdot\delta_1)^2=\frac{1}{4}t_2^{2k+2}\cdot 1_C$. These are data as in Definition 4.23 for $I_2(2k+4)$, the map $\pi_C:C\to M$ has degree 2, the map $\mathbf{q}:C\to L$ is the maximalization of L (for the missing case $I_2(4)$ compare Lemma 5.17).

(d) Generating family $F=t_1+\int_0^x(u^2-t_2)^k u\mathrm{d}u$ $(k\geq 1)$, $C=\{(x,t)\mid(t_2-x^2)x=0\}$, $\alpha_Z|_{C_{reg}}=\mathrm{d}F|_{C_{reg}}=(\mathrm{d}t_1+c\cdot x^{2k}\mathrm{d}t_2)|_{C_{reg}}$ for some $c\in\mathbb{C}-\{0\}$, $\mathbf{a}_C(\delta_2-\frac{1}{2}ct_2^k\cdot\delta_1)^2=\frac{1}{4}c^2t_2^{2k}\cdot 1_C$. These are data as in Definition 4.23 for $I_2(2k+2)$, the map $\pi_C:C\to M$ has degree 3, the map $\mathbf{q}:C\to L$ covers one component with degree 1, the other with degree 2.

Chapter 5
Singularities and Coxeter groups

In this section several families of massive F-manifolds which come from singularity theory are studied. The most important ones are the base spaces of semiuniversal unfoldings of hypersurface singularities. Three reasons for this are: (1) hypersurface singularities and their unfoldings are so universal objects; (2) their F-manifolds can be enriched to Frobenius manifolds (part 2); (3) one has a 1-1 correspondence between irreducible germs of massive F-manifolds with smooth analytic spectrum and stable right equivalence classes of singularities (Theorem 5.6). This is covered in section 5.1. The discussion of boundary singularities and their F-manifolds in section 5.2 is quite similar.

Sections 5.3 and 5.4 are devoted to finite irreducible Coxeter groups and their F-manifolds and Frobenius manifolds. The discriminant in the complex orbit space induces an F-manifold structure on the orbit space just as in Corollary 4.6. This follows independently from work of Dubrovin and from results in singularity theory by Brieskorn, Arnold, O.P. Shcherbak, Givental. We extend work of Givental in order to characterize these F-manifolds (Theorems 5.20, 5.21, 5.22) and use this to prove a conjecture of Dubrovin about the corresponding Frobenius manifolds (Theorem 5.26).

In section 5.5 other families of F-manifolds with quite different properties are constructed. A start is made on the classification of 3-dimensional germs of massive F-manifolds.

5.1 Hypersurface singularities

A distinguished class of germs of massive F-manifolds is related to isolated hypersurface singularities: the base space of a semiuniversal unfolding of an isolated hypersurface singularity is an irreducible germ of a massive F-manifold with smooth analytic spectrum (Theorem 5.3). In fact, there is a 1-1 correspondence between such germs of F-manifolds and singularities up to stable right equivalence (Theorem 5.6).

The structure of an F-manifold on the base space has excellent geometric implications and interpretations (Theorem 5.4, Remarks 5.5). Many of these have been known for a long time from different points of view. The concept of an F-manifold unifies them. On the other hand, for much of the general treatment of F-manifolds in this book the singularity case has been the model case.

An isolated hypersurface singularity is a holomorphic function germ $f : (\mathbb{C}^m, 0) \to (\mathbb{C}, 0)$ with an isolated singularity at 0. Its Milnor number $\mu \in \mathbb{N}$ is the dimension of the Jacobi algebra $\mathcal{O}_{\mathbb{C}^m,0}/(\frac{\partial f}{\partial x_1}, \ldots, \frac{\partial f}{\partial x_m}) = \mathcal{O}_{\mathbb{C}^m,0}/J_f$.

The notion of an unfolding of an isolated hypersurface singularity goes back to Thom and Mather. An *unfolding* of f is a holomorphic function germ $F : (\mathbb{C}^m \times \mathbb{C}^n, 0) \to (\mathbb{C}, 0)$ such that $F|_{\mathbb{C}^m \times \{0\}} = f$. The parameter space will be written as $(M, 0) = (\mathbb{C}^n, 0)$.

The *critical space* $(C, 0) \subset (\mathbb{C}^m \times M, 0)$ of the unfolding $F = F(x_1, \ldots, x_m, t_1, \ldots, t_n)$ is the critical space of the map $(F, pr) : (\mathbb{C}^m \times M, 0) \to (\mathbb{C} \times M, 0)$. It is the zero set of the ideal

$$J_F := \left(\frac{\partial F}{\partial x_1}, \ldots, \frac{\partial F}{\partial x_m} \right) \tag{5.1}$$

with the complex structure $\mathcal{O}_{C,0} = \mathcal{O}_{\mathbb{C}^m \times M,0}/J_F|_{(C,0)}$.

The intersection $C \cap (\mathbb{C}^m \times \{0\}) = \{0\}$ is a point and $(C, 0)$ is a complete intersection of dimension n. Therefore the projection $pr : (C, 0) \to (M, 0)$ is finite and flat with degree μ and $\mathcal{O}_{C,0}$ is a free $\mathcal{O}_{M,0}$-module of rank μ.

A kind of *Kodaira–Spencer map* is the $\mathcal{O}_{M,0}$-linear map

$$\mathbf{a}_C : \mathcal{T}_{M,0} \to \mathcal{O}_{C,0}, \quad X \mapsto \widetilde{X}(F)|_{(C,0)} \tag{5.2}$$

where $\widetilde{X}$ is any lift of $X \in \mathcal{T}_{M,0}$ to $(\mathbb{C}^m \times M, 0)$. Dividing out the submodules $\mathbf{m}_{M,0} \cdot \mathcal{T}_{M,0}$ and $\mathbf{m}_{M,0} \cdot \mathcal{O}_{C,0}$ one obtains the *reduced Kodaira–Spencer map*

$$\mathbf{a}_C|_0 : T_0M \to \mathcal{O}_{\mathbb{C}^m,0}/J_f. \tag{5.3}$$

All these objects are independent of the choice of coordinates. In fact, they even behave well with respect to morphisms of unfoldings.

There are several possibilities to define morphisms of unfoldings (cf. Remark 5.2 (iv)). We need the following.

Let $F_i : (\mathbb{C}^m \times M_i, 0) \to (\mathbb{C}, 0)$, $i = 1, 2$, be two unfoldings of f with projections $pr_i : (\mathbb{C}^m \times M_i, 0) \to (M_i, 0)$, critical spaces C_i, and Kodaira–Spencer maps $\mathbf{a}_{C_i}$. A *morphism* from F_1 to F_2 is a pair (ϕ, ϕ_{base}) of map germs such that the following diagram commutes,

$$\begin{array}{ccc} (\mathbb{C}^m \times M_1, 0) & \xrightarrow{\phi} & (\mathbb{C}^m \times M_2, 0) \\ \downarrow{\scriptstyle pr_1} & & \downarrow{\scriptstyle pr_2} \\ (M_1, 0) & \xrightarrow{\phi_{base}} & (M_2, 0), \end{array} \tag{5.4}$$

and

$$\phi|_{\mathbb{C}^m \times \{0\}} = id, \tag{5.5}$$

$$F_1 = F_2 \circ \phi \tag{5.6}$$

hold. One says that F_1 is *induced* by (ϕ, ϕ_{base}) from F_2.

The definition of critical spaces is compatible with the morphism (ϕ, ϕ_{base}), that is, $\phi^* J_{F_2} = J_{F_1}$ and $(C_1, 0) = \phi^{-1}((C_2, 0))$. Also the Kodaira–Spencer maps behave well: the $\mathcal{O}_{M_1,0}$-linear maps

$$d\phi_{base} : \mathcal{T}_{M_1,0} \to \mathcal{O}_{M_1,0} \otimes_{\mathcal{O}_{M_2,0}} \mathcal{T}_{M_2,0}, \tag{5.7}$$

$$\mathbf{a}_{C_2} : \mathcal{O}_{M_1,0} \otimes_{\mathcal{O}_{M_2,0}} \mathcal{T}_{M_2,0} \to \mathcal{O}_{M_1,0} \otimes_{\mathcal{O}_{M_2,0}} \mathcal{O}_{C_2,0}, \tag{5.8}$$

$$\phi^*|_{(C_2,0)} : \mathcal{O}_{M_1,0} \otimes_{\mathcal{O}_{M_2,0}} \mathcal{O}_{C_2,0} \to \mathcal{O}_{C_1,0} \tag{5.9}$$

are defined in the obvious way; their composition is

$$\mathbf{a}_{C_1} = \phi^*|_{(C_2,0)} \circ \mathbf{a}_{C_2} \circ d\phi_{base}. \tag{5.10}$$

Formula (5.9) restricts to the identity on the Jacobi algebra of f because of (5.5). Therefore the reduced Kodaira–Spencer maps satisfy

$$\mathbf{a}_{C_1}|_0 = \mathbf{a}_{C_2}|_0 \circ d\phi_{base}|_0. \tag{5.11}$$

An unfolding of f is *versal* if any unfolding is induced from it by a suitable morphism. A versal unfolding $F : (\mathbb{C}^m \times M, 0) \to (\mathbb{C}, 0)$ is *semiuniversal* if the dimension of the parameter space $(M, 0)$ is minimal. Semiuniversal unfoldings of an isolated hypersurface singularity exist by the work of Thom and Mather. Detailed proofs can nowadays be found at many places, e.g. [**Was**][**AGV1**, ch. 8].

Theorem 5.1 *An unfolding* $F : (\mathbb{C}^m \times M, 0) \to (\mathbb{C}, 0)$ *of an isolated hypersurface singularity* $f : (\mathbb{C}^m, 0) \to (\mathbb{C}, 0)$ *is versal if and only if the reduced Kodaira–Spencer map* $\mathbf{a}_C|_0 : T_0M \to \mathcal{O}_{\mathbb{C}^m,0}/J_f$ *is surjective. It is semiuniversal if and only if* $\mathbf{a}_C|_0$ *is an isomorphism.*

Remarks 5.2 (i) Because of the lemma of Nakayama $\mathbf{a}_C|_0$ is surjective (an isomorphism) if and only if $\mathbf{a}_C$ is surjective (an isomorphism).

(ii) A convenient choice of a semiuniversal unfolding $F : (\mathbb{C}^m \times \mathbb{C}^\mu, 0) \to (\mathbb{C}, 0)$ is $F(x_1, \ldots, x_m, t_1, \ldots, t_\mu) = f + \sum_{i=1}^{\mu} m_i t_i$, where $m_1, \ldots, m_\mu \in \mathcal{O}_{\mathbb{C}^m,0}$ represent a basis of the Jacobi algebra of f, preferably with $m_1 = 1$.

(iii) The critical space of an unfolding $F : (\mathbb{C}^m \times \mathbb{C}^n, 0) \to (\mathbb{C}, 0)$ is reduced and smooth if and only if the matrix $(\frac{\partial^2 F}{\partial x_i \partial x_j}, \frac{\partial^2 F}{\partial x_i \partial t_k})(0)$ has maximal rank m. This is satisfied for versal unfoldings.

(iv) In the literature (e.g. [**Was**]) one often finds a slightly different notion of morphisms of unfoldings: An (r)-*morphism* between unfoldings F_1 and F_2 as above is a triple $(\phi, \phi_{base}, \tau)$ of map germs ϕ and ϕ_{base} with (5.4) and (5.5) and $\tau : (M_1, 0) \to (\mathbb{C}, 0)$ with (5.6) replaced by

$$F_1 = F_2 \circ \phi + \tau. \tag{5.12}$$

The (r)-*versal* and (r)-*semiuniversal* unfoldings are defined analogously. They exist because of the following fact [**Was**]: an unfolding $F : (\mathbb{C}^m \times M, 0) \to (\mathbb{C}, 0)$ is (r)-versal ((r)-semiuniversal) if and only if the map

$$\mathbb{C} \oplus T_0 M \to \mathcal{O}_{\mathbb{C}^m,0}/J_f, \quad (c, X) \mapsto c + \mathbf{a}_C|_0(X) \tag{5.13}$$

is surjective (an isomorphism).

So one gains a bit: the base space of an (r)-semiuniversal unfolding $F^{(r)}$ has dimension $\mu - 1$; if $F^{(r)} = F^{(r)}(x_1, \ldots, x_m, t_2, \ldots, t_\mu)$ is (r)-semiuniversal then $t_1 + F^{(r)}$ is semiuniversal; between two semiunversal unfoldings $t_1 + F_1^{(r)}$ and $t_1 + F_2^{(r)}$ of this form there exist isomorphisms which come from (r)-isomorphisms of $F_1^{(r)}$ and $F_2^{(r)}$. (The relation between $F^{(r)}$ and $t_1 + F^{(r)}$ motivates the '(r)', which stands for 'restricted'.)

On the other hand, one loses (5.10). Anyway, one should keep (r)-semiuniversal unfoldings in mind. They are closely related to miniversal Lagrange maps (see the proof of Theorem 5.6 and [**AGV1**, ch. 19]).

(v) One can generalize the notion of a morphism between unfoldings if one weakens condition (5.5): Let $F_i : (\mathbb{C}^m \times M_i), 0) \to (\mathbb{C}, 0),\ i = 1, 2$, be unfoldings of two isolated hypersurface singularities f_1 and f_2. A *generalized morphism* from F_1 to F_2 is a pair (ϕ, ϕ_{base}) of map germs with a commutative diagram as in (5.4) such that (5.6) holds and $\phi|_{\mathbb{C}^m \times \{0\}}$ is a coordinate change (between f_1 and f_2).

Then f_1 and f_2 are *right equivalent*. If the generalized morphism is invertible then F_1 and F_2 are also called *right equivalent*.

Critical spaces and Kodaira–Spencer maps also behave well for generalized morphisms; (5.10) holds, in (5.11) one has to take into account the isomorphism of the Jacobi algebras of f_1 and f_2 which is induced by $\phi|_{\mathbb{C}^m\times\{0\}}$.

The multiplication on the base space of a semiuniversal unfolding was first defined by K. Saito [**SK6**, (1.5)][**SK9**, (1.3)].

Theorem 5.3 *Let $f : (\mathbb{C}^m, 0) \to (\mathbb{C}, 0)$ be an isolated hypersurface singularity and $F : (\mathbb{C}^m \times M, 0) \to (\mathbb{C}, 0)$ be a semiuniversal unfolding.*

The Kodaira–Spencer map $\mathbf{a}_C : \mathcal{T}_{M,0} \to \mathcal{O}_{C,0}$ is an isomorphism and induces a multiplication on $\mathcal{T}_{M,0}$. Then $(M, 0)$ is an irreducible germ of a massive F-manifold with smooth analytic spectrum, and $E := \mathbf{a}_C^{-1}(F|_C)$ is an Euler field of weight 1.

Proof: The map $\mathbf{a}_C : \mathcal{T}_{M,0} \to \mathcal{O}_{C,0}$ is an isomorphism because of Theorem 5.1 and Remark 5.2 (i). The critical space $(C, 0)$ is reduced and smooth. One applies Corollary 4.21 to $(Z, 0) = (\mathbb{C}^m \times M, 0)$ and $\alpha_Z = dF$. The map $\mathbf{q} : (C, 0) \to (L, \pi^{-1}(0))$ is an isomorphism, and $\pi^{-1}(0)$ is a point. Theorem 4.20 (c) shows that $F|_C \circ \mathbf{q}^{-1}$ is a holomorphic generating function. Therefore E is an Euler field of weight 1. □

Theorem 5.4 *Let $f : (\mathbb{C}^m, 0) \to (\mathbb{C}, 0)$ be an isolated hypersurface singularity and $F_i : (\mathbb{C}^m \times M_i, 0) \to (\mathbb{C}, 0)$, $i = 1, 2$, be two semiuniversal unfoldings.*

There exists a unique isomorphism $\varphi : (M_1, 0) \to (M_2, 0)$ of F-manifolds such that $\phi_{base} = \varphi$ for any isomorphism (ϕ, ϕ_{base}) of the unfoldings F_1 and F_2.

Proof: The map $\phi_{base} : (M_1, 0) \to (M_2, 0)$ is an isomorphism of F-manifolds because of (5.10).

Suppose that $F_1 = F_2$ and $(M_1, 0) = (M_2, 0)$. The tangent map of ϕ_{base} on T_0M_1 is $d\phi_{base}|_0 = id$ because of (5.11). The group of all automophisms of $(M_1, 0)$ as F-manifold is finite (Theorem 4.14). Therefore $\phi_{base} = id$. □

Remarks 5.5 (i) The rigidity of the base morphism ϕ_{base} in Theorem 5.4 is in sharp contrast to the general situation for deformations of geometric objects. Usually only a part of the base space of a miniversal deformation is rigid with respect to automorphisms of the deformation.

(ii) The reason for the rigidity is, via Theorem 4.14 and Theorem 3.19, the existence of the canonical coordinates at generic parameters. The corresponding result for singularities is that the critical values of F form coordinates on the base at generic parameters. This has been proved by Looijenga [**Lo1**].

(iii) Because of this rigidity the openness of versality (e.g. [**Te2**]) also takes a special form: For any point $t \in M$ in a representative of the base space

$(M,0) = (\mathbb{C}^\mu, 0)$ of a semiuniversal unfolding F, Theorem 2.11 yields a unique decomposition $(M,t) = \prod_{k=1}^{l}(M_k, t)$ into a product of irreducible germs of F-manifolds. These germs (M_k, t) are the base spaces of semiuniversal unfoldings of the singularities of $F|_{\mathbb{C}^m \times \{t\}}$. The multigerm of F at $\mathbb{C}^m \times \{t\} \cap C$ itself is isomorphic – in a way which can easily be made precise – to a transversal union of versal unfoldings of these singularities.

(iv) The tangent space $T_t M \cong \bigoplus_{k=1}^{l} T_t M_k$ is canonically isomorphic to the direct sum of the Jacobi algebras of singularities of $F|_{\mathbb{C}^m \times \{t\}}$. The vector in $T_t M$ of the Euler field E is mapped to the direct sum of the classes of the function $F|_{\mathbb{C}^m \times \{t\}}$ in these Jacobi algebras. A result of Scherk [**Sche2**] says:

> *The Jacobi algebra $\mathcal{O}_{\mathbb{C}^m,0}/J_f$ of an isolated hypersurface singularity $f : (\mathbb{C}^m, 0) \to (\mathbb{C}, 0)$ together with the class $[f] \in \mathcal{O}_{\mathbb{C}^m,0}/J_f$ determines f up to right equivalence.*

This result shows that the base space M as an F-manifold with Euler field E determines for each parameter $t \in M$ the singularities of $F|_{\mathbb{C}^m \times \{t\}}$ up to right equivalence and also the critical values. Theorem 5.6 will give an even stronger result.

(v) The eigenvalues of $E\circ : T_t M \to T_t M$ are the critical values of $F|_{\mathbb{C}^m \times \{t\}}$. Therefore the discriminant of the Euler field E is

$$\mathcal{D} = \{t \in M \mid (\det(E\circ))(t) = 0\} = \pi_C(C \cap F^{-1}(0)) \qquad (5.14)$$

and it coincides with the classical discriminant of the unfolding F.

All the results of section 4.1 apply to this discriminant. Of course, many of them are classic in the singularity case.

For example, Theorem 4.1 and the isomorphism $\mathbf{q} : C \to L$ from Corollary 4.21 yield an isomorphism between the development $\widetilde{\mathcal{D}} \subset \mathbb{P}T^*M$ of the discriminant and the smooth variety $C \cap F^{-1}(0)$ which has been established by Teissier [**Te2**]. Implicitly it is also in [**AGV1**, ch. 19].

The elementary way in Corollary 4.6 in which the discriminant and the unit field determine the Jacobi algebras seems to be new. But the consequence from this and Scherk's result that the discriminant and the unit field determine the singularity (up to right equivalence) is known (compare below Theorem 5.6 and Remark 5.7 (iv)).

Arnold studied the relation between singularities and Lagrange maps [**AGV1**, ch. 19]. His results (cf. also [**Hoe**], [**Ph1**, 4.7.4.1, pp. 299–301], [**Ph2**], [**Wir**, Corollary 10]) together with those of section 3.4 yield the following correspondence between unfoldings and certain germs of F-manifolds.

Theorem 5.6 *(a) Each irreducible germ of a massive F-manifold with smooth* *analytic spectrum is the base space of a semiuniversal unfolding of an isolated hypersurface singularity.*

(b) Suppose, $F_i : (\mathbb{C}^{m_i} \times M_i, 0) \to (\mathbb{C}, 0)$, $i = 1, 2$, *are semiuniversal unfoldings of singularities* $f_i : (\mathbb{C}^{m_i}, 0) \to (\mathbb{C}, 0)$ *and* $\varphi : (M_1, 0) \to (M_2, 0)$ *is an isomorphism of the base spaces as F-manifolds. Suppose that* $m_1 \leq m_2$.

Then a coordinate change $\psi : (\mathbb{C}^{m_2}, 0) \to (\mathbb{C}^{m_2}, 0)$ *exists such that*

$$f_1(x_1, \ldots, x_{m_1}) + x_{m_1+1}^2 + \cdots + x_{m_2}^2 = f_2(x_1, \ldots, x_{m_2}) \circ \psi \qquad (5.15)$$

and an isomorphism (ϕ, ϕ_{base}) *of the unfoldings* $F_1 + x_{m_1+1}^2 + \cdots + x_{m_2}^2$ *and* $F_2 \circ \psi$ *exists with*

$$F_1 + x_{m_1+1}^2 + \cdots + x_{m_2}^2 = F_2 \circ \psi \circ \phi \quad \textit{and} \quad \phi_{base} = \varphi. \qquad (5.16)$$

Proof: (a) The restricted Lagrange map of the germ of a massive F-manifold with smooth analytic spectrum is a miniversal germ of a Lagrange map with smooth Lagrange variety (section 3.4). Arnold [**AGV1**, 19.3] constructed a generating family $F^{(r)} = F^{(r)}(x, t_2, \ldots, t_\mu)$ for it. Looking at the notions of stable maps and generating families in [**AGV1**, ch. 19], one sees: $F^{(r)}$ is an (r)-semiuniversal unfolding of $F^{(r)}(x, 0)$ (cf. Remark 5.2 (iv)). The unfolding $t_1 + F^{(r)}$ is a semiuniversal unfolding of $F^{(r)}(x, 0)$. Its base space is the given germ of a massive F-manifold.

(b) The unfolding F_i is isomorphic to an unfolding $t_1 + F_i^{(r)}(x_1, \ldots, x_{m_i}, t_2, \ldots, t_\mu)$ as in Remark 5.2 (iv) over the same base. Then $F_i^{(r)}$ is an (r)-semiuniversal unfolding and a generating family for the restricted Lagrange map of the F-manifold $(M_i, 0)$.

The isomorphism $\varphi : (M_1, 0) \to (M_2, 0)$ induces an isomorphism of the restricted Lagrange maps. Then the main result in [**AGV1**, 19.4] establishes a notion of equivalence for $F_1^{(r)}$ and $F_2^{(r)}$, stable $\mathcal{R}^+$-equivalence, which yields the desired equivalence in Theorem 5.6 (b) for F_1 and F_2. □

Remarks 5.7 (i) Two isolated hypersurface singularities $f_i : (\mathbb{C}^{m_i}, 0) \to (\mathbb{C}, 0)$ with $m_1 \leq m_2$ are *stably right equivalent* if a coordinate change $\psi : (\mathbb{C}^{m_2}, 0) \to (\mathbb{C}^{m_2}, 0)$ with (5.15) exists. Furthermore they are right equivalent if $m_1 = m_2$. The splitting lemma says:

> *An isolated hypersurface singularity* $f : (\mathbb{C}^m, 0) \to (\mathbb{C}, 0)$ *with* $r := m - \text{rank}(\frac{\partial^2 f}{\partial x_i \partial x_j})(0)$ *is stably right equivalent to a singularity* $g : (\mathbb{C}^r, 0) \to (\mathbb{C}, 0)$ *with* $\text{rank}(\frac{\partial^2 g}{\partial x_i \partial x_j})(0) = 0$; *this singularity* g *is unique up to right equivalence.*

(For the existence of g see e.g. [**Sl**, (4.2) Satz], the uniqueness of g up to right equivalence follows from Theorem 5.6 or from Scherk's result (Remark 5.5 (iv)).)

(ii) Theorem 5.6 gives a 1-1 correspondence between isolated hypersurface singularities up to stable right equivalence and irreducible germs of massive F-manifolds with smooth analytic spectrum.

But the liftability of an isomorphism $\varphi : (M_1, 0) \to (M_2, 0)$ to unfoldings which is formulated in Theorem 5.6 (b) is stronger. The 1-1 correspondence itself already follows from Theorem 5.6 (a) and Scherk's result (Remark 5.5 (iv)).

(iii) The proof of Theorem 5.6 (a) is not very difficult. If $(M, 0)$ is an irreducible germ of a massive F-manifold with analytic spectrum $(L, \lambda) \subset T^*M$, then a sufficiently generic extension of a generating function on (L, λ) to a function on (T^*M, λ) is already a semiuniversal unfolding over $(M, 0)$. A version different from [**AGV1**, 19.3] of the precise construction is given by Pham [**Ph1**, 4.7.4.1, pp. 291–301], following Hörmander [**Hoe**].

(iv) Theorem 5.6 (b) follows also from [**Ph2**] (again following Hörmander) and from [**Wir**, Corollary 10]. To apply Wirthmüller's arguments one has to start with the discriminant $\mathcal{D}$ and the unit field. Pham [**Ph1**][**Ph2**] starts with the *characteristic variety*. That is the cone in T^*M of the development $\widetilde{\mathcal{D}} \subset \mathbb{P}T^*M$ of the discriminant.

A semiuniversal unfolding $F : (\mathbb{C}^m \times M, 0) \to (\mathbb{C}, 0)$ yields data as in Corollary 4.21 for the germ $(M, 0)$ of an F-manifold:

$$(Z, 0) = (\mathbb{C}^m \times M, 0), \quad \alpha_Z = \mathrm{d}F. \tag{5.17}$$

The semiuniversal unfolding F is also a generating family of $(M, 0)$ as a germ of an F-manifold in the sense of Definition 4.23.

The following observation says that these two special cases Corollary 4.21 and Definition 4.23 of the general construction of F-manifolds in Theorem 4.20 meet only in the case of unfoldings of isolated hypersurface singularities.

Lemma 5.8 *Let $Z, M, \pi_Z, C, \alpha_Z, \mathbf{a}_C$, and $F : Z \to C$ satisfy all the properties in Corollary 4.21 and Definition 4.23.*

Then C is smooth. For any point $p \in M$ the multigerm $F : (Z, C \cap \pi_Z^{-1}(p)) \to \mathbb{C}$ is isomorphic to a transversal product of versal unfoldings of the singularities of $F|\pi_Z^{-1}(p)$ (cf. Remark 5.5 (iii)). The irreducible germs (M_k, p) of F-manifolds in the decomposition $(M, p) = \prod_{k=1}^{l}(M_k, p)$ are base spaces of semiuniversal unfoldings of the singularities of $F|\pi_Z^{-1}(p)$.

Proof: The isomorphism $\mathbf{a}_C : \mathcal{T}_M \to (\pi_C)_* \mathcal{O}_C$ of Corollary 4.21 restricts at $p \in M$ to a componentwise isomorphism of algebras

$$T_pM = \bigoplus_{k=1}^{l} T_pM_k \to \bigoplus_{z \in C \cap \pi_Z^{-1}(p)} \big(\text{Jacobi algebra of } F|(\pi_Z^{-1}(p), z)\big).$$

One applies Theorem 5.1. □

5.2 Boundary singularities

The last section showed that germs of F-manifolds with smooth analytic spectrum correspond to isolated hypersurface singularities. The simplest nonsmooth germ of an analytic spectrum of dimension n is isomorphic to

$$(\{(x, y) \in \mathbb{C}^2 \mid xy = 0\}, 0) \times (\mathbb{C}^{n-1}, 0).$$

We will see that irreducible germs of massive F-manifolds with such an analytic spectrum correspond to boundary singularities (Theorem 5.14). Boundary singularities had been introduced by Arnold [**Ar2**]. Because of the similarities to hypersurface singularities we will take things forward exactly as in section 5.1.

We always consider a germ $(\mathbb{C}^{m+1}, 0)$ with coordinates $x_0, \ldots, x_m$ together with the hyperplane $H := \{x \in \mathbb{C}^{m+1} \mid x_0 = 0\}$ of the first coordinate. A *boundary singularity* (f, H) is a holomorphic function germ $f : (\mathbb{C}^{m+1}, 0) \to (\mathbb{C}, 0)$ such that f and $f|_H$ have isolated singularities at 0. It can be considered as an extension of the hypersurface singularities f and $f|_H$.

Its Jacobi algebra is

$$\mathcal{O}_{\mathbb{C}^{m+1},0}/J_{f,H} := \mathcal{O}_{\mathbb{C}^{m+1},0} \Big/ \left(x_0 \frac{\partial f}{\partial x_0}, \frac{\partial f}{\partial x_1}, \ldots, \frac{\partial f}{\partial x_m} \right), \tag{5.18}$$

and its Milnor number $\mu = \mu(f, H) := \dim \mathcal{O}_{\mathbb{C}^{m+1},0}/J_{f,H}$ satisfies ([**Ar2**, §3], [**Sz**, §2])

$$\mu = \mu(f) + \mu(f|_H). \tag{5.19}$$

An unfolding of (f, H) is simply a holomorphic function germ $F : (\mathbb{C}^{m+1} \times \mathbb{C}^n, 0) \to (\mathbb{C}, 0)$ such that $F|\mathbb{C}^{m+1} \times \{0\} = f$, that is, an unfolding of f. Again we write the parameter space as $(M, 0) = (\mathbb{C}^n, 0)$.

But the critical space $(C, 0) \subset (\mathbb{C}^{m+1} \times M, 0)$ of F as unfolding of the boundary singularity (f, H) is the zero set of the ideal

$$J_{F,H} := \left(x_0 \frac{\partial F}{\partial x_0}, \frac{\partial F}{\partial x_1}, \ldots, \frac{\partial F}{\partial x_m} \right) \tag{5.20}$$

with the complex structure $\mathcal{O}_{C,0} = \mathcal{O}_{\mathbb{C}^{m+1}\times M,0}/J_{F,H}|_{(C,0)}$ (cf. [**Sz**]). Forgetting the complex structure, $(C, 0)$ is the union of the critical sets $(C^{(1)}, 0)$ of F and $(C^{(2)}, 0)$ of $F|_{H\times M}$ as unfoldings of hypersurface singularities.

For the same reasons as in the hypersurface case the projection $pr : (C, 0) \to (M, 0)$ is finite and flat with degree μ and $\mathcal{O}_{C,0}$ is a free $\mathcal{O}_{M,0}$-module of rank μ. The 1–form

$$\alpha_Z := -\frac{\partial F}{\partial x_0}\mathrm{d}x_0 + \mathrm{d}F = \sum_{i=1}^{m}\frac{\partial F}{\partial x_i}\mathrm{d}x_i + \sum_{j=1}^{n}\frac{\partial F}{\partial t_j}\mathrm{d}t_j \tag{5.21}$$

on $(Z, 0) := (\mathbb{C}^{m+1} \times M, 0)$ gives rise to a kind of Kodaira–Spencer map

$$\mathbf{a}_C : \mathcal{T}_{M,0} \to \mathcal{O}_{C,0}, \quad X \mapsto \alpha_Z(\widetilde{X})|_{(C,0)}, \tag{5.22}$$

where $\widetilde{X}$ is any lift of $X \in \mathcal{T}_{M,0}$ to $(Z, 0)$. It induces a reduced Kodaira–Spencer map

$$\mathbf{a}_C|_0 : T_0M \to \mathcal{O}_{\mathbb{C}^{m+1},0}/J_{f,H}. \tag{5.23}$$

The ideal $J_{F,H}$ and the maps $\mathbf{a}_C$ and $\mathbf{a}_C|_0$ behave well with respect to morphisms of unfoldings, as we will see.

A morphism between two unfoldings F_1 and F_2 as in section 5.1 of a boundary singularity (f, H) is a pair (ϕ, ϕ_{base}) of a map germ with (5.4)–(5.6) and additionally

$$\phi(H \times M_1) \subset H \times M_2. \tag{5.24}$$

Then the first entry of ϕ takes the form $x_0 \cdot unit \in \mathcal{O}_{Z,0}$. Using this one can see with a bit more work than in the hypersurface case that the critical spaces behave well with respect to morphisms:

$$\phi^* J_{F_2} = J_{F_1} \quad \text{and} \quad (C_1, 0) = \phi^{-1}((C_2, 0)). \tag{5.25}$$

Also the Kodaira–Spencer maps behave as well as in the hypersurface case. The $\mathcal{O}_{M_1,0}$-linear maps $d\phi_{base}$, $\mathbf{a}_{C_2}$, and $\phi^*|_{(C_2,0)}$ are defined as in (5.7)–(5.9); again one finds

$$\mathbf{a}_{C_1} = \phi^*|_{(C_2,0)} \circ \mathbf{a}_{C_2} \circ \mathrm{d}\phi_{base} \tag{5.26}$$

and

$$\mathbf{a}_{C_1}|_0 = \mathbf{a}_{C_2}|_0 \circ \mathrm{d}\phi_{base}|_0. \tag{5.27}$$

Versal and semiuniversal unfoldings of boundary singularities are defined analogously to the hypersurface case and they exist.

Theorem 5.9 [**Ar2**] *An unfolding $F : (\mathbb{C}^{m+1} \times M, 0) \to (\mathbb{C}, 0)$ of a boundary singularity (f, H), $f : (\mathbb{C}^{m+1}, 0) \to (\mathbb{C}, 0)$, is versal if and only if the reduced Kodaira–Spencer map $\mathbf{a}_C|_0 : T_0M \to \mathcal{O}_{\mathbb{C}^{m+1},0}/J_{f,H}$ is surjective. It is semiuniversal if and only if $\mathbf{a}_C|_0$ is an isomorphism.*

Remarks 5.10 (i) The map $\mathbf{a}_C|_0$ is surjective (an isomorphism) if and only if $\mathbf{a}_C$ is surjective (an isomorphism).

(ii) The function $F(x_0, \dots, x_m, t_1, \dots, t_\mu) = f + \sum_{i=1}^{\mu} m_i t_i$ is a semiuniversal unfolding of the boundary singularity (f, H) if $m_1, \dots, m_\mu \in \mathcal{O}_{\mathbb{C}^{m+1},0}$ represent a basis of $\mathcal{O}_{\mathbb{C}^{m+1},0}/J_{f,H}$.

(iii) The critical space of an unfolding $F : (Z, 0) = (\mathbb{C}^{m+1} \times M, 0) \to (\mathbb{C}, 0)$ of a boundary singularity (f, H) is reduced and isomorphic to $(\{(x, y) \in \mathbb{C}^2 \mid xy = 0\}, 0) \times (\mathbb{C}^{n-1}, 0)$ if and only if $\frac{\partial F}{\partial x_0}, \dots, \frac{\partial F}{\partial x_m}$ represent a generating system of the vector space $\mathbf{m}_{Z,0}/((x_0) + \mathbf{m}^2_{Z,0})$. This is equivalent to the nondegeneracy condition

$$\operatorname{rank} \begin{pmatrix} \frac{\partial^2 F}{\partial x_0 \partial x_j} & \frac{\partial^2 F}{\partial x_0 \partial t_k} \\ \frac{\partial^2 F}{\partial x_i \partial x_j} & \frac{\partial^2 F}{\partial x_i \partial t_k} \end{pmatrix}_{i,j \geq 1} (0) = m + 1 \tag{5.28}$$

(cf. [**DD**]). It is satisfied for versal unfoldings.

(iv) As in Remark 5.2 (v) for hypersurface singularities, one can define generalized morphisms between unfoldings of right equivalent boundary singularities. Again the critical spaces and Kodaira–Spencer maps behave well.

Theorem 5.11 *Let $F : (\mathbb{C}^{m+1} \times M, 0) \to (\mathbb{C}, 0)$ be a semiuniversal unfolding of a boundary singularity (f, H).*

The Kodaira–Spencer map $\mathbf{a}_C : \mathcal{T}_{M,0} \to \mathcal{O}_{C,0}$ is an isomorphism and induces a multiplication on $\mathcal{T}_{M,0}$. Then $(M, 0)$ is an irreducible germ of a massive F-manifold with analytic spectrum isomorphic to $(\{(x, y) \in \mathbb{C}^2 \mid xy = 0\}, 0) \times (\mathbb{C}^{\mu-1}, 0)$. The field $E := \mathbf{a}_C(F|_C)$ is an Euler field of weight 1.

Proof: Similar to the proof of Theorem 5.3. One wants to apply Corollary 4.21 and has to show that $\alpha_Z|_{C_{reg}}$ is exact. The critical space $(C, 0)$ as a set is the union of the smooth zero sets $(C^{(1)}, 0)$ of J_F and $(C^{(2)}, 0)$ of the ideal $(x_0, \frac{\partial F}{\partial x_1}, \dots, \frac{\partial F}{\partial x_m})$. The definition (5.21) of α_Z shows

$$\alpha_Z|_{(C^{(i)},0)} = dF|_{(C^{(i)},0)} \quad \text{for } i = 1, 2. \tag{5.29}$$

Therefore $\alpha_Z|_{C_{reg}}$ is exact and $F|_C \circ \mathbf{q}^{-1}$ is a holomorphic generating function of the analytic spectrum. □

Theorem 5.12 *Let* $F_i : (\mathbb{C}^{m+1} \times M_i, 0) \to (\mathbb{C}, 0)$, $i = 1, 2$, *be two semiuniversal unfoldings of a boundary singularity* (f, H).

There exists a unique isomorphism $\varphi : (M_1, 0) \to (M_2, 0)$ *of F-manifolds such that* $\phi_{base} = \varphi$ *for any isomorphism* (ϕ, ϕ_{base}) *of the unfoldings* F_1 *and* F_2.

Proof: Similar to the proof of Theorem 5.4. □

Remarks 5.13 (i) Let $F : (\mathbb{C}^{m+1} \times M, 0) \to (\mathbb{C}, 0)$ be a semiuniversal unfolding of a boundary singularity (f, H) with critical space $(C, 0) = (C^{(1)}, 0) \cup (C^{(2)}, 0)$. For any $t \in M$ the points in $\mathbb{C}^{m+1} \times \{t\} \cap (C, 0)$ split into three classes: The hypersurface singularities of $F|\mathbb{C}^{m+1} \times \{t\}$ in $C^{(1)} - C^{(2)}$, the hypersurface singularities of $F|H \times \{t\}$ in $C^{(2)} - C^{(1)}$, and the boundary singularities of $F|\mathbb{C}^{m+1} \times \{t\}$ in $C^{(1)} \cap C^{(2)}$.

The algebra $\mathcal{O}_C|\mathbb{C}^{m+1} \times \{0\}$ is the direct sum of their Jacobi algebras. The reduced Kodaira–Spencer map at $t \in M$ is an isomorphism from $T_t M$ to this algebra.

Hence the multigerms of F at $\mathbb{C}^{m+1} \times \{t\} \cap C^{(1)}$ and of $F|H \times M$ at $H \times \{t\} \cap (C^{(2)} - C^{(1)})$ together form a transversal union of versal unfoldings of these hypersurface and boundary singularities.

The components (M_k, t) of the decomposition $(M, t) = \prod_{k=1}^{l} (M_k, t)$ into irreducible germs of F-manifolds are bases of semiuniversal unfoldings of the hypersurface and boundary singularities.

(ii) The eigenvalues of $E\circ : T_t M \to T_t M$ are by definition of E the values of F on $\mathbb{C}^{m+1} \times \{t\} \cap C$. The discriminant of the Euler field is

$$\mathcal{D} = \{t \in M \mid (\det(E\circ))(t) = 0\} = \pi_C(C \cap F^{-1}(0)). \tag{5.30}$$

This is the union of the discriminants of F and $F|H \times M$ as unfoldings of hypersurface singularities and it coincides with the classical discriminant of F as an unfolding of a boundary singularity [**Ar2**][**Sz**]. All the results of section 4.1 apply to this discriminant.

Nguyen huu Duc and Nguyen tien Dai studied the relation between boundary singularities and Lagrange maps [**DD**]. Their results together with section 3.4 yield the following correspondence between unfoldings of boundary singularities and certain germs of F-manifolds.

Theorem 5.14 *Let* $(M, 0)$ *be an irreducible germ of a massive F-manifold with analytic spectrum* (L, λ) *isomorphic to*

$$(\{(x, y) \in \mathbb{C}^2 \mid xy = 0\}, 0) \times (\mathbb{C}^{n-1}, 0)$$

and ordered components $(L^{(1)}, \lambda) \cup (L^{(2)}, \lambda) = (L, \lambda)$.

(a) There exists a semiuniversal unfolding F of a boundary singularity such that the base space is isomorphic to $(M, 0)$ as F-manifold and the isomorphism $\mathbf{q} : (C, 0) \to (L, \lambda)$ maps $C^{(i)}$ to $L^{(i)}$.

(b) Suppose, $F_i : (\mathbb{C}^{m_i+1} \times M_i, 0) \to (\mathbb{C}, 0)$, $i = 1, 2$, are semiuniversal unfoldings of boundary singularities (f_i, H_i) and $\varphi : (M_1, 0) \to (M_2, 0)$ is an isomorphism of the base spaces as F-manifolds. Suppose that $m_1 \leq m_2$.

Then a coordinate change $\psi : (\mathbb{C}^{m_2+1}, 0) \to (\mathbb{C}^{m_2+1}, 0)$ with $\psi((H_2, 0)) = (H_2, 0)$ exists such that

$$f_1(x_0, \ldots, x_{m_1}) + x_{m_1+1}^2 + \cdots + x_{m_2}^2 = f_2(x_0, \ldots, x_{m_2}) \circ \psi \qquad (5.31)$$

and an isomorphism (ϕ, ϕ_{base}) of the unfoldings $F_1 + x_{m_1+1}^2 + \cdots + x_{m_2}^2$ and $F_2 \circ \psi$ of boundary singularities exists with

$$F_1 + x_{m_1+1}^2 + \cdots + x_{m_2}^2 = F_2 \circ \psi \circ \phi \quad \text{and} \quad \phi_{base} = \varphi. \qquad (5.32)$$

Proof: (a) In [**DD**, Proposition 1] an unfolding $F : (\mathbb{C}^{m+1} \times M, 0) \to (\mathbb{C}, 0)$ with nondegeneracy condition (5.28) of a boundary singularity is constructed such that F is a generating family for $L^{(1)} \subset T^*M$ and $F|H \times M$ is a generating family for $L^{(2)}$.

One can show that there are canonical maps $C^{(i)} \to L^{(i)}$ which combine to an isomorphism $\mathbf{q} : C \to L$ with $\mathbf{a}_C = \hat{\mathbf{q}} \circ \mathbf{a}$ (as in Theorem 4.20). Then the Kodaira–Spencer map $\mathbf{a}_C : \mathcal{T}_{M,0} \to \mathcal{O}_{C,0}$ is an isomorphism and F is a semiuniversal unfolding of a boundary singularity. (Implicitly this is also contained in [**DD**, Théorème]). Because of $\mathbf{a}_C = \hat{\mathbf{q}} \circ \mathbf{a}$ its base is $(M, 0)$ as F-manifold.

(b) [**DD**, Proposition 3]. □

Remarks 5.15 (i) Two boundary singularities $f_i : (\mathbb{C}^{m_i+1}, 0) \to (\mathbb{C}, 0)$ with $m_1 \leq m_2$ are *stably right equivalent* if a coordinate change ψ as in Theorem 5.14 (b) exists. Furthermore they are right equivalent if $m_1 = m_2$. A splitting lemma for boundary singularities is formulated below in Lemma 5.16.

(ii) Theorem 5.14 gives a 1-1 correspondence between boundary singularities up to stable right equivalence and irreducible germs of massive F-manifolds with analytic spectrum $(L, \lambda) \cong (\{(x, y) \in \mathbb{C}^2 \mid xy = 0\}, 0) \times (\mathbb{C}^2, 0)$ and ordered components $(L^{(1)}, \lambda) \cup (L^{(2)}, \lambda) = (L, \lambda)$.

(iii) Interchanging the two components of (L, λ) corresponds to a duality for boundary singularities which goes much further and has been studied by I. Shcherbak, A. Szpirglas [**Sz**][**ShS1**][**ShS2**], and others.

Lemma 5.16 *(Splitting lemma for boundary singularities)*
A boundary singularity (f, H) *with* $f : (\mathbb{C}^{m+1}, 0) \to (\mathbb{C}, 0)$ *and* $H = \{x \,|\, x_0 = 0\}$ *is stably right equivalent to a boundary singularity* $g : (\mathbb{C}^{r+1}, 0) \to (\mathbb{C}, 0)$ *in*

$$r + 1 = \max\left(2; m + 1 - \operatorname{rank}\left(\frac{\partial^2 f(x_0^2, x_1, \ldots, x_m)}{\partial x_i \partial x_j}\right)(0)\right) \quad (5.33)$$

coordinates. The boundary singularity g *is unique up to right equivalence.*

Proof: Existence of g: The group $G = \mathbb{Z}_2$ acts on $(\mathbb{C}^{m+1}, 0)$ by $(x_0, x_1, \ldots, x_m) \mapsto (\pm x_0, x_1, \ldots, x_m)$. Boundary singularities on the quotient $(\mathbb{C}^{m+1}, 0)$ correspond to G-invariant singularities on the double cover, branched along H ([**AGV1**, 17.4]).

One applies an equivariant splitting lemma of Slodowy [**Sl**, (4.2) Satz] to the G-invariant singularity $f(x_0^2, x_1, \ldots, x_m)$. The nondegenerate quadratic part of the G-invariant singularity in splitted form does not contain x_0^2 because f is not smooth.

Uniqueness of g: This follows with Theorem 5.14 (b). □

The following two observations give some information on generating families in the sense of Definition 4.23 for the F-manifolds of boundary singularities. The first one gives a distinguished generating family and is essentially well known. The second one explains why $B_2 = I_2(4)$ is missing in Example 4.24 (b).

Lemma 5.17 *(a) Let* $F : (Z, 0) = (\mathbb{C}^{m+1} \times M, 0) \to (\mathbb{C}, 0)$ *be a semiuniversal unfolding of a boundary singularity* (f, H).

Then the function $\widetilde{F} : (\widetilde{Z}, 0) = (\mathbb{C}^{m+1} \times M, 0) \to (\mathbb{C}, 0)$ *with* $\widetilde{F}(x, t) = F(x_0^2, x_1, \ldots, x_m, t)$ *is a generating family for the germ* $(M, 0)$ *of an F-manifold.*

The finite map $\widetilde{\mathbf{q}} : \widetilde{C} \to L = L^{(1)} \cup L^{(2)}$ *from its critical set* $\widetilde{C}$ *to the analytic spectrum L has degree 2 on* $L^{(1)}$ *and degree 1 on* $L^{(2)}$*. The branched covering* $\widetilde{C} \to M$ *has degree* $2\mu(f) + \mu(f|_H)$.

(b) Let $(M, 0)$ *be a germ of a massive F-manifold with analytic spectrum* $(L, \lambda) \cong (\{(x, y) \in \mathbb{C}^2 \mid xy = 0\}, 0) \times (\mathbb{C}^{n-1}, 0)$*. There does not exist a generating family* $F : (Z, 0) \to (\mathbb{C}, 0)$ *with critical set C such that the canonical map* $\mathbf{q} : C \to L$ *is a homeomorphism.*

Proof: (a) Consider the branched covering $\pi_G : \widetilde{Z} \to Z$, $(x_0, \ldots, x_m, t) \mapsto (x_0^2, x_1, \ldots, x_m, t)$ which is induced by the action $(x_0, \ldots, x_m, t) \mapsto (\pm x_0, x_1, \ldots, x_m, t)$ of the group $G = \mathbb{Z}_2$ on $\widetilde{Z}$. The composition $\widetilde{F} = F \circ \pi_G$ is an unfolding of the G-invariant singularity $\widetilde{F}|(\mathbb{C}^{m+1} \times \{0\}, 0)$, infact, semiuniversal within the G-invariant unfoldings (cf. [**Sl**, (4.5)]). The ideals

$$J_{\widetilde{F}} = \left(2x_0 \cdot \left(\frac{\partial F}{\partial x_0} \circ \pi_G\right), \frac{\partial F}{\partial x_1} \circ \pi_G, \ldots, \frac{\partial F}{\partial x_m} \circ \pi_G\right) \tag{5.34}$$

and $\pi_G^* J_{F,H}$ have the same zero sets $\widetilde{C} = \pi_G^{-1}(C)$.

Comparison of $\alpha_{\widetilde{Z}} = \mathrm{d}\widetilde{F}$ on $\widetilde{Z}$ and $\alpha_Z = -\frac{\partial F}{\partial x_0}\mathrm{d}x_0 + \mathrm{d}F$ on Z (formula (5.21)) shows that the map $\mathbf{a}_{\widetilde{C}} : \mathcal{T}_{M,0} \to \mathcal{O}_{\widetilde{C},0}$ factorizes into the Kodaira–Spencer map $\mathbf{a}_C : \mathcal{T}_{M,0} \to \mathcal{O}_{C,0}$ and the map $\pi_G^*|_{(C,0)} : \mathcal{O}_{C,0} \to \mathcal{O}_{\widetilde{C},0}$.

Therefore $\mathbf{a}_{\widetilde{C}}$ is injective with multiplication invariant image and induces the correct multiplication on $\mathcal{T}_{M,0}$. The rest is clear.

(b) Assume that such a generating family F exists. The analytic spectrum (L, λ) is its own maximalization. Therefore the homeomorphism $\mathbf{q}$ is an isomorphism.

Then $\mathbf{a}_C = \hat{\mathbf{q}} \circ \mathbf{a}$ (cf. Definition 4.23 and Theorem 4.20) is an isomorphism. We are simultaneously in the special cases Definition 4.23 and Corollary 4.21 of Theorem 4.20. By Lemma 5.8 (L, λ) is smooth, a contradiction. □

5.3 Coxeter groups and F-manifolds

The complex orbit space of a finite irreducible Coxeter group is equipped with the discriminant, the image of the reflection hyperplanes, and with a certain distinguished vector field (see below), which is unique up to a scalar. Together they induce as in Corollary 4.6 the structure of an F-manifold on the complex orbit space (Theorem 5.18). This follows independently from [**Du2**][**Du3**, Lecture 4] and from [**Gi2**, Theorem 14].

In fact, both give stronger results. Dubrovin established the structure of a Frobenius manifold. This will be discussed in section 5.4. Givental proved that these F-manifolds are distinguished by certain geometric conditions (Theorem 5.21). With one additional argument we will show that the germs of these F-manifolds and their products are the only germs of simple F-manifolds whose tangent spaces are Frobenius algebras (Theorem 5.20). This complements in a nice way the relation between Coxeter groups and simple hypersurface and boundary singularities.

We will also present simple explicit formulas for these F-manifolds which are new for H_3 and H_4 (Theorem 5.22).

A finite Coxeter group is a finite group W of linear transformations of the Euclidean space $\mathbb{R}^n$ generated by reflections in hyperplanes. Each Coxeter group is the direct sum of irreducible Coxeter groups. Their classification and description can be found in [**Co**] or [**Bou**]. They are A_n $(n \geq 1)$, D_n $(n \geq 4)$, E_6, E_7,

E_8, B_n $(n \geq 2)$, F_4, G_2, H_3, H_4, $I_2(m)$ $(m \geq 3)$ with $A_2 = I_2(3)$, $B_2 = I_2(4)$, $H_2 := I_2(5)$, $G_2 = I_2(6)$.

The Coxeter group W acts on $\mathbb{C}^n = \mathbb{R}^n \otimes_{\mathbb{R}} \mathbb{C}$ and on $\mathbb{C}[x_1, \ldots, x_n]$, where $x_1, \ldots, x_n$ are the coordinates on $\mathbb{C}^n$. The ring $\mathbb{C}[x_1, \ldots, x_n]^W$ of invariant polynomials is generated by n algebraically independent homogeneous polynomials $P_1, \ldots, P_n$. Their degrees $d_i := \deg P_i$ are unique (up to ordering). The quotient $\mathbb{C}^n/W$ is isomorphic to $\mathbb{C}^n$ as an affine algebraic variety. The $\mathbb{C}^*$-action and the vector field $\sum_i x_i \frac{\partial}{\partial x_i}$ on the original $\mathbb{C}^n$ induce a $\mathbb{C}^*$-action and a vector field $\sum d_i t_i \frac{\partial}{\partial t_i}$ on the orbit space $\mathbb{C}^n/W \cong \mathbb{C}^n$. The image in the orbit space of the union of the reflection hyperplanes is the discriminant $\mathcal{D}$ of the Coxeter group.

Suppose for a moment that W is irreducible. Then there is precisely one highest degree, which is called the Coxeter number h. The degrees can be ordered to satisfy

$$d_1 = h > d_2 \geq \ldots \geq d_{n-1} > d_n = 2, \tag{5.35}$$

$$d_i + d_{n+1-i} = h + 2. \tag{5.36}$$

The vector field $e := \frac{\partial}{\partial t_1}$ is unique up to a scalar.

Theorem 5.18 *The complex orbit space $M := \mathbb{C}^n/W \cong \mathbb{C}^n$ of a finite irreducible Coxeter group W carries a unique structure of a massive F-manifold with the unit field $e = \frac{\partial}{\partial t_1}$ and the discriminant $\mathcal{D}$. The discriminant $\mathcal{D}$ corresponds to the Euler field*

$$E = \frac{1}{h} \sum_{i=1}^{n} d_i t_i \frac{\partial}{\partial t_i} \tag{5.37}$$

of weight 1.

Proof: The uniqueness follows from Corollary 4.6.

The existence follows from Dubrovin's result ([**Du2**][**Du3**, Lecture 4], cf. Theorem 5.23) or Givental's result [**Gi2**, Theorem 14] together with Theorem 3.16.

Below in Theorem 5.22 we will follow Givental and reduce it to classical results on the appearance of discriminants in singularity theory ([**Bri1**][**Ar2**][**Ly2**] [**ShO**]). □

Remarks 5.19 (i) Corollary 4.6 gives probably the most elementary way in which e and $\mathcal{D}$ determine the multiplication on the complex orbit space $M = \mathbb{C}^n/W \cong \mathbb{C}^n$, at least at a generic point: the e-orbit of a generic point $p \in M$

intersects $\mathcal{D}$ transversally in n points. One shifts the tangent spaces of $\mathcal{D}$ at these points with the flow of e to T_pM. Then there exists a basis $e_1, \ldots, e_n$ of T_pM such that $\sum_{i=1}^n e_i = e$ and such that the hyperplanes $\bigoplus_{i\neq j} \mathbb{C}\cdot e_i, j = 1, \ldots, n,$ are the shifted tangent spaces of $\mathcal{D}$. The multiplication on T_pM is given by $e_i \circ e_j = \delta_{ij} e_i$.

(ii) The unit field $e = \frac{\partial}{\partial t_1}$ is only unique up to a scalar. The flow of the Euler field respects the discriminant $\mathcal{D}$ and maps the unit field e and the multiplication to multiples, because of $\mathrm{Lie}_E(e) = -e$ and $\mathrm{Lie}_E(\circ) = \circ$.

Therefore the isomorphism class of the F-manifold $(M, \circ, e, E)$ is independent of the choice of the scalar.

(iii) The complex orbit space of a reducible Coxeter group W is isomorphic to the product of the complex orbit spaces of the irreducible subgroups. The discriminant decomposes as in Remark 4.2 (v). Now any sum of unit fields for the components yields a unit field for $\mathbb{C}^n/W$. The choices are parameterized by $(\mathbb{C}^*)^{|irr.\ subgroups|}$. But the resulting F-manifold is unique up to isomorphism. It is the product of F-manifolds for the irreducible subgroups. This F-manifold and its germ at 0 will be denoted by the same combination of letters as the Coxeter group.

Theorem 5.20 *Let $((M, p), \circ, e)$ be a germ of a massive F-manifold.*

The germ $((M, p), \circ, e)$ is simple and T_pM is a Frobenius algebra if and only if $((M, p), \circ, e)$ is isomorphic to the germ at 0 of an F-manifold of a finite Coxeter group.

This builds on the following result, which is a reformulation with section 3.4 of a theorem of Givental [**Gi2**, Theorem 14]. Theorems 5.20, 5.21, and 5.22 will be proved below in the opposite order. Some arguments on H_3 and H_4 in the proof of Theorem 5.21 will only be outlined.

Theorem 5.21 (Givental) *(a) The F-manifold of a finite irreducible Coxeter group is simple. The analytic spectrum (L, λ) of its germ at 0 is isomorphic to $(\{(x, y) \in \mathbb{C}^2 \mid x^2 = y^r\}, 0) \times (\mathbb{C}^{n-1}, 0)$ with $r = 1$ for A_n, D_n, E_n, $r = 2$ for B_n, F_4, $r = 3$ for H_3, H_4 and $r = m - 2$ for $I_2(m)$.*

(b) An irreducible germ of a simple F-manifold with analytic spectrum isomorphic to a product of germs of plane curves is isomorphic to the germ at 0 of an F-manifold of a finite irreducible Coxeter group.

Finally, we want to present the F-manifolds of the finite irreducible Coxeter groups explicitly with data as in Corollary 4.21. We will use the notations of Corollary 4.21. The following is a consequence of results in [**Bri1**][**Ar2**][**Ly2**]

[**ShO**] on the appearance of the discriminants of Coxeter groups in singularity theory.

Theorem 5.22 *(a) The germs at 0 of the F-manifolds of the Coxeter groups* A_n, D_n, E_n, B_n, F_4 *are isomorphic to the base spaces of the semiuniversal unfoldings of the corresponding simple hypersurface singularities* A_n, D_n, E_n *and simple boundary singularities* B_n *(or* C_n*) and* F_4*.*

(b) For the F-manifolds $(M, \circ, e) = (\mathbb{C}^n, \circ, e)$ *of the finite irreducible Coxeter groups, a space* Z *with projection* $\pi_Z : Z \to M$*, a subspace* $C \subset Z$ *and a 1–form* α_Z *will be given such that the map*

$$\mathbf{a}_C : \mathcal{T}_M \to (\pi_C)_*\mathcal{O}_C, \quad X \mapsto \alpha_Z(\widetilde{X})|_C \tag{5.38}$$

is welldefined and an isomorphism of $\mathcal{O}_M$*-algebras. The space* C *is isomorphic to the analytic spectrum of* $(M, \circ, e)$*. The Euler field is always* $E = \frac{1}{h}\sum_{i=1}^n d_i t_i \frac{\partial}{\partial t_i}$*. The discriminant* $\mathcal{D} \subset M$ *is* $\mathcal{D} = \pi_C(\mathbf{a}_C(E)^{-1}(0))$*.*

(i) $A_n, B_n, H_3, I_2(m)$:
$Z = \mathbb{C} \times M = \mathbb{C} \times \mathbb{C}^n$ *with coordinates* $(x, t) = (x, t_1, \ldots, t_n)$,
$\alpha_Z = \mathrm{d}t_1 + x\mathrm{d}t_2 + \cdots + x^{n-1}\mathrm{d}t_n$,
$\widetilde{t_2}(x, t) := t_2 + 2xt_3 + \cdots + (n-1)x^{n-2}t_n$,

$A_n : C = \{(x, t) \in Z \mid x^n - \widetilde{t_2} = 0\}$,
$B_n : C = \{(x, t) \in Z \mid x \cdot (x^{n-1} - \widetilde{t_2}) = 0\}$,
$H_3 : C = \{(x, t) \in Z \mid x^3 - \widetilde{t_2}^2 = 0\}$,
$I_2(m) : C = \{(x, t) \in Z \mid x^2 - \widetilde{t_2}^{m-2} = 0\}$.

(ii) D_4, F_4, H_4 :
$Z = \mathbb{C}^2 \times M = \mathbb{C}^2 \times \mathbb{C}^4$ *with coordinates* $(x, y, t) = (x, y, t_1, \ldots, t_4)$,
$\alpha_Z = \mathrm{d}t_1 + x\mathrm{d}t_2 + y\mathrm{d}t_3 + xy\mathrm{d}t_4$,
$\widetilde{t_2}(x, y, t) := t_2 + yt_4$, $\widetilde{t_3}(x, y, t) := t_3 + xt_4$,

$D_4 : C = \{(x, y, t) \in Z \mid x^2 + \widetilde{t_2} = 0,\ y^2 + \widetilde{t_3} = 0\}$,
$F_4 : C = \{(x, y, t) \in Z \mid x^2 + \widetilde{t_2} = 0,\ y^2 + \widetilde{t_3}^2 = 0\}$,
$H_4 : C = \{(x, y, t) \in Z \mid x^2 + \widetilde{t_2} = 0,\ y^2 + \widetilde{t_3}^3 = 0\}$.

(iii) D_n, E_6, E_7, E_8 :
$Z = \mathbb{C}^2 \times M = \mathbb{C}^2 \times \mathbb{C}^n$ *with coordinates* $(x, y, t) = (x, y, t_1, \ldots, t_n)$,
$F : Z \to \mathbb{C}$ *a semiuniversal unfolding of* $F|\mathbb{C}^2 \times \{0\}$,
$\alpha_Z = \sum_{i=1}^n \frac{\partial F}{\partial t_i}\mathrm{d}t_i$ *(or* $\alpha_Z = \mathrm{d}F$*)*,
$C = \{(x, y, t) \in Z \mid \frac{\partial F}{\partial x} = \frac{\partial F}{\partial y} = 0\}$,

$D_n :\ F = x^{n-1} + xy^2 + \sum_{i=1}^{[n/2]} x^{i-1}t_i + yt_{[n/2]+1} + \sum_{i=[n/2]+2}^{n} x^{i-2}t_i$,
$E_6 :\ F = x^4 + y^3 + t_1 + xt_2 + yt_3 + x^2t_4 + xyt_5 + x^2yt_6$,

$E_7: \quad F = x^3y + y^3 + t_1 + xt_2 + yt_3 + x^2t_4 + xyt_5 + x^3t_6 + x^4t_7,$
$E_8: \quad F = x^5 + y^3 + t_1 + xt_2 + yt_3 + x^2t_4 + xyt_5 + x^3t_6 + x^2yt_7 + x^3yt_8.$

Proof of Theorem 5.22: (a) One can choose a semiuniversal unfolding $F = f(x_1, \ldots, x_m) + \sum_{i=1}^n m_i t_i$ of the hypersurface or boundary singularity which is weighted homogeneous with positive degrees in all variables and parameters. There is an isomorphism from its base space $\mathbb{C}^n$ to the complex orbit space of the corresponding Coxeter group which respects the discriminant, the Euler field, and the unit field ([**Bri1**][**Ar2**]). It also respects the F-manifold structure (Corollary 4.6).

(b) Part (i) for $I_2(m)$ is Remark 4.24 (a). Part (i) for A_n, B_n and part (ii) for D_4 follow with (a), with semiuniversal unfolding as in (a) for the singularities $-\frac{1}{n+1}x^{n+1}$ (A_n), $(-\frac{1}{n}x^n + y^2, H = \{x = 0\})$ (B_n), $\frac{1}{3}x^3 + \frac{1}{3}y^3$ (D_4). Also (iii) follows with (a).

The same procedure gives for the boundary singularity F_4 with equation $(\frac{1}{2}y^2 + \frac{1}{3}x^3, H = \{y = 0\})$ the data in (ii) with critical set

$$C' = \{(x, y, t) \in Z \mid x^2 + \widetilde{t_2} = 0,\ y^2 + y\widetilde{t_3} = 0\}. \tag{5.39}$$

It is a nontrivial, but solvable, exercise to find compatible automorphisms of Z and M which map C' to C and α_Z to α_Z modulo $I_C\Omega_Z^1$. Independently of explicit calculations, the proof of Theorem 5.21 will show that the data (Z, α_Z, C) in (ii) correspond to F_4.

The data in (i) for H_3 and in (ii) for H_4 can be obtained from results of O.P. Shcherbak [**ShO**, pp. 162 and 163] (cf. also [**Gi2**, Proposition 12]) (for H_3 one could use instead [**Ly2**]). The unfoldings

$$F_{H_3} = \int_0^y (u^2 + x)^2 du + t_1 + xt_2 + x^2t_3 \tag{5.40}$$

of D_6 and

$$F_{H_4} = \int_0^y (u^2 + t_3 + xt_4)^2 du + x^3 + t_1 + xt_2 \tag{5.41}$$

of E_8 have only critical points with even Milnor number and are maximal with this property. Their discriminants are isomorphic to the discriminants of the Coxeter groups H_3 and H_4. The unfoldings are generating families in the sense of Definition 4.23 for the F-manifolds of the Coxeter groups H_3 and H_4. We will determine the data in (ii) for H_4 from F_{H_4}; the case of H_3 is similar.

Consider the map

$$\phi : \mathbb{C}^2 \times \mathbb{C}^4 \to Z = \mathbb{C}^2 \times \mathbb{C}^4, \quad (x, y, t) \mapsto (x, \widetilde{y}, t), \tag{5.42}$$

$$\widetilde{y}(x,y,t) := \int_0^y 2(u^2+t_3+xt_4)\mathrm{d}u = \frac{2}{3}y^3 + 2(t_3+xt_4)y, \tag{5.43}$$

and observe

$$\begin{aligned} \mathrm{d}F_{H_4} &= (3x^2+t_2+\widetilde{y}t_4)\mathrm{d}x + (y^2+t_3+xt_4)^2\mathrm{d}y \\ &\quad + \mathrm{d}t_1 + x\mathrm{d}t_2 + \widetilde{y}\mathrm{d}t_3 + x\widetilde{y}\mathrm{d}t_4, \end{aligned} \tag{5.44}$$

$$\frac{9}{4}\widetilde{y}^2 + 4(t_3+xt_4)^3 = (y^2+t_3+xt_4)^2(y^2+4(t_3+xt_4)). \tag{5.45}$$

Therefore

$$\phi^*(\alpha_Z) = \mathrm{d}F - \frac{\partial F}{\partial x}\mathrm{d}x - \frac{\partial F}{\partial y}\mathrm{d}y \tag{5.46}$$

and the image under ϕ of the reduced critical set C_F of F_{H_4} is

$$\phi(C_F) = \left\{(x,y,t)\in Z|3x^2+t_2+yt_4=0,\ \frac{9}{16}y^2+t_3+xt_4=0\right\}. \tag{5.47}$$

An automorphism

$$Z \to Z,\ (x,y,t_1,t_2,t_3,t_4) \mapsto (r^{-1}x, s^{-1}y, t_1, rt_2, st_3, rst_4) \tag{5.48}$$

for suitable $r, s \in \mathbb{C}^*$ maps $\phi(C_F)$ to C and respects π_Z and α_Z, together with the induced automorphism $M \to M$. □

Sketch of the proof of Theorem 5.21: (a) Consider the data in Theorem 5.22 (b). The Euler field on $M = \mathbb{C}^n$ is $E = \frac{1}{h}\sum_i d_i t_i \frac{\partial}{\partial t_i}$. The coefficients of the Lyashko–Looijenga map $\Lambda : M \to \mathbb{C}^n$ are up to a sign the symmetric polynomials in the eigenvalues of $E\circ$. Because of $\mathrm{Lie}_E(E\circ) = E\circ$, the coefficient Λ_i is weighted homogeneous of degree i with respect to the weights $(\frac{d_1}{h}, \ldots, \frac{d_n}{h})$ for $(t_1, \ldots, t_n)$. The Lyashko–Looijenga map is a branched covering of degree

$$n! \cdot \left(\prod \frac{d_i}{h}\right)^{-1} = n!h^n \cdot |W|^{-1} \tag{5.49}$$

and $(M, \circ, e)$ is simple (Corollary 4.17 (b)). The analytic spectrum is isomorphic to C.

(b) The dimension $\dim(M_k, p)$ of an irreducible germ in the decomposition $(M,p) = \prod_{k=1}^l (M_k, p)$ of a germ of a massive F-manifold is equal to the intersection multiplicity of T_p^*M with the corresponding germ (L, λ_k) of the analytic spectrum L. This number will be called the intersection multiplicity $IM(\lambda_k)$.

$(S_\mu(q), q)$ denotes for any $q \in M$ the μ-constant stratum through q (section 4.4), and $l(q)$ the number of irreducible components of (M, q). For any subvariety $S \subset L$ we have the estimates

$$\max(l(q)|q \in \pi(S)) \le n + 1 - \min(IM(\sigma)|\sigma \in S), \tag{5.50}$$

$$\max(\dim(S_\mu(q), q) \mid q \in S) \ge \dim S, \tag{5.51}$$

$$\max(\mathrm{mod}_\mu(M, q) \mid q \in S) + n + 1 - \min(IM(\sigma) \mid \sigma \in S) \ge \dim S. \tag{5.52}$$

Therefore, if M is simple then

$$\min(IM(\sigma) \mid \sigma \in S) \le n + 1 - \dim S \tag{5.53}$$

for any subvariety $S \subset L$.

Now suppose that $((M, p), \circ, e)$ is an irreducible germ of a simple F-manifold and that $\phi : (M, p) \to \prod_{i=1}^n (C_i, 0)$ is an isomorphism to a product of germs of plane curves (they are necessarily plane because of Proposition 3.7).

If at least two curve germs were not smooth, e.g. $(C_{n-1}, 0)$ and $(C_n, 0)$, then the intersection multiplicities $IM(p)$ for points p in $S_1 := \phi^{-1}(\prod_{i=1}^{n-2}(C_i, 0) \times \{0\})$ would be at least 4; but $\dim(S_1, p) = n - 2$, a contradiction to (5.53). So, at most one curve, e.g. $(C_n, 0)$, is not smooth.

The irreducible germs of F-manifolds which correspond to generic points of $\pi(S_2)$ for $S_2 := \phi^{-1}(\prod_{i=1}^{n-1}(C_i, 0) \times \{0\})$ are at most 2-dimensional because of (5.53). Therefore

$$(C_n, 0) \cong (\{(x, y) \in \mathbb{C}^2 \mid x^2 = y^r\}, 0) \tag{5.54}$$

for some $r \in \mathbb{N}$. If $r \ge 4$ and $n \ge 3$ then the set of possible intersection multiplicities for points in S_2 has a gap at 3 and a subvariety $S_3 \subset S_2$ exists with dim $S_3 = n - 2$ and $\min(IM(\sigma) \mid \sigma \in S_3) \ge 4$ [**Gi2**, p. 3266], a contradiction to (5.53). Therefore $r \in \{1, 2, 3\}$ or $n \le 2$.

If $r \in \{1, 2\}$ then (M, p) is the base space of a semiuniversal unfolding of a hypersurface singularity ($r = 1$, Theorem 5.7, [**AGV1**, ch. 19]) or boundary singularity ($r = 2$, Theorem 5.15, [**DD**]). Simplicity of their F-manifolds corresponds to simplicity of the singularities. The simple hypersurface singularities are A_n, D_n, E_6, E_7, E_8 [**AGV1**]. The simple boundary singularities are B_n, C_n, and F_4 [**Ar2**][**AGV1**]. The boundary singularities B_n and C_n are dual boundary singularities and have isomorphic discriminants and F-manifolds.

The details of the case $r = 3$ [**Gi2**, pp. 3269–3271] are difficult and will not be given here. In that case the set of possible intersection multiplicities for points in S_2 has a gap at 5. If $n \ge 6$ then a subvariety $S_4 \subset S_2$ exists with $\dim S_4 = n - 4$ and $\min(IM(\sigma) \mid \sigma \in S_4) \ge 6$, a contradiction to (5.53). The case $r = 3$ and $n = 3$ corresponds to H_3, the case $r = 3$ and $n = 4$ corresponds to H_4. □

Proof of Theorem 5.20: It is sufficient to consider an irreducible germ $((M, p), \circ, e)$. If it corresponds to a Coxeter group then it is simple (Theorem 5.21 (a)) and T_pM is a Frobenius algebra (Theorem 5.22 (b)).

Suppose that (M, p) is simple and that T_pM is a Frobenius algebra. We will show by induction on the dimension $n = \dim M$ that the analytic spectrum (L, λ) is isomorphic to $(\{(x, y) \in \mathbb{C}^2 \mid x^2 = y^r\}, 0) \times (\mathbb{C}^{n-1}, 0)$ for some $r \in \mathbb{N}$.

This is clear for $n = 2$. Suppose that $n \geq 3$. The maximal ideal of T_pM is called $\mathbf{m}$. The socle $\mathrm{Ann}_{T_pM}(\mathbf{m})$ of the Gorenstein ring T_pM has dimension 1, therefore $\mathrm{Ann}_{T_pM}(\mathbf{m}) \subsetneq \mathbf{m}$ and $\mathbf{m}^2 \neq 0$. In the equations for the analytic spectrum $(L, \lambda) \subset T_p^*M$ one can eliminate fibre coordinates which correspond to $\mathbf{m}^2 \subset T_pM$: the embedding dimension of (L, λ) is

$$\mathrm{embdim}(L, \lambda) \leq n + \dim \frac{\mathbf{m}}{\mathbf{m}^2} \leq 2n - 2 \tag{5.55}$$

(Lemma 4.22). Then $(L, \lambda) \cong (\mathbb{C}^2, 0) \times (L'', \lambda'')$ (Proposition 3.7). There exists $\lambda_2 \in L$ close to λ such that $(L, \lambda_2) \cong (L, \lambda)$ and $\pi(\lambda_2)$ is not in the e-orbit of p.

Now for all q near p, but outside of the e-orbit of p, the germ (M, q) is reducible because of $\mathrm{mod}_\mu(M, p) = 0$. For all q near p the germ (M, q) is simple and T_qM is a Frobenius algebra (Lemma 2.2).

One can apply the induction hypothesis to the irreducible component of $(M, \pi(\lambda_2))$ which corresponds to λ_2. Its analytic spectrum (L', λ') is isomorphic to a product of a smooth germ and a curve as above. Now $(L, \lambda) \cong (L, \lambda_2) \cong (\mathbb{C}^{n-\dim L'}, 0) \times (L', \lambda')$. One applies Theorem 5.21 (b). □

5.4 Coxeter groups and Frobenius manifolds

K. Saito [**SK3**] introduced a flat metric on the complex orbit space of a finite irreducible Coxeter group. Dubrovin [**Du2**][**Du3**, Lecture 4] showed that this metric and the multiplication and the Euler field from Theorem 5.18 together yield the structure of a massive Frobenius manifold on the complex orbit space (Theorem 5.23).

The Euler field has positive degrees. Dubrovin [**Du2**][**Du3**, p. 268] conjectured that these Frobenius manifolds and products of them are the only massive Frobenius manifolds with an Euler field with positive degrees.

We will prove this conjecture (Theorem 5.25). Theorem 5.20, which builds on Givental's result (Theorem 5.21, [**Gi2**, Theorem 14]), will be crucial.

We use the same notations as in section 5.3. A metric on a complex manifold is a nondegenerate complex bilinear form on the tangent bundle. The flat standard metric on $\mathbb{C}^n$ is invariant with respect to the Coxeter group W and induces a flat metric $\check{g}$ on $M - \mathcal{D}$. Dubrovin proved the following with differential geometric tools [**Du2**][**Du3**, Lecture 4 and pp. 191 and 195].

Theorem 5.23 (Dubrovin) *Let W be a finite irreducible Coxeter group with complex orbit space $M = \mathbb{C}^n / W$, Euler field E, a unit field e, and a multiplication $\circ$ on M as in Theorem 5.18.*

The metric g on $M - \mathcal{D}$ with

$$g(X, Y) := \check{g}(E \circ X, Y) \tag{5.56}$$

for any (local) vector fields X and Y extends to a flat metric on M and coincides with K. Saito's flat metric. $(M, \circ, e, E, g)$ is a Frobenius manifold. The Euler field satisfies

$$\mathrm{Lie}_E(g) = \left(1 + \frac{2}{h}\right) g. \tag{5.57}$$

There exists a basis of flat coordinates $z_1, \ldots, z_n$ on M with $z_i(0) = 0$ and $e = \frac{\partial}{\partial z_1}$ and

$$E = \frac{1}{h} \sum d_i \cdot z_i \frac{\partial}{\partial z_i}. \tag{5.58}$$

Remarks 5.24 (i) K. Saito (and also Dubrovin) introduced the flat metric g in a way different from formula (5.56): The metrics $\check{g}$ and g on $M - \mathcal{D}$ induce two isomorphisms $T(M - \mathcal{D}) \to T^*(M - \mathcal{D})$. The metrics $\check{g}$ and g are lifted with the respective isomorphisms to metrics $\check{g}^*$ and g^* on the cotangent bundle $T^*(M - \mathcal{D})$. Then

$$g^* = \mathrm{Lie}_e(\check{g}^*) \tag{5.59}$$

([**Du3**, pp. 191 and 195]). (Here $\check{g}^*$ and g^* are considered as (0, 2)-tensors.) K. Saito introduced g with the formula (5.59).

(ii) Closely related to (5.56) and (5.59) is ([**Du3**, pp. 191 and 270])

$$\sum_{i=1}^{n} \frac{\partial Q_1}{\partial x_i} \cdot \frac{\partial Q_2}{\partial x_i} = \check{g}^*(\mathrm{d}Q_1, \mathrm{d}Q_2) = i_E(\mathrm{d}Q_1 \circ \mathrm{d}Q_2). \tag{5.60}$$

Here $Q_1, Q_2 \in \mathbb{C}[x_1, \ldots, x_n]^W$ are W-invariant polynomials; $\mathrm{d}Q_1$ and $\mathrm{d}Q_2$ are interpreted as sections in T^*M; the multiplication $\circ$ is lifted to T^*M with the isomorphism $TM \to T^*M$ induced by g; i_E is the contraction of a 1–form with E.

The first equality is trivial. Equation (5.60) is related to Arnold's convolution of invariants ([**Ar3**][**Gi1**]).

(iii) A Frobenius manifold as in Theorem 5.23 for a finite irreducible Coxeter group is not unique because the unit field and the multiplication are not unique. Contrary to the F-manifold, it is not even unique up to isomorphism. There is one complex parameter between $(M, \circ, e)$ and (M, g) to be chosen: $(M, \circ, e, E, c\cdot g)$ respectively $(M, c\cdot\circ, c^{-1}\cdot e, E, g)$ is a Frobenius manifold for any $c \in \mathbb{C}^*$.

(iv) We consider only Frobenius manifolds with an Euler field which is normalized by $\mathrm{Lie}_E(\circ) = 1\cdot\circ$ (compare Remark 2.17 (c)). The product $\prod M_i$ of Frobenius manifolds $(M_i, \circ_i, e_i, E_i, g_i)$ also carries the structure of a Frobenius manifold if $\mathrm{Lie}_{E_i}(g_i) = D\cdot g_i$ holds with the same number $D \in \mathbb{C}$ for all i. This follows from Proposition 2.10, Theorem 2.15 and Remark 2.17 (c) (compare also [**Du3**, p. 136]).

(v) Especially, the complex orbit space $\mathbb{C}^n/W$ of a reducible Coxeter group can be provided with the structure of a Frobenius manifold if the irreducible Coxeter subgroups have the same Coxeter number.

The Frobenius manifold is not unique. The different choices are parameterized by $(\mathbb{C}^*)^{|irr.\ subgroups|}$ in the obvious way (cf. the Remarks 5.19 (iii) and 5.24 (iii)).

The *spectrum* of a Frobenius manifold $(M, \circ, e, E, g)$ is defined as follows (cf. Remark 9.2 e)). The Levi–Civita connection of the metric g is denoted by ∇. The operator $\nabla E : \mathcal{T}_M \to \mathcal{T}_M, X \mapsto \nabla_X E$, acts on the space of flat fields ([**Du3**, p. 132], [**Man1**, p. 24]) and coincides there with $-\mathrm{ad}\, E$. The set of its eigenvalues $\{w_1, \ldots, w_n\}$ is the spectrum ([**Man1**]). If $-\mathrm{ad}\, E$ acts semisimple on the space of flat fields then there exists locally a basis of flat coordinates $z_1, \ldots, z_n$ with

$$E = \sum_i (w_i z_i + r_i)\frac{\partial}{\partial z_i} \tag{5.61}$$

for some $r_i \in \mathbb{C}$.

The following was conjectured by Dubrovin ([**Du2**][**Du3**, p. 268]).

Theorem 5.25 *Let* $((M, p), \circ, e, E, g)$ *be the germ of a Frobenius manifold with the following properties:*
generically semisimple multiplication;
$\mathrm{Lie}_E(\circ) = 1\cdot\circ$ *and* $\mathrm{Lie}_E(g) = D\cdot g$;

$$E = \sum w_i z_i \frac{\partial}{\partial z_i} \tag{5.62}$$

for a basis of flat coordinates z_i *with* $z_i(p) = 0$;
positive spectrum $(w_1, \ldots, w_n)$, *that is,* $w_i > 0$ *for all* i.

Then (M, p) decomposes uniquely into a product of germs at 0 of Frobenius manifolds for certain irreducible Coxeter groups. The Coxeter groups have all the same Coxeter number $h = \frac{2}{D-1}$.

Proof: As in the proof of Theorem 5.21 (a), the hypotheses on the Euler field show that the Lyashko–Looijenga map $\Lambda : (M_k, p) \to \mathbb{C}^n$ is finite and that the F-manifold $(M, \circ, e)$ is simple. One applies Theorem 5.20 and Theorem 5.26. □

Theorem 5.26 *Let $((M, p), \circ, e, E, g)$ be the germ of a Frobenius manifold such that $((M, p), \circ, e, E)$ is isomorphic to the germ at 0 of the F-manifold of a finite Coxeter group with the standard Euler field.*

Then the irreducible Coxeter subgroups have the same Coxeter number and $((M, p), \circ, e, E, g)$ is isomorphic to a product of germs at 0 of Frobenius manifolds for these Coxeter groups.

Proof: First we fix notations. W is a finite Coxeter group which acts on $V = \mathbb{C}^n$ and respects the standard bilinear form. The decomposition of W into l irreducible Coxeter groups $W_1, \ldots, W_l$ corresponds to an orthogonal decomposition $V = \bigoplus_{k=1}^{l} V_k$. The choice of n algebraically independent homogeneous polynomials $P_1, \ldots, P_n \in \mathbb{C}[x_1, \ldots, x_n]^W$ identifies the quotient $M = V/W$ with $\mathbb{C}^n$. The quotient map $\psi : V \to M$ decomposes into a product of quotient maps $\psi_k : V_k \to V_k/W_k = M_k$. The F-manifold $M \cong \prod_{k=1}^{l} M_k$ is the product of the F-manifolds M_k.

Setting $\varepsilon := \sum_{i=1}^{n} x_i \frac{\partial}{\partial x_i}$ on V and $\varepsilon_k := \varepsilon|_{V_k}$, the standard Euler field E_k on M_k is $E_k = \frac{1}{h_k} \mathrm{d}\psi_k(\varepsilon_k)$. Here h_k is the Coxeter number of W_k. The Euler field on M is

$$E = \sum_{k=1}^{l} E_k = \sum_{i=1}^{n} w_i t_i \frac{\partial}{\partial t_i}, \tag{5.63}$$

$\{w_1, \ldots, w_n\}$ is the union of the invariant degrees of W_k, divided by h_k.

Now suppose that g is a flat metric on the germ $(M, 0)$ such that $((M, 0), \circ, e, E, g)$ is a germ of a Frobenius manifold with $\mathrm{Lie}_E(g) = D \cdot g$. Consider a system of flat coordinates $z_1, \ldots, z_n$ of $(M, 0)$ with $z_i(0) = 0$. The space $\bigoplus_{i=1}^{n} \mathbb{C} \cdot \frac{\partial}{\partial z_i}$ of flat fields is invariant with respect to $ad E$ ([**Du3**, p. 132], [**Man1**, p. 24]) and the space of affine linear functions $\mathbb{C} \cdot 1 \oplus \bigoplus_{i=1}^{n} \mathbb{C} \cdot z_i \subset \mathcal{O}_{M,0}$ is invariant with respect to E. Because E vanishes at 0 even the subspace $\bigoplus_{i=1}^{n} \mathbb{C} \cdot z_i$ is invariant with respect to E.

The weights $w_1, \ldots, w_n$ of E are positive. Therefore the coordinates $z_1, \ldots, z_n$ can be chosen to be weighted homogeneous polynomials in $\mathbb{C}[t_1, \ldots, t_n]$ of degree $w_1, \dots, w_n$. Thus the spectrum of the Frobenius manifold is $\{w_1, \dots, w_n\}$. It is symmetric with respect to $\frac{D}{2}$, because of $\mathrm{Lie}_E(g) = D \cdot g$; hence $1 + \frac{2}{h_k} = D$ for all $k = 1, \dots, l$. The Coxeter numbers are all equal, $h := h_1 = \dots = h_l$, the Euler field E is

$$E = \frac{1}{h} \mathrm{d}\psi(\varepsilon). \tag{5.64}$$

It remains to show that g is induced as in Theorem 5.23 from a metric on V which it the orthogonal sum of multiples of the standard metrics on the subspaces V_k.

The operator $\mathcal{U} = E\circ : \mathcal{T}_M \to \mathcal{T}_M$ is invertible on $M - \mathcal{D}$. The metric $\check{g}$ on $M - \mathcal{D}$ with

$$\check{g}(X, Y) := g(\mathcal{U}^{-1}(X), Y) \tag{5.65}$$

is flat ([**Du3**, pp. 191 and 194], [**Man1**]). It lifts to a flat metric $\widetilde{g}$ on $V - \psi^{-1}(\mathcal{D})$. We claim that $\widetilde{g}$ extends to a flat metric on the union $\psi^{-1}(\mathcal{D})$ of the reflection hyperplanes.

It is sufficient to consider a generic point p in one reflection hyperplane. Then the e-orbit of $\psi(p)$ intersects $\mathcal{D}$ in n points; there exist canonical coordinates $u_1, \dots, u_n$ in a neighbourhood of $\psi(p)$ with $e_i \circ e_j = \delta_{ij} e_i$, $g(e_i, e_j) = 0$ for $i \neq j$,

$$E = u_1 e_1 + \sum_{i=2}^{n} (u_i + r_i) e_i \quad \text{for some } r_i \in \mathbb{C}^*,$$

$$(\mathcal{D}, \psi(p)) \cong (\{u \mid u_1 = 0\}, 0).$$

The map germ $\psi : (V, p) \to (M, \psi(p))$ is a twofold covering, branched along $(\mathcal{D}, \psi(p))$, and is given by $(\widetilde{u}_1, \dots, \widetilde{u}_n) \mapsto (\widetilde{u}_1{}^2, \widetilde{u}_2, \dots, \widetilde{u}_n) = (u_1, \dots, u_n)$ for some suitable local coordinates $\widetilde{u}_1, \dots, \widetilde{u}_n$ on (V, p). Then

$$\widetilde{g}\left(\frac{\partial}{\partial \widetilde{u}_1}, \frac{\partial}{\partial \widetilde{u}_1}\right) = \check{g}(4u_1 e_1, e_1) = \check{g}(4E \circ e_1, e_1) = 4g(e_1, e_1),$$

$$\widetilde{g}\left(\frac{\partial}{\partial \widetilde{u}_i}, \frac{\partial}{\partial \widetilde{u}_i}\right) = \check{g}(e_i, e_i) = \frac{1}{u_i + r_i} g(e_i, e_i) \quad \text{for } i \geq 2,$$

$$\widetilde{g}\left(\frac{\partial}{\partial \widetilde{u}_i}, \frac{\partial}{\partial \widetilde{u}_j}\right) = 0 \quad \text{for } i \neq j.$$

So $\widetilde{g}$ extends to a nondegenerate (and then flat) metric on V.

The Coxeter group W acts as a group of isometries with respect to $\widetilde{g}$. It remains to show that the vector space structure on V which is induced by $\widetilde{g}$ (and $0 \in V$) coincides with the original vector space structure. Then $\widetilde{g}$ is an orthogonal sum of multiples of the standard metrics on the subspaces V_k, because each W-invariant quadratic form is a sum of W_k-invariant quadratic forms on the subspaces V_k and they are unique up to scalars.

Let $\widetilde{\varepsilon}$ be the vector field on V which corresponds to the $\mathbb{C}^*$-action of the vector space structure induced by $\widetilde{g}$. Then $\mathrm{Lie}_{\widetilde{\varepsilon}}\widetilde{g} = 2 \cdot \widetilde{g}$. Because of $\mathrm{Lie}_E(\mathcal{U}) = \mathcal{U}$, $\mathrm{Lie}_E(g) = (1 + \frac{2}{h})g$, and $E = \frac{1}{h}\mathrm{d}\psi(\varepsilon)$ we also have $\mathrm{Lie}_\varepsilon \widetilde{g} = 2 \cdot \widetilde{g}$ for the vector field ε, which corresponds to the $\mathbb{C}^*$-action of the old vector space structure. The difference $\varepsilon - \widetilde{\varepsilon}$ satisfies $\mathrm{Lie}_{\varepsilon - \widetilde{\varepsilon}}(\widetilde{g}) = 0$ and is a generator of a 1-parameter group of isometries. As it is also tangent to the union of reflection hyperplanes, it vanishes. The vector field $\varepsilon = \widetilde{\varepsilon}$ determines a unique space of linear functions on V and a unique vector space structure. □

5.5 3-dimensional and other F-manifolds

The F-manifolds in sections 5.1–5.3 were special in several aspects: the analytic spectrum was weighted homogeneous and a complete intersection. Therefore an Euler field of weight 1 always existed, and the tangent spaces were Frobenius algebras. Furthermore, the stratum of points with irreducible germs of dimension ≥ 3 had codimension 2.

Here we want to present examples with different properties. A partial classification of 3-dimensional germs of massive F-manifolds will show that already in dimension 3 most germs are not simple and do not even have an Euler field of weight 1. Examples of germs (M, p) of simple F-manifolds such that T_pM is not a Frobenius algebra will complement Theorem 5.20.

First, a construction which is behind the formulas for A_n, B_n, H_3, $I_2(m)$ in Theorem 5.22 (b)(i) provides many other examples.

Proposition 5.27 *Fix the following data:* $(M, 0) = (\mathbb{C}^n, 0)$,
$(Z, 0) = (\mathbb{C}, 0) \times (M, 0)$ *with coordinates* $(x, t) = (x, t_1, \ldots, t_n)$,
the projection $\pi_Z : (Z, 0) \to (M, 0)$,
the 1–form $\alpha_Z := \mathrm{d}t_1 + x\mathrm{d}t_2 + \cdots + x^{n-1}\mathrm{d}t_n$ *on* Z,
the function $\widetilde{t_2}(x, t) := t_2 + 2xt_3 + \cdots + (n-1)x^{n-2}t_n$,
an isolated plane curve singularity (or a smooth germ) $f : (\mathbb{C}^2, 0) \to (\mathbb{C}, 0)$ *with* $f(x, 0) = x^n \cdot unit \in \mathbb{C}\{x\}$,
the subvariety $C := \{(x, t) \in Z \mid f(x, \widetilde{t_2}) = 0\} \subset Z$.

(a) The map

$$\mathbf{a}_C : \mathcal{T}_{M,0} \to \mathcal{O}_{C,0}, \quad X \mapsto \alpha_Z(\widetilde{X})|_C \tag{5.66}$$

($\widetilde{X}$ is a lift of X to Z) is welldefined and an isomorphism of $\mathcal{O}_{M,0}$-modules. The germ $(M, 0)$ with the induced multiplication on $\mathcal{T}_{M,0}$ is an irreducible germ of a massive F-manifold. Its analytic spectrum is isomorphic to

$(C, 0) \cong (\mathbb{C}^{n-1}, 0) \times (f^{-1}(0), 0)$. *For each* $t \in M$ *the tangent space* $T_t M$ *is isomorphic to a product of algebras* $\mathbb{C}\{x\}/(x^k)$ *and is a Frobenius algebra.*

(b) An Euler field of weight 1 exists on (M, p) *if and only if the curve singularity* $f(x, y)$ *is weighted homogeneous.*

(c) Suppose that $\text{mult} f = n$. *Then the caustic is* $\mathcal{K} = \{t \in M \mid t_2 = 0\}$. *The germ* (M, t) *is irreducible for all* $t \in \mathcal{K}$, *so the caustic is equal to the* μ*-constant stratum of* $(M, 0)$. *The modality is* $\text{mod}_\mu(M, 0) = n - 2$ *(the maximal possible).*

Proof: (a) The 1–form α_Z is exact on C_{reg} because of $d\alpha_Z = dx d\widetilde{t_2}$. One can apply Corollary 4.21.

(b) Corollary 3.8 (b).

(c) For $t_2 = 0$ fixed we have $f(x, \widetilde{t_2}) = x^n \cdot \textit{unit} \in \mathbb{C}\{x, t_3, \ldots, t_n\}$. Thus the projection $\pi_C : C \to M$ is a branched covering of degree n, with $\pi_C^{-1}(\{t \mid t_2 = 0\}) = \{0\} \times \{t \mid t_2 = 0\}$ and unbranched outside of $\{t \mid t_2 = 0\}$. The analytic spectrum is isomorphic to C. □

Remarks 5.28 (i) The function $\widetilde{t_2}$ is part of a coordinate system on T^*M for a different Lagrange fibration: the coordinates

$$\widetilde{y_1} = y_1,\ \widetilde{y_2} = y_2,\ \widetilde{y_i} = y_i - y_2^{i-1} \quad \text{for } i \geq 3, \tag{5.67}$$

$$\widetilde{t_1} = t_1,\ \widetilde{t_2} = t_2 + 2y_2 t_3 + \cdots + (n-1) y_2^{n-2} t_n,\ \widetilde{t_i} = t_i \quad \text{for } i \geq 3$$

satisfy

$$\sum_{i=1}^{n} d\widetilde{y_i} d\widetilde{t_i} = \sum_{i=1}^{n} dy_i dt_i = d\alpha. \tag{5.68}$$

The analytic spectrum of an F-manifold as in Proposition 5.27 is

$$L = \{(y, t) \in T^*M \mid \widetilde{y_1} = 1,\ f(\widetilde{y_2}, \widetilde{t_2}) = 0,\ \widetilde{y_i} = 0 \text{ for } i \geq 3\}. \tag{5.69}$$

It is a product of Lagrange curves.

(ii) Another different Lagrange fibration is behind the formulas for D_4, F_4, H_4 in Theorem 5.22 (b)(ii). There are many possibilities to generalize the construction of the above examples.

In dimension 3, there exist up to isomorphism only two irreducible commutative and associative algebras,

$$Q^{(1)} := \mathbb{C}\{x\}/(x^3) \quad \text{and} \tag{5.70}$$

$$Q^{(2)} := \mathbb{C}\{x, y\}/(x^2, xy, y^2); \tag{5.71}$$

and $Q^{(1)}$ is a Frobenius algebra, $Q^{(2)}$ not.

Theorem 5.29 *Let (M, p) be an irreducible germ of a 3-dimensional massive F-manifold with analytic spectrum $(L, \lambda) \subset T^*M$.*

(a) Suppose $T_pM \cong Q^{(1)}$. Then (L, λ) has embedding dimension 3 or 4 and $(L, \lambda) \cong (\mathbb{C}^2, 0) \times (C', 0)$ for a plane curve $(C', 0) \subset (\mathbb{C}^2, 0)$ with $\mathrm{mult}(C', 0) \leq 3$. An Euler field of weight 1 exists if and only if $(C', 0)$ is weighted homogeneous.

(b) Suppose $T_pM \cong Q^{(1)}$ and $(L, \lambda) \cong (\mathbb{C}^2, 0) \times (C', 0)$ with $\mathrm{mult}(C', 0) < 3$. Then $((M, p), \circ, e)$ is one of the germs A_3, B_3, H_3.

(c) Suppose $T_pM \cong Q^{(1)}$ and $(L, \lambda) \cong (\mathbb{C}^2, 0) \times (C', 0)$ with $\mathrm{mult}(C', 0) = 3$. Then the caustic $\mathcal{K}$ is a smooth surface and coincides with the μ-constant stratum; that means, $T_qM \cong Q^{(1)}$ for each $q \in \mathcal{K}$. The modality is $\mathrm{mod}_\mu(M, p) = 1$ (the maximal possible).

(d) Suppose $T_pM \cong Q^{(2)}$. Then (L, λ) has embedding dimension 5 and $(L, \lambda) \cong (\mathbb{C}, 0) \times (L^{(r)}, 0)$. Here $(L^{(r)}, 0)$ is a Lagrange surface with embedding dimension 4. Its ring $\mathcal{O}_{L^{(r)},0}$ is a Cohen–Macaulay ring, but not a Gorenstein ring.

Proof: (a) One chooses coordinates (t_1, t_2, t_3) for (M, p) (as in the proof of Lemma 4.22) with $e = \frac{\partial}{\partial t_1}$ and

$$\mathbb{C} \cdot \left[\frac{\partial}{\partial t_2}\right] + \mathbb{C} \cdot \left[\frac{\partial}{\partial t_3}\right] = \mathbf{m} \subset T_pM, \tag{5.72}$$

$$\mathbb{C} \cdot \left[\frac{\partial}{\partial t_3}\right] = \mathbf{m}^2 \subset T_pM. \tag{5.73}$$

The dual coordinates on (T^*M, T_p^*M) are $y_1, \ldots, y_n$. There exist functions $a_0, a_1, a_2, b_0, b_1, b_2 \in \mathbb{C}\{t_2, t_3\}$ with

$$L = \left\{(y, t) \mid y_1 = 1, y_3 = b_2y_2^2 + b_1y_2 + b_0, \quad y_2^3 = a_2y_2^2 + a_1y_2 + a_0\right\}. \tag{5.74}$$

The Hamilton fields of the smooth functions $y_1 - 1$ and $y_3 - \sum_{i=0}^{2} b_i y_2^i$ are $\frac{\partial}{\partial t_1}$ and $\frac{\partial}{\partial t_3} + \ldots$ in T^*M. They are tangent to L. Therefore $(L, \lambda) \cong (\mathbb{C}^2, 0) \times (C', 0)$ with $(C', 0) \cong (L, \lambda) \cap T_p^*M$ (cf. Proposition 3.7). The statement on the Euler field is contained in Corollary 3.8 (b).

(b) We have $\mathrm{mult}(C', 0) \leq 2$; and the intersection multiplicity of $(C', 0)$ with a suitable smooth curve is 3. So, $(C', 0)$ is either smooth or a double point or a cusp. In the first two cases, one can apply the correspondence between F-manifolds and hypersurface or boundary singularities (Theorem 5.6 and Theorem 5.14) and the fact that A_3, B_3, and C_3 are the only hypersurface or boundary singularities with Milnor number 3.

Suppose $(C', 0)$ is a cusp and $((M, p), \circ, e)$ is not H_3. Then it is not simple because of Theorem 5.20. The μ-constant stratum $S_\mu = \{q \in M \mid T_q M \cong Q^{(1)}\}$ is more than the e-orbit of p. It can only be the image in M of the surface of cusp points of L, because at other points q close to p the germ (M, q) is A_1^3 or $A_1 \times A_2$.

So at each cusp point λ' of L the intersection multiplicity of $T^*_{\pi(\lambda')}M$ and (L, λ') is 3. This property is not preserved by small changes of the Lagrange fibration (e.g. as in Remark 5.28 (i)). But Givental proved that a versal Lagrange map is stable with respect to small changes of the Lagrange fibration [**Gi2**, p. 3251, Theorem 3 and its proof]. This together with section 3.4 yields a contradiction.

(c) In this case, at each point λ' (close to λ) of the surface of singular points of L the intersection multiplicity of $T^*_{\pi(\lambda')}M$ and (L, λ') is 3 and the map germ $\pi : (L, \lambda') \to (M, \pi(\lambda'))$ is a branched covering of degree 3. This implies all the statements.

(d) If the embedding dimension of (L, λ) is ≤ 4 then $(L, \lambda) \cong (\mathbb{C}^2, 0) \times (C', 0)$ for some plane curve $(C', 0)$ by Proposition 3.7. Then (L, λ) is a complete intersection and the tangent spaces $T_p M$ are Frobenius algebras.

So if $T_p M \cong Q^{(2)}$ then the embedding dimension of (L, λ) is 5. The ring $\mathcal{O}_{L^{(r)},0}$ is a Cohen–Macaulay ring because the projection $L^{(r)} \to M^{(r)}$ of the restricted Lagrange map is finite and flat. It is not a Gorenstein ring because $T_p M$ is not a Gorenstein ring. □

The next result provides a complete classification and normal forms for those irreducible germs (M, p) of 3-dimensional massive F-manifolds which satisfy $T_p M \cong Q^{(1)}$ and whose analytic spectrum consists of 3 components. Part (a) gives an explicit construction of all those F-manifolds.

Theorem 5.30 *(a) Choose two discrete parameters $p_2, p_3 \in \mathbb{N}$ with $p_2 \geq p_3 \geq 2$ and choose $p_3 - 1$ holomorphic parameters $(g_0, g_1, \ldots, g_{p_3-2}) \in \mathbb{C}^* \times \mathbb{C}^{p_3-2}$ with $g_0 \neq 1$ if $p_2 = p_3$.*

Define $(M, 0) := (\mathbb{C}^3, 0)$,
$(Z, 0) := (\mathbb{C}, 0) \times (M, 0)$ with coordinates $(x, t_1, t_2, t_3) = (x, t)$,
$g := \sum_{i=0}^{p_3-2} g_i t_2^i + t_2^{p_3-2} \cdot t_3$,
$f_1 := 0,\ f_2 := t_2^{p_2},\ f_3 := t_2^{p_3} \cdot g$,
$C := \bigcup_{i=1}^3 C_i := \bigcup_{i=1}^3 \{(x, t) \in Z \mid x = \frac{\partial f_i}{\partial t_2}\} \subset Z$,
$b_2(t_2, t_3) := (p_3 \cdot g + t_2 \cdot \frac{\partial g}{\partial t_2})^{-1} \cdot (p_3 \cdot g + t_2 \cdot \frac{\partial g}{\partial t_2} - p_2 \cdot t_2^{p_2-p_3})^{-1}$
(b_2 is a unit in $\mathbb{C}\{t_2, t_3\}$),
$b_1(t_2, t_3) := -b_2 \cdot p_2 \cdot t_2^{p_2-1}$,
the 1–form $\alpha_Z := dt_1 + x dt_2 + (b_2 x^2 + b_1 x) dt_3$ on Z.

(i) Then

$$\frac{\partial f_i}{\partial t_3} = b_2 \left(\frac{\partial f_i}{\partial t_2}\right)^2 + b_1 \frac{\partial f_i}{\partial t_2} \quad \textit{for } i = 1, 2, 3, \tag{5.75}$$

$$\alpha_Z|_{C_i} = \mathrm{d}(t_1 + f_i)|_{C_i} \quad \textit{for } i = 1, 2, 3. \tag{5.76}$$

(ii) The map

$$\mathbf{a}_C : \mathcal{T}_{M,0} \to \mathcal{O}_{C,0}, \quad X \mapsto \alpha_Z(\widetilde{X})|_C \tag{5.77}$$

($\widetilde{X}$ is a lift of X to Z) is welldefined and an isomorphism of $\mathcal{O}_{M,0}$-modules. The germ $(M, 0)$ with the induced multiplication on $\mathcal{T}_{M,0}$ is an irreducible germ of a massive F-manifold with $T_0M \cong Q^{(1)}$. Its analytic spectrum is isomorphic to $(C, 0) \cong (\mathbb{C}^2, 0) \times (C', 0)$, with

$$(C', 0) = (C, 0) \cap (\{(x, t) \mid t_1 = t_3 = 0\}, 0). \tag{5.78}$$

(iii) The caustic is $\mathcal{K} = \{t \in M \mid t_2 = 0\}$ and coincides with the bifurcation diagram $\mathcal{B}$ and with the μ-constant stratum.

(iv) The functions $t_1 + f_i|_{C_i}$, $i = 1, 2, 3$, combine to a function $F : C \to \mathbb{C}$ which is continuous on C and holomorphic on $C_{reg} = C \cap \{(x, t) \mid t_2 \neq 0\}$. The Euler field E on $M - \mathcal{K}$ with $\mathbf{a}_C|_{M-\mathcal{K}}(E) = F|_{M-\mathcal{K}}$ is

$$E = t_1 \frac{\partial}{\partial t_1} + \frac{1}{p_2} t_2 \frac{\partial}{\partial t_2} + \frac{1}{p_2 t_2^{p_3 - 2}} \left((p_2 - p_3)g - t_2 \frac{\partial g}{\partial t_2}\right) \cdot \frac{\partial}{\partial t_3}. \tag{5.79}$$

The following conditions are equivalent:

(α) *The function F is holomorphic on C and E is holomorphic on M,*
(β) *one has $p_3 = 2$ or ($p_2 = p_3 \geq 3$ and $g_i = 0$ for $1 \leq i < p_3 - 2$),*
(γ) *the curve $(C', 0)$ is weighted homogeneous.*

(b) Each irreducible germ (M, p) of a massive F-manifold such that $T_pM \cong Q^{(1)}$ and such that (L, λ) has 3 components is isomorphic to a finite number of normal forms as in (a). The numbers p_2 and p_3 are determined by (L, λ). The number of isomorphic normal forms is $\leq 2p_2$ if $p_2 > p_3$ and $\leq 6p_2$ if $p_2 = p_3$.

Proof: (a) (i) Direct calculation.

(ii) The map $\mathbf{a}_C$ is an isomorphism because b_2 is a unit in $\mathbb{C}\{t_2, t_3\}$. One can apply Corollary 4.21 because of (5.76). For the analytic spectrum see Theorem 5.29 (a).

(iii) The branched covering $(C, 0) \to (M, 0)$ is branched along $\{(x, t) \mid x = t_2 = 0\}$. Compare Theorem 5.29 (c). The generating function $F : C \to \mathbb{C}$ has three different values on $\pi_C^{-1}(t)$ for $t \in M$ with $t_2 \neq 0$ because of (5.76) and the definition of f_i. Therefore $\mathcal{K} = \mathcal{B}$.

(iv) Formula (5.79) can be checked by calculation. The equivalence $(\alpha) \iff (\beta)$ follows. Corollary 3.8 (b) shows $(\alpha) \iff (\gamma)$ (one can also see $(\beta) \iff (\gamma)$ directly).

(b) We start with coordinates (t_1, t_2, t_3) for (M, p) as in the proof of Theorem 5.29 (a). The proofs of Theorem 5.29 (a) and Lemma 4.22 (b) give a unique construction of data $(Z, 0) = (\mathbb{C}, 0) \times (\mathbb{C}^3, 0)$, $(C, 0) \subset (Z, 0)$ and

$$\alpha_Z = \mathrm{d}t_1 + x\mathrm{d}t_2 + (b_2(t_2, t_3)x^2 + b_1(t_2, t_3)x + b_0(t_2, t_3))\mathrm{d}t_3 \tag{5.80}$$

as in Corollary 4.21 for the germ $(M, p) \cong (\mathbb{C}^3, 0)$ of an F-manifold.

The set $(C, 0) = \bigcup_{i=1}^3 (C_i, 0)$ is the union of 3 smooth varieties, which project isomorphically to $(\mathbb{C}^3, 0)$, and is isomorphic to the product of $(\mathbb{C}^2, 0)$ and $(C, 0) \cap (\{(x, t) \mid t_1 = t_3 = 0\}, 0)$.

The components $(C_i, 0)$ can be numbered such that the intersection numbers of the curves $(C_i, 0) \cap (\{(x, t) \mid t_1 = t_3 = 0\}, 0)$ are $p_2 - 1$ for $i = 1, 2$ and $p_3 - 1$ for $i = 1, 3$ and for $i = 2, 3$. The numbers p_2 and p_3 are defined hereby and satisfy $p_2 \geq p_3 \geq 2$.

The 1–form α_Z is exact on C_{reg} and can be integrated to a continuous function $F : (C, 0) \to (\mathbb{C}, 0)$ with $F|_{C_i} = t_1 + f_i$ for a unique function $f_i \in \mathbb{C}\{t_2, t_3\}$. Then $C_i = \{(x, t) \mid x = \frac{\partial f_i}{\partial t_2}\}$ and

$$\frac{\partial f_i}{\partial t_3} = b_2 \cdot \left(\frac{\partial f_i}{\partial t_2}\right)^2 + b_1 \cdot \frac{\partial f_i}{\partial t_2} + b_0. \tag{5.81}$$

We will refine (t_1, t_2, t_3) in several steps and change Z, C, α_Z, f_1, f_2, f_3 accordingly, without explicit mentioning.

1st step: The coordinates (t_1, t_2, t_3) can be chosen such that $(C, 0) \to (\mathbb{C}^3, 0)$ is branched precisely over $\{t \in \mathbb{C}^3 \mid t_2 = 0\}$.

2nd step: The coordinate t_1 can be changed such that $f_1 = 0$.
Then $C_1 = \{(x, t) \mid x = 0\}$ and

$$C_1 \cap C_i = \left\{(x, t) \mid x = 0 = \frac{\partial f_i}{\partial t_2}\right\} = \{(x, t) \mid x = 0, t_2 = 0\}. \tag{5.82}$$

Because of $f_i|_{C_1 \cap C_i} = f_1|_{C_1 \cap C_i} = 0$, the functions f_2 and f_3 can be written uniquely as $f_2 = t_2^{\widetilde{p}_2} \cdot \widetilde{g}$, $f_3 = t_2^{\widetilde{p}_3} \cdot g$ with $\widetilde{p}_2, \widetilde{p}_3 \geq 1$, $\widetilde{g}, g \in \mathbb{C}\{t_2, t_3\} - t_2 \cdot \mathbb{C}\{t_2, t_3\}$. Now (5.82) shows $\frac{\partial f_i}{\partial t_2} = t_i^{\widetilde{p}_i - 1} \cdot \textit{unit}$ and $\widetilde{p}_i \geq 2$. Therefore $\widetilde{p}_i = p_i$ and $\widetilde{g}$ and g are units, with $g(0) \neq \widetilde{g}(0)$ if $p_2 = p_3$.

3rd step: The coordinate t_2 can be chosen such that $f_2 = t_2^{p_2}$.
(5.81) yields $b_0 = 0$ for $i = 1$ and $b_1 = -b_2 \cdot p_2 \cdot t_2^{p_2 - 1}$ for $i = 2$ and

$$t_2^{p_3} \frac{\partial g}{\partial t_3} = b_2 \cdot t_2^{p_3 - 1}\left(p_3 g + t_2 \frac{\partial g}{\partial t_2}\right) t_2^{p_3 - 1}\left(p_3 g + t_2 \frac{\partial g}{\partial t_2} - p_2 \cdot t_2^{p_2 - p_3}\right) \tag{5.83}$$

for $i = 3$. The first, third and fifth factor on the right are units, therefore we have $\frac{\partial g}{\partial t_3} = t_2^{p_3-2} \cdot unit$.

4th step: The coordinate t_3 can be changed such that $g = \sum_{i=0}^{p_3-2} g_i \cdot t_2^i + t_2^{p_3-2} \cdot t_3$.

We have brought the germ (M, p) to a normal form as in (a). The numbering of C_1, C_2, C_3 was unique up to permutation of C_1 and C_2 if $p_2 > p_3$ and arbitrary if $p_2 = p_3$. The choice of t_2 was unique up to a unit root of order p_2. Everything else was unique. □

Remark 5.31 Certain results of Givental motivate some expectations on the moduli of germs of F-manifolds, which are satisfied in the case of Theorem 5.30.

An irreducible germ (M, p) of a massive F-manifold is determined by its restricted Lagrange map $(L^{(r)}, \lambda^{(r)}) \hookrightarrow (T^*M^{(r)}, T_p^*M^{(r)}) \to (M^{(r)}, p^{(r)})$ (section 3.4). Suppose that (M, p) is 3-dimensional with $T_pM \cong Q^{(1)}$. Then $(L^{(r)}, \lambda^{(r)})$ decomposes into a product of two Lagrange curves, a smooth one and a plane curve $(C', 0)$ (Theorem 5.29 (a), Proposition 3.7).

If we fix only the topological type of the curve $(C', 0)$, we can divide the moduli for the possible germs (M, p) into three pieces:

(i) moduli for the complex structure of the germ $(C', 0)$,
(ii) moduli for the Lagrange structure of $(C', 0)$,
(iii) moduli for the Lagrange fibration in the restricted Lagrange map.

Within the μ-constant stratum $S_\mu = \{q \in M \mid T_qM \cong Q^{(1)}\}$ of a representative M, the moduli of types (i) and (ii) are not visible because the Lagrange structure of the curve $(C', 0)$ is constant along S_μ.

But the moduli for the Lagrange fibration are precisely reflected by S_μ because of a result of Givental [**Gi2**, proof of Theorem 3]: as a miniversal Lagrange map, the restricted Lagrange map is stable with respect to small changes of the Lagrange fibration which preserve the symplectic structure; that means, the germ of the Lagrange map after such a small change is the restricted Lagrange map of (M, q) for a point $q \in S_\mu$ close to p.

In view of Theorem 4.15 and Theorem 5.29 (b) and (c) there is one module of type (iii) if mult$(C', 0) = 3$ and no module of type (iii) if mult$(C', 0) = 1$ or 2.

Fixing the complex structure of the plane curve $(C', 0)$, the choice of a Lagrange structure is equivalent to the choice of a volume form. Equivalence classes of it are locally parameterized by $H^1_{Giv}((C', 0))$ ([**Gi2**, Theorem 1], [**Va5**]), so the number of moduli of type (ii) is $\mu - \tau$ (Theorem 3.5 (b)). It is equal to the number of moduli of right equivalence classes of function germs $f : (\mathbb{C}^2, 0) \to (\mathbb{C}, 0)$ with $(f^{-1}(0), 0) \cong (C', 0)$.

The μ-constant stratum of a plane curve singularity in the semiuniversal unfolding is smooth by a result of Wahl ([**Wah**], cf. also [**Matt**]) and its dimension depends only on the topological type of the curve. So one may expect that the number of moduli of types (i) and (ii) together depends only on the topological type of $(C', 0)$ and is equal to this dimension.

(But a canonical relation between the choice of a Lagrange structure and the choice of a function germ for a plane curve $(C', 0)$ is not known.)

In the case of Theorem 5.30 these expectations are met: the topological type of the plane curve is given by the intersection numbers $p_2 - 1$ and $p_3 - 1$; the last one g_{p_3-2} of the complex moduli is of type (iii), it is the module for the μ-constant stratum and for the Lagrange fibration; the other $p_3 - 2$ moduli $(g_0, \ldots, g_{p_3-3})$ are of types (i) and (ii). One can check with [**Matt**, 4.2.1] that $p_3 - 2$ is the dimension of the μ-constant stratum for such a plane curve singularity.

Finally, at least a few examples of germs (M, p) of F-manifolds with $T_0M \cong Q^{(2)}$ will be presented.

Proposition 5.32 *Consider $M = \mathbb{C}^3$ with coordinates (t_1, t_2, t_3) and T^*M with fibre coordinates y_1, y_2, y_3. Choose $p_2, p_3 \in \mathbb{N}_{\geq 2}$. Then the variety*

$$L = \{(y, t) \in T^*M \mid y_1 = 1,\ y_2(y_2 - p_2 t_2^{p_2-1}) = y_2 y_3 = y_3(y_3 - p_3 t_3^{p_3-1}) = 0\} \tag{5.84}$$

has three smooth components and is the analytic spectrum of the structure of a simple F-manifold on M with $T_0M \cong Q^{(2)}$. The field

$$E = t_1 \frac{\partial}{\partial t_1} + \frac{1}{p_2} t_2 \frac{\partial}{\partial t_2} + \frac{1}{p_3} t_3 \frac{\partial}{\partial t_3} \tag{5.85}$$

is an Euler field of weight 1.

Proof: One checks easily that $\alpha = y_1 dt_1 + y_2 dt_2 + y_3 dt_3$ is exact on the three components of L, that the map $\mathbf{a}_C : \mathcal{T}_M \to \pi_* \mathcal{O}_L$ is an isomorphism of $\mathcal{O}_M$-modules, and that E in (5.85) is an Euler field.

The weights of E are positive. This shows via the Lyashko–Looijenga map that $(M, \circ, e)$ is a simple F-manifold (cf. the proof of Theorem 5.21 (a)). □

Remark 5.33 In [**Gi2**, Theorem 15] the restricted Lagrange maps of two other series of simple F-manifolds with $M \cong \mathbb{C}^n$ and T_0M not a Frobenius manifold are given, they are the series Ξ_n ($n \geq 3$) and Ω_n ($n \geq 4$) (also $\Xi_1 = A_1$, $\Xi_2 = H_2$, $\Omega_2 = A_2$, $\Omega_3 = H_3$).

They have Euler fields of weight 1 with positive weights. The analytic spectra of Ξ_n and Ω_n are isomorphic to $\mathbb{C} \times \Sigma_{n-1}(2n-1)$ and $\mathbb{C}^2 \times \Sigma_{n-2}(2n-3)$, respectively. Here $\Sigma_k(2k+1)$ is the *open swallowtail*, the subset of polynomials in the set of polynomials $\{z^{2k+1} + a_2 z^{2k-1} + \cdots + a_{2k+1} \mid a_2, \ldots, a_{2k+1} \in \mathbb{C}\}$ which have a root of multiplicity $\geq k+1$ ([**Gi2**, p. 3256]). It has embedding dimension $2k$.

The germs $((M, 0), \circ, e)$ are irreducible for Ξ_n and Ω_n, the socle $\mathrm{Ann}_{T_0 M}(\mathbf{m})$ of $T_0 M$ is the maximal ideal $\mathbf{m} \subset T_0 M$ itself in the case of Ξ_n and has dimension $n-2$ in the case of Ω_n.

Givental [**Gi2**, Theorem 15] proved that the germs $(M, 0)$ for Ξ_n and Ω_n are the only irreducible germs of simple F-manifolds whose analytic spectra are products of smooth germs and open swallowtails. Generating functions in the sense of Definition 4.23 are due to O.P. Shcherbak and are given in [**Gi2**, Proposition 12].

Part 2

Frobenius manifolds, Gauß–Manin connections, and moduli spaces for hypersurface singularities

Chapter 6
Introduction to part 2

The notion of a Frobenius manifold was introduced by Dubrovin in 1991 [**Du1**], motivated by topological field theory. It has been studied since then by him, Manin, Kontsevich, and many others. It plays a role in quantum cohomology [**Man2**] and in mirror symmetry.

But the first big class of Frobenius manifolds had already been constructed in 1983 in singularity theory. K. Saito [**SK6**][**SK9**] studied the semiuniversal unfolding of an isolated hypersurface singularity and its Gauß–Manin connection. He was interested in period maps and defined the primitive forms as volume forms with very special properties in relation to the Gauß–Manin connection.

Any primitive form provides the base space of a semiuniversal unfolding of a singularity with the structure of a Frobenius manifold.

He proved the existence of primitive forms in special cases and M. Saito proved their existence in the general case [**SM2**][**SM3**]. Using the work of Malgrange [**Mal3**][**Mal5**] on deformations of microdifferential systems, M. Saito showed that the choice of a certain filtration on the cohomology of the Milnor fibre yields a primitive form and thus a Frobenius manifold.

This construction of Frobenius manifolds in singularity theory has been quite inaccessible to nonspecialists, because the Gauß–Manin systems are treated using the natural, though sophisticated language of algebraic analysis and especially Malgrange's results require microdifferential systems and certain Fourier–Laplace transforms. This also made it difficult to apply the construction.

The first purpose of part 2 of this book is to give a detailed account of a simplified version of the construction. This version stays largely within the framework of meromorphic connections and is sufficiently explicit to work with it.

The second purpose is to present several applications. The most difficult one is the construction of global moduli spaces for singularities in one μ-homotopy class as an analytic geometric quotient.

Outlines of the construction and the applications are offered in the following two sections. In these outlines the reader can jump straightaway to the main points. An orientation is given how all the material in the subsequent chapters is used and a motivation to study it.

The applications are given in chapters 12, 13, and 14. In chapter 12 we give a canonical complex structure on the μ-constant stratum of a singularity and an infinitesimal Torelli type result, which strengthens a result of M. Saito. In chapter 13 the global moduli spaces for singularities are constructed and symmetries of singularities are discussed, extending some work of Slodowy and Wall. In chapter 14 the G-function of a Frobenius manifold is used to study the variance of the spectral numbers of a singularity.

The construction of Frobenius manifolds in singularity theory is carried out in chapter 11. It requires the majority of the results which are presented in chapters 7 to 10. In chapters 7 and 8 a lot of material on meromorphic connections is given, most of which is known, but not presented in this form in the literature. It will be used for the discussion of the meromorphic connections in chapter 9 and 10.

In chapter 10 most of the known results on the Gauß–Manin connection for (unfoldings of) singularities are put together in a concise survey. In chapter 9 Frobenius manifolds are defined and certain meromorphic connections, which arise from them, are studied. This extends some work of K. Saito, Dubrovin, and Manin.

Sabbah generalized most of K. Saito and M. Saito's construction to the case of tame functions with isolated singularities on affine manifolds [**Sab3**][**NS**] [**Sab2**][**Sab4**]. But the details are quite different; there one uses oscillating integrals, and the results are not as complete as in the local case. The case of tame functions is important for mirror symmetry. A special case had been studied by Barannikov [**Ba3**]. All of this is discussed in sections 11.3 and 11.4.

6.1 Construction of Frobenius manifolds for singularities

A Frobenius manifold here is a complex manifold M with a multiplication $\circ$ and a metric g on the holomorphic tangent bundle TM and with two global vector fields, the unit field e and the Euler field E. The multiplication is commutative and associative on each tangent space, the metric is flat, and all the data satisfy a number of natural compatibility conditions (see Definition 9.1).

Let $f : (\mathbb{C}^{n+1}, 0) \to (\mathbb{C}, 0)$ be a holomorphic function germ with an isolated singularity at 0. The dimension of its Jacobi algebra $\mathcal{O}_{\mathbb{C}^{n+1},0}/(\frac{\partial f}{\partial x_0}, \dots, \frac{\partial f}{\partial x_n}) =: \mathcal{O}/J_f$ is the Milnor number μ. A semiuniversal unfolding is a function germ

$F : (\mathbb{C}^{n+1} \times \mathbb{C}^{\mu}, 0) \to (\mathbb{C}, 0)$ with $F|(\mathbb{C}^{n+1} \times \{0\}, 0) = f$ and such that the derivatives $\frac{\partial F}{\partial t_i}|(\mathbb{C}^{n+1} \times \{0\}, 0)$, $i = 1, \ldots, \mu$, represent a basis of the Jacobi algebra; here $(x, t) = (x_0, \ldots, x_n, t_1, \ldots, t_\mu) \in \mathbb{C}^{n+1} \times \mathbb{C}^{\mu}$.

One can choose a representative $F : \mathcal{X} \to \Delta$ with $\Delta = B^1_\delta \subset \mathbb{C}$, $M = B^{\mu}_{\theta} \subset \mathbb{C}^{\mu}$ and $\mathcal{X} = F^{-1}(\Delta) \cap (B^{n+1}_{\varepsilon} \times M) \subset \mathbb{C}^{n+1} \times \mathbb{C}^{\mu}$ for suitable small $\varepsilon, \delta, \theta > 0$. One should see it as a family of functions $F_t : \mathcal{X} \cap (B^{n+1}_{\varepsilon} \times \{t\}) \to \Delta$, parameterized by $t \in M$, with $F_0 = f$. The manifold M is the candidate for a Frobenius manifold.

Its tangent bundle carries a canonical multiplication: The critical space $C \subset \mathcal{X}$ of the unfolding is defined by the ideal $J_F = (\frac{\partial F}{\partial x_0}, \ldots, \frac{\partial F}{\partial x_n})$. Its projection $pr_{C,M} : C \to M$ is finite and flat of degree μ. There is a canonical isomorphism

$$\mathbf{a} : \mathcal{T}_M \to (pr_{C,M})_* \mathcal{O}_C, \quad \frac{\partial}{\partial t_i} \mapsto \left. \frac{\partial F}{\partial t_i} \right|_C \tag{6.1}$$

of free $\mathcal{O}_M$-modules of rank μ, the Kodaira–Spencer map. Here $\mathcal{T}_M$ is the holomorphic tangent sheaf of M. The natural multiplication on the right hand side induces a multiplication $\circ$ on $\mathcal{T}_M$. This was first observed by K. Saito **[SK6][SK9]**. With $e = \mathbf{a}^{-1}(1|_C)$ as the unit field and $E := \mathbf{a}^{-1}(F|_C)$ as the Euler field, M carries a canonical structure of an F-manifold $(M, \circ, e, E)$ with Euler field (Definition 2.8 and Theorem 5.3). The Kodaira–Spencer map gives for each tangent space $T_t M$ an isomorphism

$$(T_t M, \circ, E|_t) \cong \left(\bigoplus_{x \in Sing\ F_t} \text{Jacobi algebra of } (F_t, x), \text{mult.}, [F_t] \right). \tag{6.2}$$

In order to find a metric one may look for a similar isomorphism as in (6.1) from the tangent sheaf to a sheaf with a nondegenerate symmetric bilinear form. Such a sheaf exists, the sheaf

$$\Omega_F := (pr_M)_* \Omega^{n+1}_{\mathcal{X}/\Delta \times M} \tag{6.3}$$

of relative differential forms with respect to the map φ in (6.5) is a free $\mathcal{O}_M$-module of rank μ and is equipped with the Grothendieck residue pairing $J_F : \Omega_F \times \Omega_F \to \mathcal{O}_M$ (see section 10.4)

It is also a free $(pr_{C,M})_* \mathcal{O}_C$-module of rank 1; generators are represented by suitable volume forms $unit(x, t) \mathrm{d}x_0 \ldots \mathrm{d}x_n$. The choice of a generator induces isomorphisms

$$\mathcal{T}_M \to (pr_{C,M})_* \mathcal{O}_C \to \Omega_F, \tag{6.4}$$

and the pairing J_F yields a metric on TM. But one needs special volume forms, the primitive forms of K. Saito, in order to obtain flat metrics and Frobenius manifold structures on M. Their construction uses the Gauß–Manin connection.

In the following we intend to give an idea of this construction. The details can be found in section 11.1. The map

$$\varphi : \mathcal{X} \to \Delta \times M, \quad (x, t) \mapsto (F(x, t), t) \tag{6.5}$$

is a C^∞-fibration of Milnor fibres outside of the discriminant $\check{\mathcal{D}} := \varphi(C) \subset \Delta \times M$. The cohomology bundle

$$H^n = \bigcup_{(z,t)\in\Delta\times M-\check{\mathcal{D}}} H^n(\varphi^{-1}(z, t), \mathbb{C}) \tag{6.6}$$

has rank μ and is flat. There is a distinguished extension $\mathcal{H}^{(0)}$ of its sheaf of holomorphic sections over the discriminant.

The sheaf $\mathcal{H}^{(0)}$ can be defined in two ways. In terms close to the relative de Rham cohomology it is

$$\mathcal{H}^{(0)} = \varphi_*\Omega^{n+1}_{\mathcal{X}/M}/\mathrm{d}F \wedge \mathrm{d}\varphi_*\Omega^{n-1}_{\mathcal{X}/M}. \tag{6.7}$$

It is a coherent and even free $\mathcal{O}_{\Delta\times M}$-module of rank μ [**Gre**]. It has a logarithmic pole along $\check{\mathcal{D}}$, that means, the derivatives of sections in $\mathcal{H}^{(0)}$ by logarithmic vector fields are still in $\mathcal{H}^{(0)}$. The residue endomorphism along $\check{\mathcal{D}}_{reg}$ has eigenvalues $(\frac{n-1}{2}, 0, \ldots, 0)$ and is (for $n \geq 2$) semisimple. The second description of $\mathcal{H}^{(0)}$ is this: it is the maximal coherent extension of the sheaf of holomorphic sections of H^n over $\check{\mathcal{D}}$ with these properties along $\check{\mathcal{D}}$ (see Lemma 10.2 and section 10.3).

One can extend the cohomology bundle H^n uniquely to a flat bundle over $\mathbb{C} \times M - \check{\mathcal{D}}$ and the sheaf $\mathcal{H}^{(0)}$ to $\mathbb{C} \times M$. Now the key point is to look for extensions $\overline{\mathcal{H}^{(0)}}$ of $\mathcal{H}^{(0)}$ over $\mathbb{P}^1 \times M$ which are free $\mathcal{O}_{\mathbb{P}^1\times M}$-modules and which have a logarithmic pole along $\{\infty\} \times M$. Figure 6.1 may help to visualize the situation.

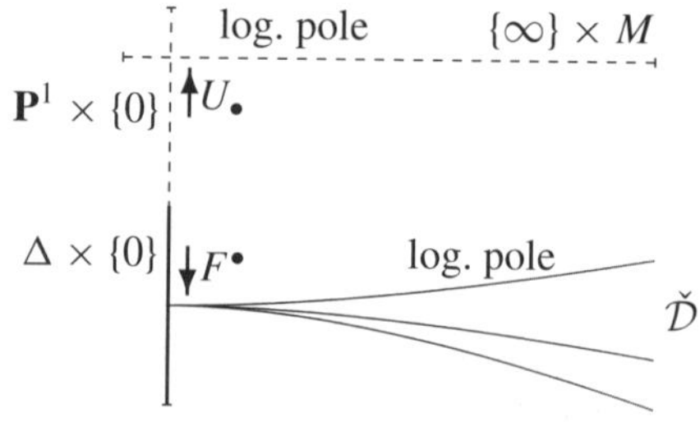

Figure 6.1

Extensions with a logarithmic pole are not difficult to obtain. Denote by H^∞ the μ-dimensional space of the global flat manyvalued sections in $(\mathbb{C}-\Delta)\times M$ of the extended cohomology bundle (there is only the monodromy around $\{\infty\}\times M$). This space H^∞ is equipped with a monodromy operator. There is a one-to-one correspondence between locally free extensions $\overline{\mathcal{H}^{(0)}}$ with a logarithmic pole along $\{\infty\}\times M$ and monodromy invariant (increasing exhaustive) filtrations $U_\bullet$ on H^∞ (see sections 7.3 and 8.2).

But for a free $\mathcal{O}_{\mathbb{P}^1\times M}$-module $\overline{\mathcal{H}^{(0)}}$ one needs special filtrations $U_\bullet$.

First one observes that it is sufficient to show that the restrictions of the sections to $\mathbb{P}^1\times\{0\}$ yield a free $\mathcal{O}_{\mathbb{P}^1}$-module. A classical theorem (cf. for example [**Sab4**, I 5.b], [**Mal4**, §4]) on families of vector bundles over $\mathbb{P}^1$ then asserts that the sheaf over $\mathbb{P}^1\times M$ is a free module (for M sufficiently small).

The germs in $(\Delta\times\{0\},0)$ of the sections in $\mathcal{H}^{(0)}$ form the Brieskorn lattice H_0'' [**Bri2**], a free $\mathbb{C}\{z\}$-module of rank μ. In general it does not have a logarithmic pole at 0, but its sections have moderate growth, so it is regular singular. Varchenko showed that the principal parts of its sections give rise to a Hodge filtration on H^∞ (see section 10.6). His construction was modified to obtain Steenbrink's Hodge filtration $F^\bullet$ on H^∞. Here H^∞ is canonically identified with the space of the global flat multivalued sections on the cohomology bundle over $\Delta^*\times\{0\}$.

Now M. Saito found that an opposite filtration $U_\bullet$ to this Hodge filtration (see Definition 10.19) is what one needs. In fact, he did not look for extensions to ∞, but he constructed from $U_\bullet$ a basis for H_0'' with properties such that he could apply Malgrange's results. The fact that an opposite filtration $U_\bullet$ gives rise to an $\mathcal{O}_{\mathbb{P}^1}$-free extension to $\mathbb{P}^1$ of the Brieskorn lattice H_0'' with a logarithmic pole at ∞ is a solution of a Riemann–Hilbert–Birkhoff problem and is discussed in section 7.4.

The existence of opposite filtrations $U_\bullet$ to $F^\bullet$ follows from properties of mixed Hodge structures. In general $U_\bullet$ is not unique.

Now fix a choice of $U_\bullet$ and the corresponding extension $\overline{\mathcal{H}^{(0)}}$. Denote by $\pi:\mathbb{P}^1\times M\to M$ the projection. The sheaf $\pi_*\overline{\mathcal{H}^{(0)}}$ of fibrewise global sections is a free $\mathcal{O}_M$-module of rank μ. It contains the μ-dimensional subspace of sections whose restrictions to $\{\infty\}\times M$ are flat with respect to the residual connection along $\{\infty\}\times M$ (see section 8.2).

The residue endomorphism along $\{\infty\}\times M$ acts on this space. It turns out that it acts semisimple with eigenvalues $-\alpha_1,\ldots,-\alpha_\mu$ (the rational numbers $\alpha_1,\ldots,\alpha_\mu$ are the spectral numbers of H_0'', see sections 7.2 and 10.6). The smallest spectral number α_1 has multiplicity 1 (Theorem 10.33).

Let v_1 be a global section on $\overline{\mathcal{H}^{(0)}}$ which is flat along $\{\infty\}\times M$ with respect to the residual connection there and which is an eigenvector with eigenvalue

$-\alpha_1$ of the residue endomorphism. It is uniquely determined up to a scalar. It turns out that it is a primitive form in the sense of K. Saito.

It yields a period map and an isomorphism $\pi_*\overline{\mathcal{H}^{(0)}} \cong \mathcal{T}_M$. The residual connection along $\{\infty\} \times M$ induces a flat connection on $\mathcal{T}_M$. In order to see that the isomorphism (6.4) for v_1 as volume form gives a flat metric on $\mathcal{T}_M$ one needs two things: K. Saito's higher residue pairings (see section 10.4) and the fact that the Hodge filtration $F^\bullet$ on H^∞ is part of a *polarized* mixed Hodge structure. There is a polarizing form S on H^∞ which also has to be respected by the opposite filtration $U_\bullet$. It is related to the higher residue pairings (see section 10.6). Altogether one obtains the following (see Theorem 11.1).

Theorem 6.1 *A monodromy invariant opposite filtration $U_\bullet$ to the Hodge filtration $F^\bullet$ on H^∞ induces a flat metric g up to a scalar on M such that $(M, \circ, e, E, g)$ is a Frobenius manifold.*

All the material in Sabbah's book [**Sab4**] and the second structure connections in Manin's book [**Man2**, II§2] have been very helpful for carving the above version of the construction of Frobenius manifolds in singularity theory.

The second structure connections in [**Man2**] are a family $\check{\nabla}^{(s)}$, $s \in \mathbb{C}$, of meromorphic connections over $\mathbb{P}^1 \times M$ for a semisimple Frobenius manifold. The definition generalizes to arbitrary Frobenius manifolds (chapter 9). In the case of singularities the connection $\check{\nabla}^{(-\frac{n}{2})}$ turns out to be isomorphic to the (extended) Gauß–Manin connection with the sheaf $\overline{\mathcal{H}^{(0)}}$. The connection $\check{\nabla}^{(0)}$ was also defined by Dubrovin [**Du3**]. In the singularity case, K. Saito had already defined the germs at 0 of all the connections $\check{\nabla}^{(s)}$ in a different way [**SK9**, §5].

6.2 Moduli spaces and other applications

Singularities which are contained in a μ-constant family have isomorphic Milnor lattices and for $n \neq 2$ even the same topological type. They are called μ-homotopic. Singularities which differ only by a local coordinate change are called right equivalent and should be considered as isomorphic. One may ask about the moduli space of right equivalence classes of singularities in one μ-homotopy class.

Let us fix a μ-homotopy class $\mathcal{E} \subset \mathbf{m}^2 \subset \mathcal{O}_{\mathbb{C}^{n+1},0}$ of singularities and an integer $k \geq \mu + 1$. The right equivalence class of a singularity $f \in \mathcal{E}$ is determined by its k-jet $j_k f \in \mathbf{m}^2/\mathbf{m}^{k+1}$. The set $j_k\mathcal{E} \subset \mathbf{m}^2/\mathbf{m}^{k+1}$ is a quasiaffine variety (possibly reducible as a variety, but connected as a topological space). The algebraic group $j_k\mathcal{R}$ of k-jets of coordinate changes acts on it. The quotient $j_k\mathcal{E}/j_k\mathcal{R}$ parametrizes the right equivalence classes of singularities in the

μ-homotopy class $\mathcal{E}$. One knows that the orbits all have the same dimension, but a priori not much more about the group action. We can prove the following (see Theorem 13.15).

Theorem 6.2 *The quotient $j_k\mathcal{E}/j_k\mathcal{R}$ is an analytic geometric quotient.*

A priori this is a global statement. But with the construction of unfoldings and with some results of Gabrielov and Teissier one can translate it into statements on semiuniversal unfoldings of singularities in $\mathcal{E}$. Then one can use the rich structure of their base spaces as Frobenius manifolds.

Theorem 6.2 includes the claim that the quotient topology on $j_k\mathcal{E}/j_k\mathcal{R}$ is Hausdorff. I now want to sketch the proof of that part in one page.

Consider a singularity $f \in \mathcal{E}$ and a semiuniversal unfolding F with base space M as in section 6.1. The μ-constant stratum in M is $S_\mu = \{t \in M \mid \text{Sing } F_t = \{x\} \text{ and } F_t(x) = 0\}$.

A result of Scherk [**Sche2**] says that for any $t \in M$ the datum in (6.2) determines the right equivalence classes of the germs (F_t, x) for $x \in \text{Sing } F_t$. So the base space as F-manifold $(M, \circ, e, E)$ with Euler field knows the right equivalence classes of all the singularities above it. There is a related result of Arnold, Hörmander, and others on Lagrange maps and generating families. It implies (Theorem 5.6) that the germ $((M, t), \circ, e, E)$ of an F-manifold for $t \in S_\mu$ determines the germ of the unfolding at the singular point of F_t up to right equivalence of unfoldings.

Now consider for f, F, M, and $S_\mu(f)$ as above a sequence $(t_i)_{i\in\mathbb{N}}$ with $t_i \in S_\mu(f)$ and $t_i \to 0$ for $i \to \infty$ and suppose that there is a second singularity $\widetilde{f}$ with $\widetilde{F}$, $\widetilde{M}$, $S_\mu(\widetilde{f})$ and $(\widetilde{t}_i)_i$ defined analogously and with a sequence of coordinate changes φ_i such that $F_{t_i} = \widetilde{F}_{\widetilde{t}_i} \circ \varphi_i$. One has to show that then f and $\widetilde{f}$ are also right equivalent. This will imply that the quotient topology is Hausdorff. Figure 6.2 illustrates the situation.

There is no possibility of controlling the coordinate changes φ_i and finding a limit coordinate change. But it turns out that they induce unique isomorphisms

$$\varphi_{i,M} : ((M, t_i), \circ, e, E) \to ((\widetilde{M}, \widetilde{t}_i), \circ, \widetilde{e}, \widetilde{E}) \tag{6.8}$$

of the germs of F-manifolds. With the construction of Frobenius manifolds we can control these and show that there is a subsequence which gets stationary for large i and extends to an isomorphism $(M, 0) \to (\widetilde{M}, 0)$. Then one can apply Scherk's or Arnold's and Hörmander's result and see that f and $\widetilde{f}$ are right equivalent.

Essential for controlling the sequence $\varphi_{i,M}$ is the strong link between flat structures on M and filtrations on H^∞ by Theorem 6.1. The coordinate changes

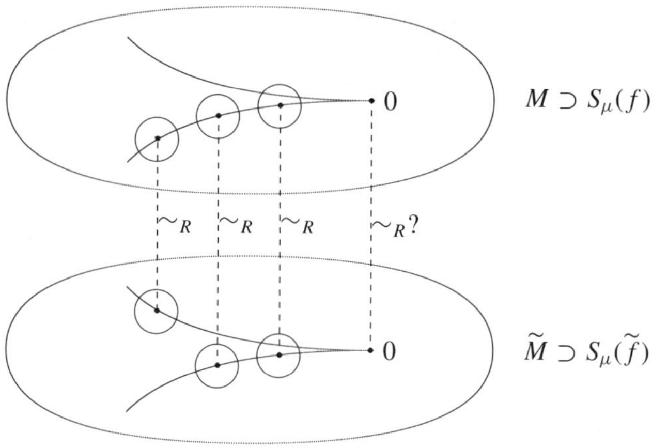

Figure 6.2

φ_i also induce isomorphisms

$$\varphi_{i,coh} : \left(H^\infty, H^\infty_{\mathbb{Z}}, h, S, F^\bullet(t_i)\right) \to \left(\widetilde{H^\infty}, \widetilde{H^\infty_{\mathbb{Z}}}, \widetilde{h}, \widetilde{S}, \widetilde{F}^\bullet(\widetilde{t}_i)\right); \qquad (6.9)$$

here H^∞ is the space from section 6.1 with canonical lattice $H^\infty_{\mathbb{Z}}$, monodromy operator h, polarizing form S and Hodge filtration $F^\bullet(t_i)$ from the singularity F_{t_i}.

The data $(H^\infty, H^\infty_{\mathbb{Z}}, h, S, F^\bullet(t))$ encode polarized mixed Hodge structures. There exists a classifying space D_{PMHS} for them and a period map

$$S_\mu \to D_{PMHS}, \quad t \mapsto F^\bullet(t). \qquad (6.10)$$

The discrete group $G_{\mathbb{Z}} = \mathrm{Aut}(H^\infty, H^\infty_{\mathbb{Z}}, h, S)$ acts properly discontinuously on D_{PMHS} (because the mixed Hodge structures are polarized). This implies that the isomorphisms $\varphi_{i,coh}$ are all contained in a finite set. Via the construction of flat structures on M the same then holds for the isomorphisms $\varphi_{i,M}$ (for more details see section 13.3).

One can be more precise about the local structure of the moduli space $j_k\mathcal{E}/j_k\mathcal{R}$. The group of automorphisms $\mathrm{Aut}((M, 0), \circ, e, E)$ is in fact finite and acts on the μ-constant stratum $S_\mu \subset M$. Similar arguments as above yield the following for a germ of the moduli space (see Theorem 13.15).

Theorem 6.3

$$(j_k\mathcal{E}/j_k\mathcal{R}, [j_k f]) \cong (S_\mu, 0)/\mathrm{Aut}((M, 0), \circ, e, E).$$

We can equip the μ-constant stratum with a canonical complex structure. An opposite filtration $U_\bullet$ as in Theorem 6.1 induces a flat metric up to a scalar on

M and a unique flat structure. By construction there exists a basis of flat vector fields $\delta_1, \dots, \delta_\mu$ with

$$[\delta_i, E] = (1 + \alpha_1 - \alpha_i) \cdot \delta_i. \tag{6.11}$$

The coefficients $\varepsilon_{ij} \in \mathcal{O}_{M,0}$ in

$$\delta_i \circ E = \sum_j \varepsilon_{ij} \cdot \delta_j \tag{6.12}$$

are close to a part of Dubrovin's deformed flat coordinates (see section 11.2) and they determine the μ-constant stratum (see section 12.1).

Theorem 6.4 *The μ-constant stratum $(S_\mu, 0)$ is the zero set of the ideal $(\varepsilon_{ij} \mid \alpha_j - 1 - \alpha_i < 0)$. This ideal is independent of the choice of $\delta_1, \dots, \delta_\mu$ and even of the choice of the opposite filtration $U_\bullet$.*

I expect that this ideal provides in general a nonreduced complex structure. But it is very difficult to compute. I do not have examples. I also expect that the induced complex structure on the moduli space $j_k\mathcal{E}/j_k\mathcal{R}$ is a good candidate for a coarse moduli space with respect to some functor of μ-constant deformations over arbitrary bases.

In [**He1**][**He2**][**He3**] Torelli type questions for hypersurface singularities were studied. There is a datum which is even finer than the polarized mixed Hodge structure $(H^\infty, H^\infty_\mathbb{Z}, h, S, F^\bullet)$ of a singularity, the datum $(H^\infty, H^\infty_\mathbb{Z}, h, S, H_0'')$, the Brieskorn lattice together with topological information. A classifying space D_{BL} for such data was constructed in [**He4**]. It is a fibre bundle over the classifying space D_{PMHS}. The group $G_\mathbb{Z}$ also acts properly discontinuously on D_{BL}. There is a local period map

$$S_\mu \to D_{BL}, \quad t \mapsto H_0''(t) \tag{6.13}$$

for the μ-constant stratum of a singularity and a global period map

$$j_k\mathcal{E}/j_k\mathcal{R} \to D_{BL}/G_\mathbb{Z} \tag{6.14}$$

for a μ-homotopy class of singularities. A global Torelli type conjecture [**He2**] asks whether (6.14) is always injective. In all known examples it is true, but a general answer is still unknown. Now Theorem 6.2 shows at least that the moduli space on the left is an analytic variety. The map (6.14) is now a morphism between varieties.

It is easy to see that (6.13) is injective (for small S_μ) if S_μ is smooth. M. Saito [**SM4**] used this fact to show that for general S_μ it is finite-to-one. (He did not have the classifying space D_{BL}, but considered a period map to a bigger space, a subset of a certain flag manifold.) With the flat coordinates of a Frobenius manifold structure on the base space M we can show a stronger infinitesimal Torelli type result (see Theorem 12.8).

Theorem 6.5 *The period map $S_\mu \to D_{BL}$ is an embedding for small S_μ.*

All these applications made use of the flat structure on M which is induced by an opposite filtration, but not of the metric itself and not much use of its other properties, the multiplication invariance and the potentiality (see Definition 9.1). These give rise to an extremely rich hidden structure on Frobenius manifolds which has been uncovered by Dubrovin and Zhang [**DuZ1**][**DuZ2**] and Givental [**Gi8**]. Exploiting this for singularities is a big task for the future. We can present one surprising application.

There is a function $G(t)$ associated to each semisimple Frobenius manifold ([**DuZ1**][**DuZ2**][**Gi7**], cf. section 14.2). Following a suggestion of Givental we show (Theorem 14.6) that it extends in the singularity case to a holomorphic function on the whole base space. It has the stunning property

$$E\, G(t) = -\frac{1}{4}\sum_{i=1}^{\mu}\left(\alpha_i - \frac{n-1}{2}\right)^2 + \frac{\mu(\alpha_\mu - \alpha_1)}{48} =: \gamma. \tag{6.15}$$

The spectral numbers α_i satisfy the symmetry $\alpha_i + \alpha_{\mu+1-i} = n-1$. So one can consider $\frac{n-1}{2}$ as their *expectation value*. Then $\frac{1}{\mu}\sum_{i=1}^{\mu}(\alpha_i - \frac{n-1}{2})^2$ is their *variance*. The G-function gives a grip at this variance. In the case of a quasihomogeneous singularity f, one has $f \in J_f$ and $E|_0 = 0$ because of (6.2). This shows the following (Theorem 14.9).

Theorem 6.6 *If f is a quasihomogeneous singularity then $\gamma = 0$. The variance of the spectral numbers is*

$$\frac{1}{\mu}\sum_{i=1}^{\mu}\left(\alpha_i - \frac{n-1}{2}\right)^2 = \frac{\alpha_\mu - \alpha_1}{12}. \tag{6.16}$$

It is part of the motivation for the following conjecture.

Conjecture 6.7 *For any isolated hypersurface singularity, the variance of the spectral numbers satisfies*

$$\frac{1}{\mu}\sum_{i=1}^{\mu}\left(\alpha_i - \frac{n-1}{2}\right)^2 \leq \frac{\alpha_\mu - \alpha_1}{12}. \tag{6.17}$$

Chapter 7

Connections over the punctured plane

The only initial datum in section 7.1 is a monodromy operator. For the corresponding flat vector bundle over the punctured plane $\mathbb{C}^*$ notions such as the elementary sections and the V-filtration are introduced. In section 7.2 $\mathcal{O}_\mathbb{C}$-free extensions over 0 with regular singularity at 0 of the sheaf of holomorphic sections of the vector bundle are discussed. Comparison with the V-filtration leads to the spectral numbers and certain filtrations. Sections 7.1 and 7.2 are elementary and classical.

The subject of section 7.3 is those extensions over 0 which not only have a regular singularity at 0, but also a logarithmic pole. There is a correspondence between such extensions and certain filtrations, which is not so well known. It has a generalization in section 8.2. It is used in section 7.4 for the solution of a Riemann–Hilbert–Birkhoff problem. This is based on ideas of M. Saito. It is central to the construction of Frobenius manifolds in section 11.1.

In section 7.5 a formula is given for the sum of the spectral numbers in a global situation, when one has on a compact Riemann surface a locally free sheaf and a flat connection with several singularities as above. It is useful in the case of $\mathbb{P}^1$ for the Riemann–Hilbert–Birkhoff problem.

7.1 Flat vector bundles on the punctured plane

Let us fix a holomorphic vector bundle $H \to \mathbb{C}^*$ of rank $\mu \geq 1$ with a flat connection ∇ on the punctured plane $\mathbb{C}^* = \mathbb{C} - \{0\}$. We want to discuss special sections in H and extensions of the sheaf $\mathcal{H}$ of its holomorphic sections over $0 \in \mathbb{C}$.

Of course this is classical and has been done in many ways (e.g. [**De1**], [**AGV2**], [**SM3**], [**He1**], [**Ku**], [**Sab4**]). But we have to establish comfortable notations in order to discuss later the information which is contained in certain extensions of $\mathcal{H}$ over 0, for example in the Brieskorn lattice (section 10.6). We more or less follow [**SM3**][**He1**].

A positively oriented loop around 0 induces a monodromy h on each fibre H_z, $z \in \mathbb{C}^*$, of the bundle. The monodromy determines the bundle uniquely up to isomorphism. Let $h = h_s \cdot h_u = h_u \cdot h_s$ be the decomposition into semisimple and unipotent parts, $N := \log h_u$ the nilpotent part, and

$$H_z = \bigoplus_\lambda H_{z,\lambda}, \quad H_{z,\lambda} := \ker((h_s - \lambda) : H_z \to H_z), \tag{7.1}$$

the decomposition into generalized eigenspaces. We will use the universal covering

$$e : \mathbb{C} \to \mathbb{C}^*, \quad \zeta \mapsto e^{2\pi i \zeta}. \tag{7.2}$$

Global flat sections A of the bundle $e^*H \to \mathbb{C}$ induce via the projection $pr : e^*H \to H$ maps $pr \circ A : \mathbb{C} \to H$, which are called *global flat multivalued sections*. The space of these global flat multivalued sections is denoted by H^∞. It is canonically isomorphic to each fibre $(e^*H)_\zeta$, $\zeta \in \mathbb{C}$. The monodromy h acts on it with eigenspace decomposition $H^\infty = \bigoplus_\lambda H^\infty_\lambda$ and with

$$h\ A(\zeta) = A(\zeta + 1) \tag{7.3}$$

for any $A \in H^\infty$.

Now we can define some special global sections in H. Fix $A \in H^\infty_\lambda$ and $\alpha \in \mathbb{C}$ with $e^{-2\pi i \alpha} = \lambda$. The map

$$\mathbb{C} \to H, \quad \zeta \mapsto e(\alpha\zeta)\exp(-\zeta N)A(\zeta),$$

is invariant with respect to the shift $\zeta \mapsto \zeta + 1$ and therefore induces a holomorphic section

$$es(A, \alpha) : \mathbb{C}^* \to H, \quad z \mapsto e(\alpha\zeta)\exp(-\zeta N)A(\zeta) \quad \text{for } e(\zeta) = z \tag{7.4}$$

of the bundle H. It is called an *elementary section* [**AGV2**] and is usually denoted informally as

$$z \mapsto z^\alpha \exp\left(-\log z \cdot \frac{N}{2\pi i}\right) A, \tag{7.5}$$

α is called its order. It is nowhere vanishing if $A \neq 0$ because the twist with $e(\alpha\zeta)\exp(-\zeta N)$ is invertible.

The symbol C^α denotes the space of all elementary sections $es(A, \alpha)$, $A \in H^\infty_\lambda$, with a fixed order α. The map

$$\psi_\alpha : H^\infty_\lambda \to C^\alpha, \quad A \mapsto es(A, \alpha) \tag{7.6}$$

is an isomorphism of vector spaces. By definition one has

$$z \cdot es(A,\alpha) = es(A,\alpha+1),\ z \circ \psi_\alpha = \psi_{\alpha+1}, \tag{7.7}$$

$$\nabla_{\partial_z} es(A,\alpha) = \alpha \cdot es(A,\alpha-1) + es\left(\frac{-N}{2\pi i}A, \alpha-1\right),$$

$$(z\nabla_{\partial_z} - \alpha) es(A,\alpha) = es\left(-\frac{N}{2\pi i}A, \alpha\right) = -\frac{N}{2\pi i} es(A,\alpha),$$

$$\begin{aligned} z : C^\alpha &\to C^{\alpha+1} \quad \text{bijective,} \\ \nabla_{\partial_z} : C^\alpha &\to C^{\alpha-1} \quad \text{bijective iff } \alpha \neq 0, \\ z\nabla_{\partial_z} - \alpha = -\frac{N}{2\pi i} : C^\alpha &\to C^\alpha \quad \text{nilpotent,} \end{aligned} \tag{7.8}$$

so C^α is a generalized eigenspace of $z\nabla_{\partial_z}$. We simply call these spaces C^α *eigenspaces*.

To obtain a filtration for these eigenspaces we fix a total order $\prec$ on the set $\{\alpha \mid e^{-2\pi i\alpha} \text{ eigenvalue of } h\} \cup \mathbb{Z}$, which satisfies

$$\begin{aligned} &\alpha \prec \alpha+1, \\ &\alpha \prec \beta \text{ iff } \alpha+1 \prec \beta+1, \\ \forall\, \alpha,\beta\ \exists\, m \in \mathbb{Z} \qquad &\alpha \prec \beta + m. \end{aligned} \tag{7.9}$$

Later h will be quasiunipotent and, if not said otherwise, the order will be the natural order $<$ on $\mathbb{Q}$. But different orders can also be interesting (cf. Remarks 11.7). To simplify notations, we will write the usual symbol $<$ for the order $\prec$. It should be clear when the usual order $<$ on $\mathbb{R}$ is meant and when the order $\prec$ on $\{\alpha \mid e^{-2\pi i\alpha} \text{ eigenvalue of } h\} \cup \mathbb{Z}$ is meant.

From now on we will concentrate on germs at 0 of sections in H, that means, on the stalk $(i_*\mathcal{H})_0$ at 0 of the sheaf $i_*\mathcal{H}$, where $i : \mathbb{C}^* \hookrightarrow \mathbb{C}$ is the inclusion. The space of function germs $(i_*\mathcal{O}_{\mathbb{C}^*})_0$ and the operator ∇_{∂_z} act on this stalk. The eigenspaces C^α and the elementary sections are identified with their images in $(i_*\mathcal{H})_0$. Then C^α is characterized as

$$C^\alpha = \ker((z\nabla_{\partial_z} - \alpha)^m : (i_*\mathcal{H})_0 \to (i_*\mathcal{H})_0) \tag{7.10}$$

for some $m \gg 0$. A basis of elementary sections in $\bigoplus_{-1<\alpha\leq 0} C^\alpha$ induces a basis of each fibre $H_z,\ z \in \mathbb{C}^*$. Therefore

$$(i_*\mathcal{H})_0 = \bigoplus_{-1<\alpha\leq 0} (i_*\mathcal{O}_{\mathbb{C}^*})_0 C^\alpha \tag{7.11}$$

is an $(i_*\mathcal{O}_{\mathbb{C}^*})_0$-vector space of dimension $\mu = \dim_{\mathbb{C}} H_z\ (z \in \mathbb{C}^*)$.

The space of all germs at 0 of sections of moderate growth (in the sense of Deligne [**De1**, II §§1–4], comparing them with the flat multivalued sections) is

$$V^{>-\infty} := \bigoplus_{-1<\alpha\leq 0} \mathbb{C}\{z\}[z^{-1}]C^\alpha \subset (i_*\mathcal{H})_0. \tag{7.12}$$

The space $V^{>-\infty}$ is a $\mathbb{C}\{z\}[z^{-1}]$-vector space of dimension μ. It is a regular holonomic $\mathcal{D}_{\mathbb{C},0}$-module of meromorphic type [**SM3**, §1]. Obviously it decomposes into $\mathcal{D}_{\mathbb{C},0}$- and $\mathbb{C}\{z\}[z^{-1}]$-submodules which correspond to a Jordan block decomposition of H^∞.

Within this space one has the Kashiwara–Malgrange V-filtration, an exhaustive decreasing filtration, indexed by $(\{\alpha \mid e^{-2\pi i\alpha}$ eigenvalue of $h\}, <)$ and defined by

$$V^\alpha := \sum_{\beta\geq\alpha} \mathbb{C}\{z\}C^\beta = \bigoplus_{\alpha\leq\beta<\alpha+1} \mathbb{C}\{z\}C^\beta. \tag{7.13}$$

One supplements this with

$$V^{>\alpha} := \sum_{\beta>\alpha} \mathbb{C}\{z\}C^\beta = \bigoplus_{\alpha<\beta\leq\alpha+1} \mathbb{C}\{z\}C^\beta. \tag{7.14}$$

Then V^α and $V^{>\alpha}$ are free $\mathbb{C}\{z\}$-modules of rank μ, satisfying

$$z : V^\alpha \to V^{\alpha+1} \quad \text{bijective,} \tag{7.15}$$

$$\nabla_{\partial_z} : V^\alpha \to V^{\alpha-1} \quad \text{bijective if } \alpha > 0,$$

$$\nabla_{\partial_z} : V^{>0} \to V^{>-1} \quad \text{bijective,}$$

$$Gr_V^\alpha := V^\alpha / V^{>\alpha} \cong C^\alpha \quad \text{canonically isomorphic.} \tag{7.16}$$

Any section $\omega \in V^{>-\infty}$ is a sum (often infinite) of unique elementary sections,

$$\omega = \sum_\alpha s(\omega, \alpha), \quad s(\omega, \alpha) \in C^\alpha, \tag{7.17}$$

whose orders are bounded from below by some number

$$\alpha(\omega) := \max(\alpha \mid \omega \in V^\alpha) = \min(\alpha \mid s(\omega, \alpha) \neq 0), \tag{7.18}$$

which is called the order of ω. The elementary section $s(\omega, \alpha(\omega))$ is the *principal part* of ω. All the sections $s(\omega, \alpha)$ are called the *elementary parts* of ω.

Finally, we will need the ring

$$\mathbb{C}\{\{\partial_z^{-1}\}\} = \left\{ \sum_{i\geq 0} a_i \partial_z^{-i} \;\middle|\; \sum_{i\geq 0} \frac{a_i}{i!} z^i \in \mathbb{C}\{z\} \right\} \tag{7.19}$$

of microdifferential operators with constant coefficients [**Ph1**, part 2]. Just as $\mathbb{C}\{z\}$, it is a discrete valuation ring and a principal ideal domain. In view of the action $\partial_z^{-1} z^k = \frac{1}{k+1} z^{k+1}$ of ∂_z^{-1} on $\mathbb{C}\{z\}$, the ring $\mathbb{C}\{\{\partial_z^{-1}\}\}$ is designed to act on $\mathbb{C}\{z\}$ such that $\mathbb{C}\{z\}$ is a free $\mathbb{C}\{\{\partial_z^{-1}\}\}$-module of rank 1 with generator 1. It is well known ([**Mal2**, 4.1]) (and not hard to prove elementarily) that this generalizes as follows.

Lemma 7.1 *The map* $\nabla_{\partial_z} : \mathbb{C}\{z\}C^{\alpha+1} \to \mathbb{C}\{z\}C^\alpha$ *for* $\alpha \notin \mathbb{Z}_{<0}$ *is bijective, the inverse extends to an action of* $\mathbb{C}\{\{\partial_z^{-1}\}\}$ *such that* $\mathbb{C}\{z\}C^\alpha$ *is a free* $\mathbb{C}\{\{\partial_z^{-1}\}\}$*-module of rank* $\dim C^\alpha$.

Especially, $V^{>-1}$ *and all* V^α, $V^{>\alpha}$ *for* $\alpha > -1$ *are free* $\mathbb{C}\{\{\partial_z^{-1}\}\}$*-modules of rank* μ, *with* $\nabla_{\partial_z}^{-1} : V^{>-1} \to V^{>0}$.

7.2 Lattices

We stay in the situation of section 7.1. Up to now the only initial datum was the monodromy h. It determined the flat bundle $H \to \mathbb{C}^*$ up to isomorphy. Everything in section 7.1 was developed from this.

But usually one has another ingredient, a $\mathbb{C}\{z\}$-module in $V^{>-\infty}$, which contains additional information (e.g. the Brieskorn lattice, cf. section 10.6). One wants to understand this information by comparing with the V-filtration and the elementary sections.

We first discuss $\mathbb{C}\{z\}$-modules of $V^{>-\infty}$, most of the discussion also applies to $\mathbb{C}\{\{\partial_z^{-1}\}\}$-submodules of $V^{>-1}$.

A finitely generated $\mathbb{C}\{z\}$-submodule of $V^{>-\infty}$ is free, as $\mathbb{C}\{z\}$ is a principal ideal domain. The name $\mathbb{C}\{z\}$-*lattice* will be reserved for free $\mathbb{C}\{z\}$-modules of the maximal rank μ. A $\mathbb{C}\{z\}$-lattice $\mathcal{L}_0$ can be extended uniquely to an $\mathcal{O}_\mathbb{C}$-free subsheaf of rank μ in $i_*\mathcal{H}$. The correspondence between $\mathbb{C}\{z\}$-lattices $\mathcal{L}_0 \subset V^{>-\infty}$ and $\mathcal{O}_\mathbb{C}$-free subsheaves of rank μ of $i_*\mathcal{H}$ whose sections have moderate growth is one-to-one and justifies focusing on the stalks at 0.

A $\mathbb{C}\{z\}$-lattice $\mathcal{L}_0$ satisfies $\mathbb{C}\{z\}[z^{-1}]\mathcal{L}_0 = V^{>-\infty}$, just as do the $\mathbb{C}\{z\}$-lattices V^α. Therefore there exist α' and α'' with $V^{\alpha'} \supset \mathcal{L}_0 \supset V^{\alpha''}$. The principal parts of the sections of $\mathcal{L}_0$ are placed together in the subspaces

$$\mathrm{Gr}_V^\alpha \mathcal{L}_0 = (V^\alpha \cap \mathcal{L}_0 + V^{>\alpha})/V^{>\alpha} \subset \mathrm{Gr}_V^\alpha = C^\alpha. \tag{7.20}$$

One can visualize them in figure 7.1.

Obviously $z\mathrm{Gr}_V^{\alpha-1}\mathcal{L}_0 = \mathrm{Gr}_V^\alpha z\mathcal{L}_0 \subset \mathrm{Gr}_V^\alpha \mathcal{L}_0$. The dimensions of the quotients are

$$\dim\left(\mathrm{Gr}_V^\alpha \mathcal{L}_0 / z\mathrm{Gr}_V^{\alpha-1}\mathcal{L}_0\right) = \dim \mathrm{Gr}_V^\alpha \mathcal{L}_0 - \dim \mathrm{Gr}_V^{\alpha-1}\mathcal{L}_0 =: d(\alpha) \tag{7.21}$$

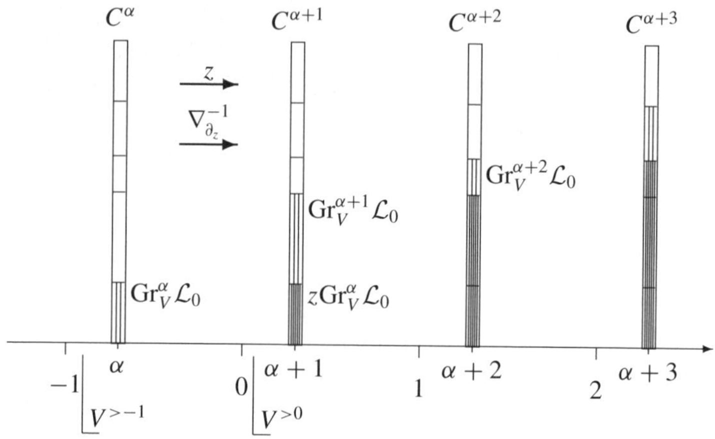

Figure 7.1

and satisfy

$$\sum_{k\in\mathbb{Z}} d(\alpha+k) = \dim C^{\alpha}. \tag{7.22}$$

They give rise to the *spectral numbers* $(\alpha_1, \ldots, \alpha_\mu) = Sp(\mathcal{L}_0)$ of the lattice $\mathcal{L}_0$ [**AGV2**], defined by

$$\sharp(i \mid \alpha_i = \alpha) = d(\alpha), \tag{7.23}$$

and ordered by $\alpha_1 \leq \cdots \leq \alpha_\mu$. Via the isomorphisms $\psi_\alpha : H_\lambda^\infty \to C^\alpha$ (cf. (7.6)), the subspaces $\mathrm{Gr}_{\mathrm{V}}^{\alpha}\mathcal{L}_0$ induce an increasing exhaustive h_s-invariant filtration $F_\bullet$ on H^∞,

$$\begin{aligned} F_p H_\lambda^\infty &:= \psi_{\alpha+p}^{-1}\mathrm{Gr}_{\mathrm{V}}^{\alpha+\mathrm{p}}\mathcal{L}_0 \quad \text{for } \lambda = \mathrm{e}^{-2\pi i\alpha},\ -1<\alpha\leq 0, \\ &= \psi_\alpha^{-1} z^{-p}\mathrm{Gr}_{\mathrm{V}}^{\alpha+\mathrm{p}}\mathcal{L}_0, \end{aligned} \tag{7.24}$$

$$F_p H^\infty := \bigoplus_\lambda F_p H_\lambda^\infty. \tag{7.25}$$

Remarks 7.2 (a) In the V-filtration, in $\alpha_1 \leq \cdots \leq \alpha_\mu$, and in $-1 < \alpha \leq 0$ the order $\prec$ from (7.9) is used.

(b) In the case of the Brieskorn lattice of an isolated hypersurface singularity $f(x_0, \ldots, x_n)$, the filtration $F^\bullet = F_{n-\bullet}$ is essentially Varchenko's Hodge filtration ([**Va1**], cf. also section 10.6). It reflects the information contained in the principal parts of the sections in the Brieskorn lattice and is already highly trancendental. The question of how to treat the higher elementary parts of the sections and the whole information in the Brieskorn lattice leads to M. Saito's work [**SM3**, §3], which will be taken up in section 7.4, and to my Torelli type results ([**He1**]–[**He4**] and section 12.2).

Lemma 7.3 *A $\mathbb{C}\{z\}$-lattice $\mathcal{L}_0 \subset V^{>-\infty}$ with spectral numbers $\alpha_1 \leq \cdots \leq \alpha_\mu$ satisfies*

$$V^{\alpha_1} \supset \mathcal{L}_0 \supset V^{>\alpha_\mu - 1}. \tag{7.26}$$

Elements $\omega_1, \ldots, \omega_\mu \in \mathcal{L}_0$ whose principal parts represent a basis of the space $\bigoplus_\alpha \mathrm{Gr}_V^\alpha \mathcal{L}_0 / z\mathrm{Gr}_V^{\alpha-1}\mathcal{L}_0$ form a $\mathbb{C}\{z\}$-basis of $\mathcal{L}_0$.

Proof: We may suppose $\alpha(\omega_i) = \alpha_i$. The elements $\omega_1, \ldots, \omega_\mu$ are $\mathbb{C}\{z\}$-linearly independent because in any linear combination with nonvanishing coefficients at least one of the principal parts is not cancelled by anything. They generate a $\mathbb{C}\{z\}$-lattice $\mathcal{L}_0' \subset \mathcal{L}_0$. Because of $\mathrm{Gr}_V^\alpha \mathcal{L}_0' = \mathrm{Gr}_V^\alpha \mathcal{L}_0$ for all α, one can enlarge the order of elements of $\mathcal{L}_0$ arbitrarily by adding elements of $\mathcal{L}_0'$. So $\mathcal{L}_0 \subset \mathcal{L}_0' + V^\beta$ for some large β and then $\mathcal{L}_0 = \mathcal{L}_0'$.

In the same way one obtains $V^{>\alpha_\mu - 1} \subset \mathcal{L}_0' + V^\beta = \mathcal{L}_0' = \mathcal{L}_0$. The inclusion $V^{\alpha_1} \supset \mathcal{L}_0$ is obvious. □

A free $\mathbb{C}\{\{\partial_z^{-1}\}\}$-module $\mathcal{L}_0 \subset V^{>-1}$ of maximal rank μ is called a $\mathbb{C}\{\{\partial_z^{-1}\}\}$-lattice. Spectral numbers $\alpha_1, \ldots, \alpha_\mu$ are defined as in (7.21) and (7.23). One obtains an increasing exhaustive h_s-invariant filtration $F_\bullet^{alg}$ on H^∞ by

$$F_p^{alg} H_\lambda^\infty = \psi_\alpha^{-1} \nabla_{\partial_z}^p \mathrm{Gr}_V^{\alpha+p} \mathcal{L}_0 \quad \text{for } \lambda = e^{-2\pi i\alpha},\ -1 < \alpha \leq 0, \tag{7.27}$$

$$F_p^{alg} H^\infty = \bigoplus_\lambda F_p^{alg} H_\lambda^\infty. \tag{7.28}$$

Lemma 7.3 holds analogously for $\mathbb{C}\{\{\partial_z^{-1}\}\}$-lattices.

Often a subspace $\mathcal{L}_0 \subset V^{>-1}$ is given which is a $\mathbb{C}\{z\}$- and a $\mathbb{C}\{\{\partial_z^{-1}\}\}$-lattice. Then the definitions (7.21) and (7.23) give the same set of spectral numbers when one replaces z in (7.21) by ∂_z^{-1}. But the filtrations $F_\bullet$ and $F_\bullet^{alg}$ may differ.

Lemma 7.4 *(a) A $\mathbb{C}\{z\}$-lattice $\mathcal{L}_0 \subset V^{>-1}$ is a $\mathbb{C}\{\{\partial_z^{-1}\}\}$-lattice if and only if $\nabla_{\partial_z}^{-1} \mathcal{L}_0 \subset \mathcal{L}_0$. A $\mathbb{C}\{\{\partial_z^{-1}\}\}$-lattice $\mathcal{L}_0 \subset V^{>-1}$ is a $\mathbb{C}\{z\}$-lattice if and only if $z\mathcal{L}_0 \subset \mathcal{L}_0$.*

(b) The filtrations $F_\bullet$ and $F_\bullet^{alg}$ of a $\mathbb{C}\{z\}$- and $\mathbb{C}\{\{\partial_z^{-1}\}\}$-lattice $\mathcal{L}_0 \subset V^{>-1}$ satisfy $F_{-1}^{alg} = F_{-1} = 0$ and are related for $p \geq 0$ by the formula

$$F_p^{alg} H_\lambda^\infty = \prod_{k=1}^{p} \left(-\frac{N}{2\pi i} + k + \alpha \right) F_p H_\lambda^\infty \tag{7.29}$$

where $\lambda = e^{-2\pi i\alpha}$, $-1 < \alpha \leq 0$.

Proof: (a) The inclusion $\mathcal{L}_0 \supset V^{\alpha'}$ for some α', Lemma 7.1, and $\nabla_{\partial_z}^{-1}\mathcal{L}_0 \subset \mathcal{L}_0$ imply that $\mathcal{L}_0$ is a $\mathbb{C}\{\{\partial_z^{-1}\}\}$-lattice, analogously when $z\mathcal{L}_0 \subset \mathcal{L}_0$ for $\mathbb{C}\{z\}$.

(b) This follows from the definitions of $F_\bullet$ and $F_\bullet^{alg}$ and from the formula $z\nabla_{\partial_z} - \alpha = -\frac{N}{2\pi i}$ on C^α (cf. (7.8)). □

Remarks 7.5 (a) The filtration F_p^{alg} has the index 'alg' because in the case of a hypersurface singularity $f(x_0, \dots, x_n)$ the operator ∇_{∂_z} has a more algebraic flavour than the operator z (cf. sections 10.3 and 10.6).

(b) The Brieskorn lattice is a $\mathbb{C}\{\{\partial_z^{-1}\}\}$-lattice, and $F_{n-\bullet}^{alg}$ is the Hodge filtration of Steenbrink [**Stn**] (and Scherk [**SchSt**], M. Saito [**SM1**], Pham [**Ph3**]). Because of (7.29) the filtrations $F_{n-\bullet}^{alg}$ and $F_{n-\bullet}$ coincide on the quotients of the weight filtration (which is defined via N, cf. [**Schm**][**AGV2**] and section 10.5) and are both Hodge filtrations of mixed Hodge structures. But $F_{n-\bullet}^{alg}$ behaves better with respect to a polarizing form and is part of a *polarized* mixed Hodge structure ([**He4**], cf. sections 10.5 and 10.6).

(c) This polarizing form is, after the monodromy, a second topological ingredient and can be married to the structure $(V^{-1}, C^\alpha, z, \partial_z^{-1})$ giving a $\mathbb{C}\{\{\partial_z^{-1}\}\}$-sesquilinear form on $V^{>-1}$, which in fact coincides with the restriction to $V^{>-1}$ of K. Saito's higher residue pairings and which fits together with the Brieskorn lattice [**He4**]. This will be discussed in section 10.6.

7.3 Saturated lattices

We stay in the situation of section 7.1. A *saturated lattice* $\mathcal{L}_0$ is a free $\mathbb{C}\{z\}$-module $\mathcal{L}_0 \subset (i_*\mathcal{H})_0$ of rank μ with $z\nabla_{\partial_z}\mathcal{L}_0 \subset \mathcal{L}_0$. Then the germs of sections in $\mathcal{L}_0$ have moderate growth, that means, $\mathcal{L}_0 \subset V^{>-\infty}$. This follows from the classical theorem of Sauvage, that the solutions of a system of linear differential equations with simple pole have moderate growth (cf. for example [**De1**, II.1] or [**Sab4**, II.2]).

We will see that there are correspondences

$$\begin{aligned}&\text{saturated lattices}\\ &\quad\leftrightarrow \text{filtrations on } H \to \mathbb{C}^* \text{ by flat subbundles}\\ &\quad\leftrightarrow \text{monodromy invariant filtrations on } H^\infty\end{aligned} \tag{7.30}$$

and that the saturated lattices and these correspondences are invariant under change of the coordinate z. They are even independent of the choice of the order $<$ in (7.9). A result related to these correspondences has been given by Sabbah [**Sab4**, III 1.1].

The structure of a saturated lattice is summarized in the following lemma.

Lemma 7.6 *(a) Let $\mathcal{L}_0 \subset V^{>-\infty}$ be a saturated lattice with spectral numbers $\alpha_1, \ldots, \alpha_\mu$ (cf. (7.23)) and filtration $F_\bullet$ on H^∞ (cf. (7.24) and (7.25)). Then*

$$\mathcal{L}_0 = \left(\bigoplus_{\alpha_1 \le \alpha \le \alpha_\mu - 1} \mathcal{L}_0 \cap C^\alpha \right) \oplus V^{>\alpha_\mu - 1}, \tag{7.31}$$

so $\mathcal{L}_0$ contains all the elementary parts $s(\omega, \alpha)$ of a section $\omega \in \mathcal{L}_0$. The spaces $\mathcal{L}_0 \cap C^\alpha \subset C^\alpha$ and $F_p \subset H^\infty$ are N-invariant and thus monodromy invariant.

(b) Any increasing exhaustive monodromy invariant filtration $F_\bullet$ on H^∞ induces a saturated lattice in the following way.

The filtration $F_\bullet$ on H^∞ induces a filtration of $H \to \mathbb{C}^*$ by flat subbundles $F_pH \to \mathbb{C}^*$; the sheaf of holomorphic sections in the subbundle F_pH is denoted by $F_p\mathcal{H}$. Then

$$\mathcal{L}_0 := \bigoplus_{-1<\alpha\le 0} \sum_{p\in\mathbb{Z}} (i_* F_p \mathcal{H})_0 \cap \mathbb{C}\{z\} C^{\alpha+p} \tag{7.32}$$

is a saturated lattice and $F_\bullet$ on H^∞ is the filtration of $\mathcal{L}_0$ defined in (7.24) and (7.25).

Proof: (a) Lemma 7.3 yields the inclusions $V^{\alpha_1} \supset \mathcal{L}_0 \supset V^{>\alpha_\mu - 1}$. The operator $z\nabla_{\partial_z}$ acts on the space $V^{\alpha_1}/V^{>\alpha_\mu-1} = \oplus_{\alpha_1 \le \alpha \le \alpha_\mu - 1} C^\alpha$, its generalized eigenspaces are the spaces C^α. The subspace $\mathcal{L}_0/V^{>\alpha_\mu-1}$ is invariant under $z\nabla_{\partial_z}$. This implies (7.31) and $z\nabla_{\partial_z}(\mathcal{L}_0 \cap C^\alpha) \subset \mathcal{L}_0 \cap C^\alpha$. The formula $z\nabla_{\partial_z} - \alpha = -\frac{N}{2\pi i}$ on C^α (cf. (7.8)) gives the N-invariance of $\mathcal{L}_0 \cap C^\alpha$ and of $F_p \subset H^\infty$.

(b) Exhaustive means that there exist integers $a < b$ with $0 = F_a \subset F_b = H^\infty$. Then $V^{>a} \supset \mathcal{L}_0 \supset V^{>b-1}$. The stalk $(i_* F_p \mathcal{H})_0$ is an $(i_* \mathcal{O}_{\mathbb{C}^*})_0$-module and invariant under ∇_{∂_z}. The $\mathbb{C}\{z\}$-module $\mathbb{C}\{z\}C^{\alpha+p}$ is invariant under $z\nabla_{\partial_z}$. Therefore $\mathcal{L}_0$ is a saturated lattice. The equalities

$$\mathrm{Gr}_V^{\alpha+p} \mathcal{L}_0 = \mathcal{L}_0 \cap \mathrm{C}^{\alpha+\mathrm{p}} = (\mathrm{i}_* \mathrm{F}_\mathrm{p} \mathcal{H})_0 \cap \mathrm{C}^{\alpha+\mathrm{p}} \tag{7.33}$$

for $-1 < \alpha \le 0$ show that (7.24) and (7.25) give the filtration $F_\bullet$ on H^∞. □

If $\varphi : (\mathbb{C}, 0) \to (\mathbb{C}, 0)$ is an isomorphism of germs then $\varphi(z)$ itself can be considered as a new coordinate on $(\mathbb{C}, 0)$. One can lift φ to an automorphism Φ of the flat bundle $H \to \mathbb{C}^*$ (in a neighbourhood of 0) such that $\Phi : H_z \to H_{\varphi(z)}$ is an isomorphism obtained from the flat structure by some path from z to $\varphi(z)$. The isomorphism Φ is not unique, only unique up to a power of the monodromy. The isomorphism Φ maps $(i_*\mathcal{H})_0$ to itself and $\mathbb{C}\{z\}$-lattices to $\mathbb{C}\{z\}$-lattices.

But $\Phi(C^\alpha)$ does not usually coincide with C^α: for the definition of elementary sections the choice of a coordinate z was crucial. Also many $\mathbb{C}\{z\}$-lattices are not invariant under Φ.

Lemma 7.7 *Let φ and Φ be as above. Then the sets $\mathbb{C}\{z\}C^\alpha$ and all saturated lattices are invariant under Φ.*

Proof: One can use (7.5) to see that an elementary section is mapped to a section (not necessarily elementary) with the same order. Thus $\Phi(C^\alpha) \subset \mathbb{C}\{z\}C^\alpha$, and Φ^{-1} gives $\Phi(\mathbb{C}\{z\}C^\alpha) = \mathbb{C}\{z\}C^\alpha$.

Now formula (7.32) shows that any saturated lattice is invariant under Φ, because the flat subbundles are invariant by definition of Φ. □

The subspaces V^α of the Kashiwara–Malgrange V-filtration are saturated lattices. Their independence of coordinates is of course known. We could have considered the spaces $\mathbb{C}\{z\}C^\alpha$ as the subspaces of the V-filtration for the flat eigenspace subbundle $\ker(h_s - e^{-2\pi i\alpha}) \to \mathbb{C}^*$ with eigenvalue $e^{-2\pi i\alpha}$.

Now, for a coordinate free reformulation of the previous discussion, let M be a 1-dimensional manifold, $q \in M$ a point, $i : M - \{q\} \hookrightarrow M$ the inclusion, and $H \to M - \{q\}$ a flat vector bundle with sheaf $\mathcal{H}$ of holomorphic sections.

The $\mathcal{O}_{M,q}$-modules $\mathcal{L}_q \subset (i_*\mathcal{H})_q$ of rank μ (= rank of $H \to M - \{q\}$) are called $\mathcal{O}_{M,q}$-lattices. They are the germs in q of $\mathcal{O}_M$-locally free extensions $\mathcal{L}$ of $\mathcal{H}$ over q. If z is a coordinate around q then the vector field $z\partial_z$ generates the logarithmic vector fields in a neighbourhood of q (cf. section 8.2). We can rewrite the above definition of a saturated lattice coordinate freely.

Definition 7.8 An $\mathcal{O}_{M,q}$-lattice $\mathcal{L}_q \subset (i_*\mathcal{H})_q$ is *saturated* if and only if it is invariant under the logarithmic vector fields along $\{q\} \subset M$ (cf. section 8.2). An $\mathcal{O}_M$-locally free extension $\mathcal{L}$ of $\mathcal{H}$ over q has a *logarithmic pole* at q if and only if the lattice $\mathcal{L}_q$ is saturated.

The Lemmata 7.6 and 7.7 give the following.

Theorem 7.9 *The saturated lattices correspond one-to-one to the increasing exhaustive filtrations by flat subbundles of the restriction of H to some disc around q.*

A saturated lattice $\mathcal{L}_q \subset (i_*\mathcal{H})_0$ is equipped with a *residue endomorphism*

$$\mathrm{Res}_q = z\nabla_{\partial_z} : \mathcal{L}_q/z\mathcal{L}_q \to \mathcal{L}_q/z\mathcal{L}_q, \tag{7.34}$$

here $z \in \mathcal{O}_{M,q}$ is a coordinate with $z(q) = 0$. The residue Res_q is independent of the coordinate: if $\widetilde{z}$ is any other coordinate with $\widetilde{z}(q) = 0$ then $\widetilde{z}\partial_{\widetilde{z}} = u(z)z\partial_z$ with $u \in \mathcal{O}_{M,q}$, $u(0) = 1$.

In fact, the vector space $\mathcal{L}_q/z\mathcal{L}_q$ can be identified with the fibre at q of a vector bundle on M which extends H and has $\mathcal{L}_q$ as the space of germs of holomorphic sections. Then the monodromy automorphism h of the bundle H (in a neighbourhood of q) extends to this fibre and coincides there with $e^{-2\pi i \mathrm{Res}_q}$ [**De1**, II 1.17]. The endomorphism $e^{-2\pi i \mathrm{Res}_q}$ has the same eigenvalues as h, but may have a simpler Jordan block structure. The following more precise (and well known) statements will be useful later. We again fix a coordinate z, identifying (M, q) and $(\mathbb{C}, 0)$ and having the spaces C^α of elementary sections at our disposal.

Theorem 7.10 *Let $\mathcal{L}_0 \subset (i_*\mathcal{H})_0$ be a saturated lattice and $F_\bullet$ the corresponding (mondromy invariant) filtration on H^∞.*

(a) The coordinate z induces an isomorphism

$$\mathcal{L}_0/z\mathcal{L}_0 \to \bigoplus_p \mathrm{Gr}_\mathrm{F}^\mathrm{p}\mathrm{H}^\infty \tag{7.35}$$

which identifies the actions of $e^{-2\pi i \mathrm{Res}_0}$ on the left hand side and of h on the right hand side.

(b) The eigenvalues of the residue endomorphism Res_0 *are the spectral numbers $\alpha_1, \ldots, \alpha_\mu$ of $\mathcal{L}_0$.*

(c) The endomorphism Res_0 *is semisimple if and only if $N(F_p) \subset F_{p-1}$ for all p.*

(d) The endomorphism $e^{-2\pi i \mathrm{Res}_0}$ has the same Jordan normal form as h if no two spectral numbers differ by a nonzero integer (nonresonance condition).

Proof: (a) The coordinate z provides the spaces C^α of elementary sections. Because of (7.31) and (7.24) this yields isomorphisms

$$\begin{aligned} \mathcal{L}_0/z\mathcal{L}_0 &\cong \bigoplus_{-1<\alpha\le 0} \bigoplus_p \mathcal{L}_0 \cap C^{\alpha+p}/(z\mathcal{L}_0 \cap C^{\alpha+p}) \\ &\cong \bigoplus_{-1<\alpha\le 0} \bigoplus_p \mathrm{Gr}_\mathrm{F}^\mathrm{p}\mathrm{H}^\infty_{\mathrm{e}^{-2\pi i\alpha}} \\ &\cong \bigoplus_p \mathrm{Gr}_\mathrm{F}^\mathrm{p}\mathrm{H}^\infty. \end{aligned} \tag{7.36}$$

On each subspace C^α the monodromy h acts as $\exp(-2\pi i z\nabla_{\partial_z})$ (cf. (7.8)). This shows (a).

(b) and (c) follow immediately from (a) and its proof.

(d) Under the nonresonance condition there exists for each eigenvalue λ of h an index $p(\lambda)$ with $\mathrm{Gr}_F^{p(\lambda)} H_\lambda^\infty = H_\lambda^\infty$. Now (d) also follows from (a). □

Example 7.11 Suppose that the residue endomorphism Res_0 of a saturated lattice $\mathcal{L}_0 \subset (i_*\mathcal{H})_0$ is semisimple with eigenvalues $(s, 0, \ldots, 0)$ for some $s \in \mathbb{C}$.

If $s \in \mathbb{C} - (\mathbb{Z} - \{0\})$ then the monodromy is semisimple with eigenvalues $(e^{-2\pi is}, 1, \ldots, 1)$ and the lattice is $\mathcal{L}_0 = \mathbb{C}\{z\}C^0 \oplus \mathbb{C}\{z\}C^s$.

If $s \in \mathbb{Z} - \{0\}$ there are two possibilities:

(a) Either the monodromy is the identity; then the saturated lattice can be any one in a family parameterized by $\mathbb{P}\,C^0$ for $s < 0$ and by $\mathbb{P}Hom(C^0, \mathbb{C})$ for $s > 0$.

(b) Or there is one 2×2 Jordan block; then the saturated lattice is unique; for $s > 0$ it is $\mathbb{C}\{z\}\ker(N : C^0 \to C^0) + V^s$, for $s < 0$ it is $\mathbb{C}\{z\}z^s \cdot N(C^0) + V^0$.

7.4 Riemann–Hilbert–Birkhoff problem

The most recent references in book form on the Riemann–Hilbert problem and the Birkhoff problem are [**AB**] and [**Sab4**]. In chapter IV in [**Sab4**] Sabbah discusses several versions of the problem. One unifying general version can be stated as follows.

Hypotheses: $\Sigma_1 \subset \mathbb{P}^1$ and $\Sigma_2 \subset \mathbb{P}^1$ are two disjoint finite sets, $H \to \mathbb{P}^1 - (\Sigma_1 \cup \Sigma_2)$ is a flat vector bundle of rank $\mu \geq 1$ and with sheaf $\mathcal{H}$ of holomorphic sections, and $\mathcal{L}$ is a free $\mathcal{O}_{\mathbb{P}^1 - \Sigma_2}$-module of rank μ with an isomorphism $\mathcal{L}|\mathbb{P}^1 - (\Sigma_1 \cup \Sigma_2) = \mathcal{H}$.

Problem: Does there exist an extension of $\mathcal{L}$ to a free $\mathcal{O}_{\mathbb{P}^1}$-module $\overline{\mathcal{L}}$ which has logarithmic poles along the points of Σ_2?

Because of Theorem 7.9 one can extend $\mathcal{L}$ to a locally free $\mathcal{O}_{\mathbb{P}^1}$-module with logarithmic poles along Σ_2 without any problem. The requirement that $\overline{\mathcal{L}}$ shall be a free $\mathcal{O}_{\mathbb{P}^1}$-module makes the problem difficult.

Often the problem is formulated in terms of the (trivial) vector bundles which correspond to the sheaves $\mathcal{L}$ and $\overline{\mathcal{L}}$.

Usually one makes additional assumptions on the poles along Σ_1. At least one supposes that for $q \in \Sigma_1$ the coefficients of the connection matrix with respect to a basis of $\mathcal{L}_q$ are meromorphic. A much stronger assumption would be that the sections in $\mathcal{L}_q$, $q \in \Sigma_1$, are of bounded growth, that means, the connection is regular singular there.

The classical Riemann–Hilbert problem is the case $\Sigma_1 = \emptyset$, the classical Birkhoff problem is the case $\Sigma_1 = \{0\}$, $\Sigma_2 = \{\infty\}$, with the assumption that the connection matrix with respect to a basis of $\mathcal{L}_0$ has a pole of order ≤ 2. Sabbah [**Sab4**] calls the case $\Sigma_1 = \{0\}$, $\Sigma_2 = \{\infty\}$ without special assumptions on $\mathcal{L}_0$ the Riemann–Hilbert–Birkhoff problem.

A particular case of it was treated implicitly by M. Saito [**SM3**, §3]. He gave (implicitly) a correspondence between certain solutions and certain filtrations on the flat bundle $H \to \mathbb{C}^*$. The purpose of this section is to resume and generalize this correspondence. In the case of hypersurface singularities such filtrations exist because of properties of mixed Hodge structures (cf. section 10.5).

Let us stick to the situation and notations in section 7.1. Again $H \to \mathbb{C}^*$ is a flat vector bundle of rank μ and with monodromy $h = h_s \cdot h_u = h_u \cdot h_s$, $N = \log h_u$, and $\mathcal{H}$, H^∞, C^α, ψ_α, V^α, $V^{>\alpha}$, $\mathbb{C}\{\{\partial_z^{-1}\}\}$ as in section 7.1. Again $\mathcal{L}_0 \subset V^{>-\infty}$ will be a $\mathbb{C}\{z\}$-lattice (of rank μ), and $\mathcal{L}$ is the $\mathcal{O}_\mathbb{C}$-free extension of $\mathcal{H}$ to 0 with germ $\mathcal{L}_0$ at 0.

Because of Theorem 7.9 the extensions of $\mathcal{L}$ to *locally* free $\mathcal{O}_{\mathbb{P}^1}$-modules $\overline{\mathcal{L}}$ with a logarithmic pole at ∞ correspond one-to-one to the monodromy invariant finite increasing filtrations $U_\bullet$ on H^∞. Theorems 7.16 and 7.17 will describe distinguished extensions to *free* $\mathcal{O}_{\mathbb{P}^1}$-modules.

The objects in section 7.1 for the point ∞ instead of 0 will all be equipped with a tilde: the coordinate $\widetilde{z} = \frac{1}{z}$, satisfying

$$\frac{d\widetilde{z}}{\widetilde{z}} = -\frac{dz}{z}, \qquad \widetilde{z}\partial_{\widetilde{z}} = -z\partial_z, \tag{7.37}$$

the spaces $\widetilde{C}^\alpha = C^{-\alpha}$, $\widetilde{V}^{>\alpha} \subset \widetilde{V}^\alpha \subset \widetilde{V}^{>-\infty} \subset (\widetilde{i}_*\mathcal{H})_\infty$, the spectral numbers $\widetilde{\alpha}_1, \ldots, \widetilde{\alpha}_\mu$ for a $\mathbb{C}\{\widetilde{z}\}$-lattice $\widetilde{\mathcal{L}}_\infty \subset \widetilde{V}^{>-\infty}$.

Example 7.12 Let $H \to \mathbb{C}^*$ be the trivial bundle of rank 2 with basis e_1, e_2 of flat sections. The following is a 1-parameter family of free $\mathcal{O}_{\mathbb{P}^1}$-modules $\mathcal{L}(r)$, $r \in \mathbb{C}$, extending $\mathcal{H}$ to $\mathbb{P}^1$,

$$\mathcal{L}(r) = \mathcal{O}_{\mathbb{P}^1} \cdot (re_1 + ze_2) \oplus \mathcal{O}_{\mathbb{P}^1} \cdot ze_1. \tag{7.38}$$

The lattice $\mathcal{L}_\infty(r) = \widetilde{V}^{-1}$ is constant and saturated with spectral numbers $(\widetilde{\alpha}_1, \widetilde{\alpha}_2) = (-1, -1)$. The lattice $\mathcal{L}_0(r) \subset V^0$ is not constant and saturated only for $r = 0$; even the spectral numbers jump, $(\alpha_1, \alpha_2)(0) = (1, 1)$ and $(\alpha_1, \alpha_2) = (0, 2)$ for $r \neq 0$. A $\mathbb{C}\{z\}$-basis of $\mathcal{L}_0(r)$ for $r \neq 0$ whose principal parts represent a basis of the space $\bigoplus_\alpha \mathrm{Gr}_V^\alpha \mathcal{L}_0(\mathrm{r})/\mathrm{Gr}_V^\alpha \mathrm{z}\mathcal{L}_0(\mathrm{r})$ (cf. Lemma 7.3) is given by

$$\mathcal{L}_0(r) = \mathbb{C}\{z\} \cdot (re_1 + ze_2) \oplus \mathbb{C}\{z\} \cdot (z^2 e_2), \; r \neq 0. \tag{7.39}$$

It does not extend to a basis at $r = 0$.

In the example $\sum_i \widetilde{\alpha}_i + \sum_i \alpha_i = 0$. Theorem 7.20 will show that this holds in general. We will be interested in free $\mathcal{O}_{\mathbb{P}^1}$-modules with a stronger property, which includes $\widetilde{\alpha}_i = -\alpha_{\mu+1-i}$. We resume two definitions from [**SM3**, §3].

Definition 7.13 (a) Let $\mathcal{L}_0 \subset V^{>-\infty}$ be a $\mathbb{C}\{z\}$-lattice. A μ-dimensional subspace $W \subset \mathcal{L}_0$ such that the projection $pr : W \to \mathcal{L}_0/z\mathcal{L}_0$ is an isomorphism is the image of a unique section $v : \mathcal{L}_0/z\mathcal{L}_0 \to \mathcal{L}_0$ with $pr \circ v = id$. The space W is called a good $\mathcal{L}_0/z\mathcal{L}_0$-section if the following two equivalent conditions hold:

(i) The filtrations $pr(V^\bullet \cap W)$ and $(V^\bullet \cap \mathcal{L}_0 + z\mathcal{L}_0)/z\mathcal{L}_0$ on $\mathcal{L}_0/z\mathcal{L}_0$ coincide.
(ii) The space W has a basis whose principal parts represent a basis of the space $\bigoplus_\alpha \mathrm{Gr}_V^\alpha \mathcal{L}_0/\mathrm{z}\mathrm{Gr}_V^{\alpha-1}\mathcal{L}_0$.

(b) If $\mathcal{L}_0 \subset V^{>-1}$ is a $\mathbb{C}\{\{\partial_z^{-1}\}\}$-lattice, good $\mathcal{L}_0/\nabla_{\partial_z}^{-1}\mathcal{L}_0$-sections are defined analogously.

Remarks 7.14 ([**SM3**, §3]) A basis of the image Im v of any $\mathbb{C}$-linear section $v : \mathcal{L}_0/z\mathcal{L}_0 \to \mathcal{L}_0$ of a $\mathbb{C}\{z\}$-lattice $\mathcal{L}_0 \subset V^{>-\infty}$ is a $\mathbb{C}\{z\}$-basis of $\mathcal{L}_0$, by the lemma of Nakayama. Always $pr(V^\alpha \cap W) \subset (V^\alpha \cap \mathcal{L}_0 + z\mathcal{L}_0)/z\mathcal{L}_0$.

In Example 7.12, the space $W = \mathbb{C} \cdot (re_1 + ze_2) + \mathbb{C} \cdot ze_1 \subset \mathcal{L}_0(r)$ is not a good $\mathcal{L}_0/z\mathcal{L}_0$-section for $r \neq 0$.

The interplay between two filtrations on a finite dimensional space is discussed in [**De2**]. If $F_\bullet$ and $U_\bullet$ are two increasing filtrations on a finite dimensional vector space then one has canonical isomorphisms

$$\begin{aligned}\mathrm{Gr}_p^F \mathrm{Gr}_q^U &\cong F_p \cap U_q/(F_p \cap U_{q-1} + F_{p-1} \cap U_q)\\ &\cong \mathrm{Gr}_q^U \mathrm{Gr}_p^F.\end{aligned}$$

This helps to explain the equivalences in the following definition (cf. [**De2**], [**SM3**, §3]).

Definition 7.15 Two increasing filtrations $F_\bullet$ and $U_\bullet$ on a finite dimensional vector space are called opposite to one another if the following three equivalent conditions hold:

(i) If $\mathrm{Gr}_p^F \mathrm{Gr}_q^U \neq 0$ then $p + q = 0$.
(ii) The vector space splits into a direct sum $\bigoplus_p F_p \cap U_{-p}$.
(iii) One has decompositions $F_p = \bigoplus_{q \le p} F_q \cap U_{-q}$ and $U_p = \bigoplus_{q \le p} F_{-q} \cap U_q$ for all p.

Theorem 7.16 *Let $\mathcal{L}$ be an $\mathcal{O}_{\mathbb{C}}$-free extension over 0 of the sheaf $\mathcal{H}$ of sections of the flat vector bundle $H \to \mathbb{C}^*$ with $\mathcal{L}_0 \subset V^{>-\infty}$. The filtration which $\mathcal{L}_0$ induces on H^∞ by (7.25) is denoted by $F_\bullet$.*

There is a one-to-one correspondence between the two sets of data:

(i) Extensions of $\mathcal{L}$ to $\mathcal{O}_{\mathbb{P}^1}$-free modules $\overline{\mathcal{L}}$ which have a logarithmic pole at ∞ and satisfy: the μ-dimensional space $\overline{\mathcal{L}}(\mathbb{P}^1)$ of global sections is a good $\mathcal{L}_0/z\mathcal{L}_0$-section.
(ii) Monodromy invariant increasing exhaustive filtrations $U_\bullet$ on H^∞ such that $F_\bullet H_1^\infty$ and $U_\bullet H_1^\infty$ are opposite and, for $\lambda \neq 1$, $F_\bullet H_\lambda^\infty$ and $U_{\bullet+1} H_\lambda^\infty$ are opposite.

The spectral numbers $(\alpha_1, \ldots, \alpha_\mu)$ at 0 and $(\widetilde{\alpha}_1, \ldots, \widetilde{\alpha}_\mu)$ at ∞ of such an extension $\overline{\mathcal{L}}$ are related by $\widetilde{\alpha}_i = -\alpha_{\mu+1-i}$.

Proof: Let $U_\bullet$ be a filtration as in (ii). The following explicit construction of a basis of global sections of $\overline{\mathcal{L}}$ from $U_\bullet$ is a key idea in [**SM3**, §3] (cf. also [**He4**, chapter 5]). The space H^∞ decomposes into

$$H^\infty = \bigoplus_p H_1^\infty \cap F_p \cap U_{-p} \oplus \bigoplus_{\lambda \neq 1} \bigoplus_p H_\lambda^\infty \cap F_p \cap U_{1-p}. \tag{7.40}$$

These subspaces lead to distinguished spaces of elementary sections

$$G^{\alpha+p} := z^p \psi_\alpha \left(H^\infty_{e^{-2\pi i \alpha}} \cap F_p \cap U_{(0\ or\ 1)-p} \right) \subset C^{\alpha+p} \tag{7.41}$$

for $-1 < \alpha \le 0$, $p \in \mathbb{Z}$, with the properties

$$C^\alpha = \bigoplus_{p \in \mathbb{Z}} z^{-p} G^{\alpha+p} \supset \mathrm{Gr}_V^\alpha \mathcal{L}_0 = \bigoplus_{p \le 0} z^{-p} G^{\alpha+p} = G^\alpha \oplus z \mathrm{Gr}_V^{\alpha-1} \mathcal{L}_0, \tag{7.42}$$

$$N G^\alpha = \left(z \nabla_{\partial_z} - \alpha \right) G^\alpha \subset \bigoplus_{p \ge 0} z^{-p} G^{\alpha+p}. \tag{7.43}$$

Formula (7.43) follows from the monodromy invariance of $U_\bullet$ and from $z\nabla_{\partial_z} - \alpha = -\frac{N}{2\pi i}$ on C^α (7.8). The purpose of the filtration $U_\bullet$ is really the splitting in (7.42) of the filtration of C^α by $\mathrm{Gr}_V^\alpha z^\bullet \mathcal{L}_0$.

One can choose a basis of elementary sections $s_1, \ldots, s_\mu$ of $\bigoplus_\alpha G^\alpha$ with orders $\alpha(s_i) = \alpha_i$. They form a basis of H_z for each $z \in \mathbb{C}^*$. They are principal parts of germs in $\mathcal{L}_0$. The point now is that there exist unique germs v_i in $\mathcal{L}_0$

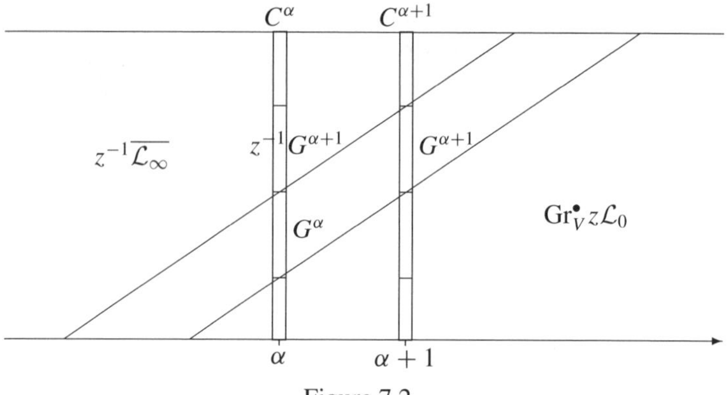

Figure 7.2

whose principal parts are s_i and whose higher elementary parts are contained in $\bigoplus_{\beta>\alpha_i}\bigoplus_{p>0} z^{-p}G^{\beta+p}$ for each $i = 1, \dots, \mu$: starting with some sections whose principal parts are s_i, one can inductively for increasing i and increasing orders erase all higher elementary parts in $\bigoplus_{\beta>\alpha_i}\bigoplus_{p\le 0} z^{-p}G^{\beta+p}$ by adding elements of $\mathcal{L}_0$. The uniqueness of the section

$$v_i \in \mathcal{L}_0 \cap \left(s_i + \bigoplus_{\beta>\alpha_i}\bigoplus_{p>0} z^{-p}G^{\beta+p} \right) \tag{7.44}$$

is clear because the difference of two such sections would have an impossible principal part.

Figure 7.2 shall illustrate the construction. The eigenspaces C^α should be imagined as columns above the αs; the picture does not take into account the different dimensions of the C^α and the discreteness of the values α.

A space G^β is nonzero if and only if $\beta \in \{\alpha_1, \dots, \alpha_\mu\}$. Therefore the sections v_i are sums of finitely many elementary parts, so they are global sections in $\mathcal{H}(\mathbb{C})$. They form a basis of H_z for each $z \in \mathbb{C}^*$ because the s_i do and (7.44) gives a triangular coefficient matrix.

The lattice at ∞ of the $\mathcal{O}_{\mathbb{P}^1}$-module $\overline{\mathcal{L}} := \bigoplus_i \mathcal{O}_{\mathbb{P}^1} \cdot v_i$ is

$$\overline{\mathcal{L}}_\infty = \bigoplus_i \mathbb{C}\{\widetilde{z}\} \cdot v_i = \bigoplus_i \mathbb{C}\{\widetilde{z}\} s_i = \bigoplus_\alpha \mathbb{C}\{\widetilde{z}\} G^\alpha. \tag{7.45}$$

Formula (7.43) shows that it is saturated. Going through (7.24) and Lemma 7.6 (b) one checks that the filtration on H^∞ which corresponds to $\overline{\mathcal{L}}_\infty$ via (7.30) is precisely $U_\bullet$. Here one needs $\widetilde{z}\partial_{\widetilde{z}} = -z\partial_z$ and $\widetilde{C}^\alpha = C^{-\alpha}$ (7.37). Obviously, the spectral numbers $(\widetilde{\alpha}_1, \dots, \widetilde{\alpha}_\mu)$ at infinity are $\widetilde{\alpha}_i = -\alpha_{\mu+1-i}$.

It remains to go from (i) to (ii).

Let $\overline{\mathcal{L}}$ be as in (i). One can choose a basis $v_1, \dots, v_\mu$ of global sections in $\overline{\mathcal{L}}(\mathbb{P}^1)$ whose principal parts $s_1, \dots, s_\mu$ represent a basis of the space

$\bigoplus_\alpha \mathrm{Gr}_V^\alpha \overline{\mathcal{L}}_0/\mathrm{Gr}_V^\alpha z\overline{\mathcal{L}}_0$ and have orders α_i. Then these principal parts $s_1, \ldots, s_\mu$ also form a global basis of the bundle $H \to \mathbb{C}^*$, because they form a $\mathbb{C}\{z\}[z^{-1}]$-basis of $V^{>-\infty}$ and because they are elementary sections.

They generate a $\mathbb{C}\{\widetilde{z}\}$-lattice $\mathcal{L}'_\infty$ at infinity whose spectral numbers are $-\alpha_1, \ldots, -\alpha_\mu$. But the lattice $\overline{\mathcal{L}}_\infty$ is saturated and contains all elementary parts of the v_i (Lemma 7.6 (a)). Therefore $\mathcal{L}'_\infty \subset \overline{\mathcal{L}}_\infty$. If $\overline{\mathcal{L}}_\infty$ were bigger than $\mathcal{L}'_\infty$, the spectral numbers $\widetilde{\alpha}_1, \ldots, \widetilde{\alpha}_\mu$ of $\overline{\mathcal{L}}_\infty$would satisfy $\sum \widetilde{\alpha}_i + \sum \alpha_i < 0$. But Theorem 7.20 will show

$$\sum \widetilde{\alpha}_i + \sum \alpha_i = 0. \tag{7.46}$$

Therefore $\mathcal{L}'_\infty = \overline{\mathcal{L}}_\infty$. Now one defines

$$G^\alpha := \bigoplus_{\alpha_i = \alpha} \mathbb{C} \cdot s_i \subset C^\alpha. \tag{7.47}$$

The identity (7.42) holds. The identity (7.43) holds because $\overline{\mathcal{L}}_\infty$ is saturated by assumption. Reading (7.41) and (7.40) backwards one obtains a monodromy invariant increasing exhaustive filtration $U_\bullet$ on H^∞ which satisfies the properties in (ii) and which is just the filtration corresponding to $\overline{\mathcal{L}}_\infty$ via (7.30). □

In the singularity case one is interested in $\mathbb{C}\{z\}$-lattices $\mathcal{L}_0 \subset V^{>-1}$ which also are $\mathbb{C}\{\{\partial_z^{-1}\}\}$-lattices. Then a stronger result holds.

Theorem 7.17 *Let $\mathcal{L}$ be an $\mathcal{O}_\mathbb{C}$-free extension over 0 of the sheaf $\mathcal{H}$ of sections of the flat vector bundle $H \to \mathbb{C}^*$ with $\mathcal{L}_0 \subset V^{>-1}$, and suppose that $\mathcal{L}_0$ also is a $\mathbb{C}\{\{\partial_z^{-1}\}\}$-lattice. The filtrations which $\mathcal{L}_0$ induces on H^∞ by* (7.25) *and* (7.28) *are denoted by $F_\bullet$ and $F_\bullet^{alg}$.*

(a) The two conditions for a monodromy invariant increasing exhaustive filtration $U_\bullet$ on H^∞ are equivalent:

(i) The filtrations $F_\bullet$ and $U_\bullet$ are opposite in H_1^∞; the filtrations $F_\bullet$ and $U_{\bullet+1}$ are opposite in H_λ^∞ for $\lambda \neq 1$.
(ii) The filtrations $F_\bullet^{alg}$ and $U_\bullet$ are opposite in H_1^∞; the filtrations $F_\bullet^{alg}$ and $U_{\bullet+1}$ are opposite in H_λ^∞ for $\lambda \neq 1$.

(b) There is a one-to-one correspondence between the two sets of data:

(i) Extensions of $\mathcal{L}$ to $\mathcal{O}_{\mathbb{P}^1}$-free modules $\overline{\mathcal{L}}$ which have a logarithmic pole at ∞ and which satisfy: the μ-dimensional space $\overline{\mathcal{L}}(\mathbb{P}^1)$ of global sections is a good $\mathcal{L}_0/\nabla_{\partial_z}^{-1}\mathcal{L}_0$-section.
(ii) Filtrations as in (a) (i) and (ii).

Then $\overline{\mathcal{L}}(\mathbb{P}^1)$ *is also a good* $\mathcal{L}_0/z\mathcal{L}_0$*-section; a basis of it is a* $\mathbb{C}\{z\}$*- and* $\mathbb{C}\{\{\partial_z^{-1}\}\}$*-basis of* $\mathcal{L}_0$*. The spectral numbers at 0 and* ∞ *satisfy* $\widetilde{\alpha}_i = -\alpha_{\mu+1-i}$.

Proof: (a) We will prove that condition (i) for H_1^∞ implies condition (ii) for H_1^∞. Everything else is analogous.

We have $F_{-1}^{alg} = F_{-1} = 0$ because of $\mathcal{L}_0 \subset V^{>-1}$. For any $p \geq 0$ the automorphism $\varphi_p := \prod_{k=1}^p (\frac{-N}{2\pi i} + k)$ of H_1^∞ in (7.29) maps $F_p H_1^\infty$ to $F_p^{alg} H_1^\infty$ and respects $U_\bullet H_1^\infty$. Therefore on H_1^∞

$$\varphi_p(F_p \cap U_{-p}) \subset F_p^{alg} \cap U_{-p}, \tag{7.48}$$

$$(F_p \cap U_{-p}) \oplus U_{-p-1} = U_{-p} = \varphi_p(F_p \cap U_{-p}) \oplus U_{-p-1}, \tag{7.49}$$

$$H_1^\infty = \bigoplus_p \varphi_p(F_p \cap U_{-p}). \tag{7.50}$$

The equality $\dim F_p = \dim F_p^{alg}$ now implies

$$F_p^{alg} = \bigoplus_{q \leq p} \varphi_p(F_q \cap U_{-q}), \tag{7.51}$$

$$F_p^{alg} \cap U_{-p} = \varphi_p(F_p \cap U_{-p}). \tag{7.52}$$

(b) First we go from (ii) to (i). Suppose $U_\bullet$ satisfies the properties in (a). Theorem 7.16 and its proof yield an extension $\overline{\mathcal{L}}$ and global sections v_i with principal parts s_i and spaces $G^\alpha = \bigoplus_{\alpha_i = \alpha} \mathbb{C} \cdot s_i \subset C^\alpha$. It remains to show $\mathrm{Gr}_V^\alpha \mathcal{L}_0 = G^\alpha \oplus \nabla_{\partial_z}^{-1} \mathrm{Gr}_V^{\alpha-1} \mathcal{L}_0$. We need the formula

$$z^p \psi_\alpha\big(H_{e^{-2\pi i\alpha}}^\infty \cap U_{-q}\big) = \nabla_{\partial_z}^{-p} \psi_\alpha\big(H_{e^{-2\pi i\alpha}}^\infty \cap U_{-q}\big) \tag{7.53}$$

for $-1 < \alpha \leq 0,\ p \geq 0,\ q \in \mathbb{Z}$. This follows from the monodromy invariance of U_{-q} and from

$$z = \nabla_{\partial_z}^{-1} \left(\alpha + 1 - \frac{N}{2\pi i}\right) : C^\alpha \to C^{\alpha+1} \quad \text{for } \alpha > -1 \tag{7.54}$$

with $(\alpha + 1 - \frac{N}{2\pi i})$ being invertible.

The formulas (7.41), (7.53), and the definitions of $F_\bullet$ and $F_\bullet^{alg}$ show

$$\begin{aligned} G^{\alpha+p} &= \big(\mathrm{Gr}_V^{\alpha+p} \mathcal{L}_0\big) \cap \big(z^p \psi_\alpha\big(H_{e^{-2\pi i\alpha}}^\infty \cap U_{(0 \text{ or } 1)-p}\big) \\ &= \nabla_{\partial_z}^{-p} \psi_\alpha\big(H_{e^{-2\pi i\alpha}}^\infty \cap F_p^{alg} \cap U_{(0 \text{ or } 1)-p}\big) \end{aligned} \tag{7.55}$$

for $-1 < \alpha \leq 0,\ p \geq 0$. This formula together with the analogon of (7.40) for $F_\bullet^{alg}$ shows that (7.42) is also valid for $\alpha > -1$ if one replaces z by $\nabla_{\partial_z}^{-1}$,

$$C^\alpha = \bigoplus_{p \in \mathbb{Z}} \nabla_{\partial_z}^p G^{\alpha+p} \supset \mathrm{Gr}_V^\alpha \mathcal{L}_0 = \bigoplus_{p \leq 0} \nabla_{\partial_z}^p G^{\alpha+p}. \tag{7.56}$$

This implies that $\overline{\mathcal{L}}(\mathbb{P}^1)$ is a good $\mathcal{L}_0/\nabla_{\partial_z}^{-1}\mathcal{L}_0$-section.

It remains to go from (i) to (ii).

Let $\overline{\mathcal{L}}$ be as in (i) and $v_1, \dots, v_\mu$ be a basis of global sections in $\overline{\mathcal{L}}(\mathbb{P}^1)$ whose principal parts $s_1, \dots, s_\mu$ at 0 represent a basis of the space $\bigoplus_\alpha \mathrm{Gr}_V^\alpha \overline{\mathcal{L}}_0/\nabla_{\partial_z}^{-1} \mathrm{Gr}_V^{\alpha-1}\overline{\mathcal{L}}_0$. It is sufficient to show that they also represent a basis of $\bigoplus_\alpha \mathrm{Gr}_V^\alpha \overline{\mathcal{L}}_0/z\mathrm{Gr}_V^{\alpha-1}\overline{\mathcal{L}}_0$. Then one can apply Theorem 7.16.

We may choose $v_1, \dots, v_\mu$ such that $\alpha_1, \dots, \alpha_\mu$ are their orders. Define

$$G^\alpha := \bigoplus_{\alpha_i=\alpha} \mathbb{C}\cdot s_i \subset C^\alpha. \tag{7.57}$$

Then

$$C^\alpha = \bigoplus_{p\in\mathbb{Z}} \nabla_{\partial_z}^p G^{\alpha+p} \qquad \text{for } \alpha > -1. \tag{7.58}$$

The lattice $\overline{\mathcal{L}}_\infty$ is saturated and contains with v_i also its principal part s_i (Lemma 7.6 (a)). Because of $\partial_z = -\widetilde{z}^2\partial_{\widetilde{z}}$ then

$$\overline{\mathcal{L}}_\infty \supset \bigoplus_{-1<\alpha\le 0}\ \bigoplus_{p\ge 0} \nabla_{\partial_z}^p G^{\alpha+p} \supset \bigoplus_{-1<\alpha\le 0} C^\alpha. \tag{7.59}$$

The sum in the middle generates a $\mathbb{C}\{\widetilde{z}\}$-lattice $\mathcal{L}'_\infty$ whose spectrum is already $-\alpha_1, \dots, -\alpha_\mu$. Theorem 7.20 implies in the same way as in the proof of Theorem 7.16 that $\overline{\mathcal{L}}_\infty = \mathcal{L}'_\infty$. Now $\partial_z = \widetilde{z}\cdot(-\widetilde{z}\partial_{\widetilde{z}})$ again shows

$$\overline{\mathcal{L}}_\infty \cap C^\alpha = G^\alpha \oplus (\widetilde{z}\overline{\mathcal{L}}_\infty \cap C^\alpha) \quad \text{for } \alpha > -1 \tag{7.60}$$

and then

$$C^\alpha = \bigoplus_{p\in\mathbb{Z}} z^{-p} G^{\alpha+p} \qquad \text{for } \alpha > -1, \tag{7.61}$$

$$\mathrm{Gr}_V^\alpha \mathcal{L}_0 = G^\alpha \oplus z\mathrm{Gr}_V^{\alpha-1}\mathcal{L}_0 \qquad \text{for } \alpha > -1. \tag{7.62}$$

Now one can apply Theorem 7.16. □

Remarks 7.18 (a) The main result Theorem 3.6 in [**SM3**, §3] can be stated as follows: Under the same assumptions as in Theorem 7.17 there is a one-to-one correspondence between filtrations $U_\bullet$ as in Theorem 7.17 (a) (ii) and good $\mathcal{L}_0/\nabla_{\partial_z}^{-1}\mathcal{L}_0$-sections $W \subset \mathcal{L}_0$ with

$$z\,W \subset W + \nabla_{\partial_z}^{-1} W. \tag{7.63}$$

Comparing the proofs of Theorem 7.17 and [**SM3**, Theorem 3.6], one finds that these μ-dimensional spaces W are precisely the spaces $\overline{\mathcal{L}}(\mathbb{P}^1)$ of global sections of extensions $\overline{\mathcal{L}}$ as in Theorem 7.17 (b) (i). But this is quite nontrivial.

M. Saito does not consider the questions of whether a basis of germs in W is extendable to a basis of global sections in the bundle $H \to \mathbb{C}^*$ and whether they give a logarithmic pole at ∞. He does not formulate Theorems 7.17 and 7.16 and does not use the filtration $F_\bullet$.

(b) The main application of Theorem 7.17 and of [**SM3**, Theorem 3.6] is the same: to find a distinguished basis of sections in the Gauß–Manin connection of a versal unfolding of a singularity which induces a Frobenius manifold structure on the base.

In section 11.1 we will use Theorem 7.17 and Theorem 8.7 in order to extend the Gauß–Manin connection in an explicit way to ∞. M. Saito uses (7.63) in order to apply a result of Malgrange on deformations of microdifferential systems ([**Mal3**, §5], [**Mal5**]). In the proof of Malgrange's result there also is an extension of a connection along ∞, but other ingredients are a Fourier transformation and microlocal aspects.

(c) One can separate the proof of [**SM3**, Theorem 3.6] into four pieces:

(i) to construct from a filtration $U_\bullet$ as in Theorem 7.17 (a) (ii) a good $\mathcal{L}_0/\nabla_{\partial_z}^{-1}\mathcal{L}_0$-section $W \subset \mathcal{L}_0$ ([**SM3**, Proposition 3.4], similar to the first part in the proof of Theorem 7.16 with $\nabla_{\partial_z}^{-1}$ instead of z),

(ii) to show that this section satisfies (7.63) (a discussion of principal parts, this is not so difficult),

(iii) to recover a filtration $U_\bullet$ from a good $\mathcal{L}_0/\nabla_{\partial_z}^{-1}\mathcal{L}_0$-section $W \subset \mathcal{L}_0$ with (7.63) ([**SM3**, Proposition 3.5]),

(iv) to show that this section is the unique section constructed from $U_\bullet$ in (i) (this is the most difficult piece; it is not even clear a priori that the elements of a section W with (7.63) consist of finitely many elementary sections).

(d) We will meet in the singularity case an extension of (7.63) to the Gauß–Manin connection of a versal unfolding, cf. chapter 11, (11.20) and (11.54). The coefficients will carry precious structure.

7.5 Spectral numbers globally

The results in this section are inspired by and partly due to [**Sab4**, IV 1.10].

We stay in the situation of section 7.1. The spectral numbers $(\alpha_1, \ldots, \alpha_\mu)$ of a $\mathbb{C}\{z\}$-lattice $\mathcal{L}_0 \subset V^{>-\infty}$ had been defined by

$$\sharp(i \mid \alpha_i = \alpha) = d(\alpha) := \dim \mathrm{Gr}_V^\alpha \mathcal{L}_0 - \dim \mathrm{Gr}_V^{\alpha-1} \mathcal{L}_0, \tag{7.64}$$

cf. (7.21) and (7.23), and were ordered by $\alpha_1 \leq \ldots \leq \alpha_\mu$.

The V-filtration is independent of the coordinate z in $\mathbb{C}$ (Lemma 7.7), so the spectral numbers of a lattice are also coordinate independent. But in (7.9) a total order $<$ had been chosen. A different order may give a different V-filtration and other spectral numbers. An example due to M. Saito [**SM3**, (4.4)] will be presented in Remark 11.7.

The following result shows that the sum $\sum_i \alpha_i$ is even independent of the order $<$.

The determinant bundle $\det H = H^{\wedge\mu} \to \mathbb{C}^*$ is a line bundle and carries an induced flat connection. The monodromy is given by its eigenvalue $\det h \in \mathbb{C}^*$. The determinant sheaf $\mathcal{L}^{\wedge\mu}$ of an $\mathcal{O}_{\mathbb{C}}$-free extension $\mathcal{L}$ to 0 of the sheaf $\mathcal{H}$ of sections of $H \to \mathbb{C}^*$ is an $\mathcal{O}_{\mathbb{C}}$-free extension of the sheaf $\mathcal{H}^{\wedge\mu}$ of sections of $H^{\wedge\mu} \to \mathbb{C}^*$. If $\mathcal{L}_0 \subset V^{>-\infty}$ then the sections of $\mathcal{L}^{\wedge\mu}$ also are of bounded growth at 0. More precisely, one has the following.

Theorem 7.19 *Let $\mathcal{L}$ be as above with $\mathcal{L}_0 \subset V^{>-\infty}$ and with spectral numbers $\alpha_1, \dots, \alpha_\mu$ at 0. Then the germ $\mathcal{L}_0^{\wedge\mu}$ of $\mathcal{L}^{\wedge\mu}$ is a saturated lattice in $(i_*\mathcal{H}^{\wedge\mu})_0$ with residue eigenvalue $\sum_{i=1}^{\mu} \alpha_i$.*

Proof: One chooses a $\mathbb{C}\{z\}$-basis $\omega_1, \dots, \omega_\mu$ of $\mathcal{L}_0$ whose principal parts $s_1, \dots, s_\mu$ represent a basis of $\bigoplus_\alpha \mathrm{Gr}_V^\alpha \mathcal{L}_0 / z\mathrm{Gr}_V^{\alpha-1}\mathcal{L}_0$ and have orders $\alpha_1, \dots, \alpha_\mu$. Then $\omega_1, \dots, \omega_\mu$ generates $\mathcal{L}_0^{\wedge\mu}$ over $\mathbb{C}\{z\}$. It is sufficient to show

$$\omega_1 \wedge \dots \wedge \omega_\mu = u(z) s_1 \wedge \dots \wedge s_\mu \tag{7.65}$$

with $u(z) \in \mathbb{C}\{z\}$, $u(0) = 1$. Then $\mathcal{L}_0^{\wedge\mu} = \mathbb{C}\{z\}\omega_1 \wedge \dots \wedge \omega_\mu = \mathbb{C}\{z\}s_1 \wedge \dots \wedge s_\mu$ has residue eigenvalue $\sum \alpha_i$ (in the rank 1 case a lattice with sections of bounded growth is automatically saturated).

The principal parts $s_1, \dots, s_\mu$ form a $\mathbb{C}\{z\}[z^{-1}]$-basis of $V^{>-\infty}$, so

$$\omega_i = \sum_j \sum_{k \in \mathbb{Z}} a_{ij}^{(k)} z^k s_j, \tag{7.66}$$

with unique $a_{ij}^{(k)} \in \mathbb{C}$, which satisfy $a_{ii}^{(0)} = 1$ and

$$a_{ij}^{(k)} \neq 0 \quad \text{only for } (j,k) = (i,0) \text{ or } \alpha_i < \alpha_j + k. \tag{7.67}$$

Expanding $\omega_1 \wedge \dots \wedge \omega_\mu$ with (7.66) one has to see that all combinations of summands except for the combinations of the principal parts are contained in $z\mathbb{C}\{z\}s_1 \wedge \dots \wedge s_\mu$. This is elementary; it follows from (7.67) and the properties in (7.9) of the transitive order $<$. □

Theorem 7.20 *Let the following be given: a holomorphic vector bundle $L \to M$ of rank μ on a compact Riemann surface M with sheaf $\mathcal{L}$ of holomorphic sections, a finite set $\Sigma \subset M$, a flat connection on $L|_{M-\Sigma}$ such that the sections of $\mathcal{L}$ have bounded growth everywhere. Let $(\alpha_1(q), \dots, \alpha_\mu(q))$ denote the spectral* *numbers of the lattice $\mathcal{L}_q$ for $q \in \Sigma$. The line bundle $L^{\wedge\mu}$ is the determinant bundle of L. Then*

$$\sum_q \sum_i \alpha_i(q) = -\deg L^{\wedge\mu}. \tag{7.68}$$

Proof: The restriction $L^{\wedge\mu}|_{M-\Sigma}$ is equipped with the induced flat connection. The sheaf $\mathcal{L}^{\wedge\mu}$ has logarithmic poles with residue eigenvalues $\sum_i \alpha_i(q)$ at all points $q \in \Sigma$, because of Theorem 7.19.

One can choose any nonvanishing meromorphic section on $L^{\wedge\mu}$ with zeros and poles in a set Σ' with $\Sigma' \cap \Sigma = \emptyset$. It yields a meromorphic connection 1-form ω with simple poles in $\Sigma' \cup \Sigma$. The sum of its residues vanishes.

$$\sum_{q\in\Sigma} \sum_i \alpha_i(q) = \sum_{q\in\Sigma} res_q\omega = -\sum_{q\in\Sigma'} res_q\omega = -\deg L^{\wedge\mu}. \tag{7.69}$$

□

Chapter 8

Meromorphic connections

Section 8.1 is a reminder of logarithmic vector fields and differential forms and some other classical facts. In sections 8.2–8.4 extensions with logarithmic poles along a divisor $\mathcal{D} \subset M$ of the sheaf of holomorphic sections of a flat vector bundle on $M - \mathcal{D}$ are discussed. In the case of a smooth divisor $\mathcal{D}$ in section 3.2, there are three important tools for working with such extensions: the correspondence to certain filtrations, the classical residue endomorphism along $\mathcal{D}$, and the (less familiar) residual connections along $\mathcal{D}$, whose definitions require the choice of a transversal coordinate.

Extensions to singular divisors $\mathcal{D}$ are treated in section 8.3 in greater generality than in the literature. If an (automatically locally free) extension to $\mathcal{D}_{reg}$ with logarithmic pole is given, then there exists a unique maximal coherent extension to $\mathcal{D}$. It is locally free only under special circumstances. The case of a normal crossing divisor is discussed in section 8.3, the Gauß–Manin connections for singularities provide other very instructive examples (Theorem 10.3, Theorem 10.7 (b)). In Section 8.4 only some remarks on regular singularities are made.

8.1 Logarithmic vector fields and differential forms

For the reader's convenience we put together some definitions and results from **[SK4][De1][Ser]**, which will be useful in sections 8.2 and 8.3.

Let M be an m-dimensional complex manifold, $\mathcal{D} \subset M$ a hypersurface, and $g : M \to \mathbb{C}$ a holomorphic function such that $\mathcal{D} = g^{-1}(0)$ and the ideal sheaf $(g) \subset \mathcal{O}_M$ is everywhere reduced. A (locally defined) holomorphic vector field $X \in \mathcal{T}_M$ is *logarithmic* if $X\, g \in (g)$.

A (locally defined) meromorphic q-form ω with poles at most on $\mathcal{D}$ is *logarithmic* if $g\,\omega$ and $g\, d\omega$ are holomorphic q-forms. These definitions are obviously independent of the choice of the defining function g for $\mathcal{D}$. The sheaves of all logarithmic vector fields and differential q-forms are denoted $\mathrm{Der}_M(\log \mathcal{D})$ and $\Omega^q_M(\log \mathcal{D})$, respectively.

Example 8.1 (Smooth divisor $\mathcal{D}$.) Suppose $M \subset \mathbb{C}^m$ and $g = z_1$. Then these sheaves are free $\mathcal{O}_M$-modules,

$$\mathrm{Der}_M(\log \mathcal{D}) = \mathcal{O}_M \cdot z_1 \frac{\partial}{\partial z_1} \oplus \bigoplus_{i=2}^m \mathcal{O}_M \cdot \frac{\partial}{\partial z_i}, \tag{8.1}$$

$$\Omega^1_M(\log \mathcal{D}) = \mathcal{O}_M \cdot \frac{\mathrm{d}z_1}{z_1} \oplus \bigoplus_{i=2}^m \mathcal{O}_M \cdot \mathrm{d}z_i, \tag{8.2}$$

$$\Omega^q_M(\log \mathcal{D}) = \frac{\mathrm{d}z_1}{z_1} \wedge \Omega^{q-1}_M + \Omega^q_M, \tag{8.3}$$

$$\Omega^q_M(\log \mathcal{D}) = \bigwedge^q \Omega^1_M(\log \mathcal{D}). \tag{8.4}$$

The logarithmic vector fields are the vector fields tangent to $\mathcal{D}$.

This case of a smooth divisor $\mathcal{D}$ is essential for the general case because of some codimension 2 arguments. The singular part $\mathcal{D}_{sing} \subset M$ of the divisor $\mathcal{D}$ has codimension ≥ 2 in M. Let $i : M - \mathcal{D}_{sing} \hookrightarrow M$ denote the inclusion. Then $i_*\mathcal{O}_{M-\mathcal{D}_{sing}} = \mathcal{O}_M$ by the Riemann extension theorem, and also $i_*((g)|_{M-\mathcal{D}_{sing}}) = (g)$. The above definitions show

$$\mathrm{Der}_M(\log \mathcal{D}) = i_*\big(\mathrm{Der}_M(\log \mathcal{D})|_{M-\mathcal{D}_{sing}}\big), \tag{8.5}$$

$$\Omega^q_M(\log \mathcal{D}) = i_*\big(\Omega^q_M(\log \mathcal{D})|_{M-\mathcal{D}_{sing}}\big). \tag{8.6}$$

So the logarithmic vector fields are the vector fields tangent to $\mathcal{D}_{reg}$. Formula (8.4) also holds for general M and $\mathcal{D}$.

Example 8.2 ([**De1**, II §3], the normal crossing case.) Suppose $M = \mathbb{C}^m$ and $g = z_1 \cdot z_2 \cdot \ldots \cdot z_k$ for some k with $1 \leq k \leq m$, so $\mathcal{D} = \bigcup_{i=1}^k \mathbb{C}^{i-1} \times \{0\} \times \mathbb{C}^{m-i}$. Then the sheaves $\mathrm{Der}_M(\log \mathcal{D})$ and $\Omega^q_M(\log \mathcal{D})$ are free $\mathcal{O}_M$-modules,

$$\mathrm{Der}_M(\log \mathcal{D}) = \bigoplus_{i=1}^k \mathcal{O}_M \cdot z_i \frac{\partial}{\partial z_i} \oplus \bigoplus_{i=k+1}^m \mathcal{O}_M \cdot \frac{\partial}{\partial z_i}, \tag{8.7}$$

$$\Omega^1_M(\log \mathcal{D}) = \bigoplus_{i=1}^k \mathcal{O}_M \cdot \frac{\mathrm{d}z_i}{z_i} \oplus \bigoplus_{i=k+1}^m \mathcal{O}_M \cdot \mathrm{d}z_i, \tag{8.8}$$

and the sheaves $\Omega^q_M(\log \mathcal{D})$ for $q > 1$ are determined by (8.4). Here equality for the restrictions to $M - \mathcal{D}_{sing}$ follows from Example 8.1. The sheaves on the right are free $\mathcal{O}_M$-modules. Then equality follows from (8.5), (8.6) and $i_*\mathcal{O}_{M-\mathcal{D}_{sing}} = \mathcal{O}_M$.

The sheaves of logarithmic vector fields and differential forms are coherent. This is clear for $\mathrm{Der}_M(\log \mathcal{D})$ because that is the kernel of the map $\mathcal{T}_M \to \mathcal{O}_M/(g)$, $X \mapsto X\, g$. But it is not so clear for the differential forms. It will follow from (8.9), but first we will see it in a way which will also be useful in section 8.3. The proof uses the following result of Serre.

Theorem 8.3 [**Ser**, Theorem 1 and Proposition 7] *Let X be a normal variety, $Y \subset X$ a subvariety of codimension ≥ 2, and $\mathcal{F}$ an $\mathcal{O}_{X-Y}$-coherent sheaf on $X - Y$ without torsion. Let $i : X - Y \hookrightarrow X$ denote the inclusion.*

(a) Then $i_*\mathcal{F}$ is $\mathcal{O}_X$-coherent if and only if there exists some $\mathcal{O}_X$-coherent extension of $\mathcal{F}$.

(b) (i) If $\mathcal{F}$ is reflexive and $i_*\mathcal{F}$ is $\mathcal{O}_X$-coherent then $i_*\mathcal{F}$ is reflexive and it is the only reflexive extension of $\mathcal{F}$.

(ii) Especially, if there exists some locally free extension of $\mathcal{F}$ then $i_*\mathcal{F}$ is this locally free extension.

A sheaf is *reflexive* if the map to its bidual is an isomorphism. The statement in Theorem 8.3 (b) (ii) is elementary and was already used in Example 8.2.

Arguments of the following type can be found in [**De1**, II 5.7] and [**Mal7**]. They will again be used in section 8.3.

Theorem 8.4 *Let M be a manifold and $\mathcal{D}$ a hypersurface. The sheaves of logarithmic vector fields and differential forms are coherent and reflexive $\mathcal{O}_M$-modules.*

Proof: Following Hironaka, there exists a resolution $f : (\widetilde{M}, \widetilde{\mathcal{D}}) \to (M, \mathcal{D})$, that is, a manifold $\widetilde{M}$, a normal crossing divisor $\widetilde{\mathcal{D}} \subset \widetilde{M}$, and a proper holomorphic map $f : \widetilde{M} \to M$ with $F^{-1}(\mathcal{D}) = \widetilde{\mathcal{D}}$ and such that $f : \widetilde{M} - \widetilde{\mathcal{D}} \to M - \mathcal{D}$ and $f : \widetilde{M} - F^{-1}(\mathcal{D}_{sing}) \to M - \mathcal{D}_{sing}$ are biholomorphic.

The sheaf $\mathrm{Der}_{\widetilde{M}}(\log \widetilde{D})$ on $\widetilde{M}$ is locally free because of Example 8.2. By Grauert's coherence theorem for proper maps (cf. for example [**Fi**, 1.17] the sheaf $f_*\mathrm{Der}_{\widetilde{M}}(\log \widetilde{\mathcal{D}})$ is $\mathcal{O}_M$-coherent. It is a coherent extension to M of Der_M $(\log \mathcal{D})|_{M-\mathcal{D}_{sing}}$. By Theorem 8.3 (a) and (8.5), then $\mathrm{Der}_M(\log \mathcal{D})$ is coherent and reflexive. The same applies to the sheaves of logarithmic differential forms. □

Reflexiveness of $\mathrm{Der}_M(\log \mathcal{D})$ and $\Omega^1_M(\log \mathcal{D})$ also follows from the fact that these sheaves are dual $\mathcal{O}_M$-modules by the inner product

$$\mathrm{Der}_M(\log \mathcal{D}) \times \Omega^1_M(\log \mathcal{D}) \to \mathcal{O}_M, \ (X, \omega) \mapsto \omega(X) \tag{8.9}$$

([**SK4**, (1.6) and (1.7)]). This duality is a consequence of Example 8.1 and (8.5), (8.6).

In general, $\mathrm{Der}_M(\log \mathcal{D})$ and $\Omega^1_M(\log \mathcal{D})$ are not necessarily locally free $\mathcal{O}_M$-modules. If they are, then $\mathcal{D} \subset M$ is called a *free divisor*. Several criteria for a divisor being free are given in [**SK4**]. In the case of singularities or, more generally, massive F-manifolds with Euler fields, the discriminants are always free divisors, often the bifurcation diagrams (section 4.3).

8.2 Logarithmic pole along a smooth divisor

Let $M, \mathcal{D}, g$ be as in section 8.1, M a complex manifold, $\mathcal{D} \subset M$ a hypersurface, and $g : M \to \mathbb{C}$ holomorphic with $\mathcal{D} = g^{-1}(0)$ and everywhere reduced ideal sheaf $(g) \subset \mathcal{O}_M$.

Let $H \to M - \mathcal{D}$ be a flat vector bundle with connection ∇ and sheaf $\mathcal{H}$ of holomorphic sections.

Deligne [**De1**] gave a precise meaning to the notion *regular singular* for (meromorphic) equivalence classes of $\mathcal{O}_M$-coherent extensions of $\mathcal{H}$ and showed that there is always precisely one regular singular equivalence class of extensions. We will come back to this in section 8.4

K. Saito [**SK2**] emphasized that one should look not just at this equivalence class of extensions, but at specific extensions of $\mathcal{H}$ which are usually given naturally. Here and in section 8.3 $\mathcal{O}_M$-coherent extensions with logarithmic poles will be discussed; the case of a smooth divisor here, the general case in section 8.3.

An important fact and tool is that in the case of a smooth divisor $\mathcal{D}$ the correspondence (7.30) between extensions with logarithmic poles and some filtrations extends to the higher dimensional case. Two other tools in this case are the *residue endomorphism* and a family of *residual connections*.

Definition 8.5 Let $M, \mathcal{D}, g$ be as above and $\mathcal{L}$ be an $\mathcal{O}_M$-coherent extension of $\mathcal{H}$ to M. The pair $(\mathcal{L}, \nabla)$ (or simply $\mathcal{L}$, if the connection is unambiguous) has a *pole of order* $\leq r + 1$ (or of *Poincaré rank* r) along $\mathcal{D}$ for some $r \in \mathbb{Z}_{\geq 0}$ if

$$\nabla : \mathcal{L} \to \frac{1}{g^r}\Omega^1_M(\log \mathcal{D}) \otimes \mathcal{L}. \tag{8.10}$$

The pair $(\mathcal{L}, \nabla)$ has a *logarithmic pole* along $\mathcal{D}$ if $r = 0$.

Remarks 8.6 (a) We are interested here only in the case $r = 0$.

(b) This definition is standard at least in the case $r = 0$, $\mathcal{L}, \mathcal{D}$ arbitrary and in the case r arbitrary, $\mathcal{L}$ locally free, $\mathcal{D}$ smooth. In the second case it is usually formulated in terms of vector bundles [**Sab4**][**Man2**].

(c) The definition is independent of the choice of g.

(d) Because of the duality (8.9) between logarithmic vector fields and 1-forms, $(\mathcal{L}, \nabla)$ has a logarithmic pole if and only if ∇_X maps $\mathcal{L}$ to itself for any logarithmic vector field X.

Let $(M, \mathcal{D})$ be isomorphic to $(\Delta^m, \{0\} \times \Delta^{m-1})$, where $\Delta = \{z \in \mathbb{C} \mid |z| < 1\} \subset \mathbb{C}$ is the unit disc, and let $g : M \to \mathbb{C}$ be holomorphic with $\mathcal{D} = g^{-1}(0)$ and $(g) \subset \mathcal{O}_M$ reduced. As above, $H \to M - \mathcal{D}$ is a flat vector bundle of rank μ with connection ∇ and sheaf $\mathcal{H}$ of flat sections; $i : M - \mathcal{D} \to M$ is the inclusion. The aim of this section is to make the following statements precise and explain them.

Theorem 8.7 *(a) Any $\mathcal{O}_M$-coherent extension $\mathcal{L}$ of $\mathcal{H}$ to M which has a logarithmic pole along $\mathcal{D}$ is a locally free $\mathcal{O}_M$-module.*

(b) There is a one-to-one correspondence between sheaves $\mathcal{L}$ on M as in (a) and increasing exhaustive filtrations of $H \to M - \mathcal{D}$ by flat subbundles $F_pH \to M - \mathcal{D}$, $p \in \mathbb{Z}$.

(c) There is a unique $\mathcal{O}_\mathcal{D}$-linear residue endomorphism Res *on the locally free $\mathcal{O}_\mathcal{D}$-module $(\mathcal{L}/g \cdot \mathcal{L})|_\mathcal{D}$. It is independent of g.*

(d) The function g induces a flat residual connection $\nabla^{res,g}$ on $(\mathcal{L}/g\mathcal{L})|_\mathcal{D}$. The residue endomorphism Res *is flat with respect to this residual connection.*

Two functions g and $\widetilde{g}$ with $\widetilde{g}/g \in \mathcal{O}_M^*$ induce the same residual connection if $(\widetilde{g}/g)|_\mathcal{D} \equiv$ constant.

Remarks 8.8 (i) Often one considers a priori only extensions of the bundle $H \to M - \mathcal{D}$ to a vector bundle on M with a logarithmic pole along $\mathcal{D}$. This is justified by (a). But in the case of a nonsmooth divisor $\mathcal{D}$ this approach would be too narrow.

(ii) Some particular filtrations of $H \to M - \mathcal{D}$ are the filtrations which are trivial on each subbundle of generalized eigenspaces of the monodromy, that means, $0 = F_{p-1} \subset F_p =$ *total subbundle* for some $p \in \mathbb{Z}$. They correspond to the sheaves $\mathcal{L}$ as in (a) whose residue endomorphisms satisfy the *nonresonance* condition: no two eigenvalues of Res differ by a nonzero integer.

That these extensions $\mathcal{L}$ are uniquely determined by the eigenvalues of their residue endomorphisms is classical ([**De1**, II §5], [**Man1**]).

(iii) A point of view which together with Theorem 7.9 gives Theorem 8.7(b) is formulated in [**Mal4**, p. 404] (cf. also [**Sab4**, III 1.20]): A locally free $\mathcal{O}_{\Delta \times \{0\}}$-module $\mathcal{L}'$ on $\Delta \times \{0\} \subset \Delta^m \cong M$ which is an extension of the sheaf of sections of $H|_{\Delta^* \times \{0\}}$ with a logarithmic pole along $\{0\}$ can be extended uniquely to a sheaf $\mathcal{L}$ on M as in (a).

(iv) The definition and the uniqueness of the residue endomorphism Res are classical. Res and $\nabla^{res,g}$ are presented together in [**Man2**, II 2.1] and [**Sab4**, 0.14.(b)], except for the precise dependence of $\nabla^{res,g}$ on g.

(v) Although $\nabla^{res,g}$ depends on g and is thus less canonical than Res, it will be extremely useful in the case of bundles on $M = \mathbb{P}^1 \times M'$ where the coordinate z on $\mathbb{P}^1$ or $\frac{1}{z}$ serve as g.

First we choose coordinates and set $M = \Delta^m$, $\mathcal{D} = \{0\} \times \Delta^{m-1}$, $g = z_1$. The dependence on the choices will be discussed later.

We have to repeat the construction in section 7.1 of the elementary sections with respect to z_1. Again $h = h_s \cdot h_u = h_u \cdot h_s$ denotes the monodromy of the bundle $H \to M - \mathcal{D}$, with semisimple part h_s, unipotent part h_u, and nilpotent part $N = \log h_u$. The fibres H_z, $z \in M - \mathcal{D}$ decompose into generalized eigenspaces, $H_z = \bigoplus_\lambda H_{z,\lambda}$ with $H_{z,\lambda} = \ker(h_s - \lambda)$.

These subspaces form the flat subbundles $H_{(\lambda)} = \bigcup_{z \in M-\mathcal{D}} H_{z,\lambda}$ with sheaves $\mathcal{H}_{(\lambda)}$ of holomorphic sections. Again

$$e : \mathbb{H} \times \Delta^{m-1} \to M - \mathcal{D}, \quad (\zeta, z_2, \ldots, z_m) \mapsto (\mathrm{e}^{2\pi i \zeta}, z_2, \ldots, z_m) \tag{8.11}$$

is a universal covering, and the space

$$H^\infty := \{pr \circ A : \mathbb{H} \times \Delta^{m-1} \to H \mid A \text{ is a global flat section of } e^* H\} \tag{8.12}$$

is the μ-dimensional space of global multivalued flat sections. Here $pr : e^*H \to H$ is the projection. The monodromy acts on it, $H^\infty = \bigoplus_\lambda H^\infty_\lambda$ with $H^\infty_\lambda = \ker(h_s - \lambda)$.

Then for any $A \in H^\infty_\lambda$ and $\alpha \in \mathbb{C}$ with $e^{-2\pi i \alpha} = \lambda$ there is an *elementary section*

$$\begin{aligned} es(A, \alpha) \colon M - \mathcal{D} &\to H, \qquad\qquad (8.13) \\ z &\mapsto \exp(2\pi i \alpha \zeta) \exp(-\zeta N) A(\zeta, z_2, \ldots, z_m) \\ \text{‘} &= z_1^\alpha \exp\left(-\log z_1 \frac{N}{2\pi i}\right) A\,', \end{aligned}$$

here $e(\zeta, z_2, \ldots, z_m) = z$. Formulas analogous to (7.7) hold, especially

$$\left(z_1 \nabla_{\partial_{z_1}} - \alpha\right) es(A, \alpha) = es\left(-\frac{N}{2\pi i} A, \alpha\right) = -\frac{N}{2\pi i} es(A, \alpha), \tag{8.14}$$

$$\nabla_{\partial_{z_i}} es(A, \alpha) = 0 \quad \text{for } i \geq 2. \tag{8.15}$$

Again C^α denotes the space of all elementary sections $es(A, \alpha)$, $A \in H^\infty_\lambda$, with a fixed order α. A basis of C^α is a basis of sections of the subbundle $H_{(\lambda)}$.

Now let $F_pH \to M - \mathcal{D}$, $p \in \mathbb{Z}$, be flat subbundles which form an increasing exhaustive filtration. We construct from this filtration an $\mathcal{O}_M$-locally free extension $\mathcal{L}$ of $\mathcal{H}$ with a logarithmic pole along $\mathcal{D}$ in two ways; the first way is more explicit; the independence of choices is easier to see in the second way.

1st way: The filtration $F_\bullet H$ induces a monodromy invariant filtration $F_\bullet H^\infty$. One may choose a basis $A_j \in F_{p(j)}H^\infty_{\lambda(j)}$, $i = 1, \dots, \mu$, of H^∞ which splits this filtration and the eigenspace decomposition. Then

$$\mathcal{L} := \bigoplus_{j=1}^{\mu} \mathcal{O}_M \cdot es(A_j, \alpha(j) + p(j)) \tag{8.16}$$

for $e^{-2\pi i\alpha(j)} = \lambda(j)$, $-1 < \text{Re}(\alpha(j)) \le 0$. Obviously these elementary sections form a basis of the bundle $H \to M - \mathcal{D}$. The extension $\mathcal{L}$ of $\mathcal{H}$ to M has logarithmic pole along $\mathcal{D}$ because of (8.14), (8.15) and the monodromy invariance of $F_\bullet H^\infty$.

2nd way: Let $F_p\mathcal{H}_{(\lambda)}$ denote the sheaf of holomorphic sections of $F_pH_{(\lambda)}$.

$$\mathcal{L} := \bigoplus_{\lambda} \sum_{p\in\mathbb{Z}} i_* F_p\mathcal{H}_{(\lambda)} \cap \mathcal{O}_M \cdot C^{\alpha+p} \tag{8.17}$$

for $e^{-2\pi i\alpha} = \lambda$, $-1 < \text{Re}(\alpha) \le 0$, and $i : M - \mathcal{D} \hookrightarrow M$ the inclusion. One sees that (8.16) and (8.17) give the same extension $\mathcal{L}$.

The *residue endomorphism* for this extension $\mathcal{L}$ is the $\mathcal{O}_\mathcal{D}$-linear map

$$\text{Res} = z_1\nabla_{\partial_{z_1}} : \mathcal{L}/z_1\mathcal{L}|_\mathcal{D} \to \mathcal{L}/z_1\mathcal{L}|_\mathcal{D} \tag{8.18}$$

on the locally free $\mathcal{O}_\mathcal{D}$-module $\mathcal{L}/z_1\mathcal{L}|_\mathcal{D}$ of rank μ.

The *residual connection* ∇^{res,z_1} on $\mathcal{L}/z_1\mathcal{L}|_\mathcal{D}$ is the flat connection whose flat sections are generated by the classes in $\mathcal{L}/z_1\mathcal{L}|_\mathcal{D}$ of the elementary sections $es(A_j, \alpha(j)+p(j))$ from (8.16). Obviously Res is ∇^{res,z_1}-flat. The eigenvalues of the residue endomorphism are the numbers $\alpha(j) + p(j)$, $j = 1, \dots, \mu$.

Theorem 8.7 follows from the next lemma.

Lemma 8.9 *(a) The construction in* (8.16) *and* (8.17) *gives all $\mathcal{O}_M$-coherent extensions $\mathcal{L}$ of $\mathcal{H}$ with a logarithmic pole along $\mathcal{D}$.*

(b) Let $\sigma_1, \dots, \sigma_\mu$ be an $\mathcal{O}_M$-basis of an extension $\mathcal{L}$ as in (8.16) *and* (8.17) *and $A_1(z)\frac{dz_1}{z_1} + \sum_{i\ge2} A_i(z)dz_i$ be the connection matrix with respect to it.*

Then $A_1(0, z_2, \dots, z_m)$ and $\sum_{i\ge2} A_i(0, z_2, \dots, z_m)dz_i$ are the matrices for the residue endomorphism Res *and the residual connection ∇^{res,z_1} with respect to the basis $\sigma_1|_\mathcal{D}, \dots, \sigma_\mu|_\mathcal{D}$ of $\mathcal{L}/z_1\mathcal{L}|_\mathcal{D}$.*

(c) The residue endomorphism Res *is independent of the choice of coordinates. The residual connection ∇^{res,z_1} depends only on z_1, not on $z_2, \dots, z_m$, and $\nabla^{res,z_1} = \nabla^{res,\widetilde{z}_1}$ if $(\widetilde{z}_1/z_1)|_\mathcal{D} = $ constant.*

(d) The $\mathcal{O}_M$-locally free extension $\mathcal{O}_M \cdot C^\alpha \subset i_*\mathcal{H}_{(\lambda)}$ of $\mathcal{H}_{(\lambda)}$ depends only on α, not on $(z_1, \dots, z_m)$. It is the unique extension with logarithmic pole along $\mathcal{D}$ and with α as the only eigenvalue of its residue endomorphism.

(e) The construction of $\mathcal{L}$ in (8.17) *from the filtration $F_\bullet H$ is independent of the choice of coordinates. It gives a one-to-one correspondence between increasing exhaustive filtrations of $H \to M - \mathcal{D}$ by flat subbundles $F_p H \to M - \mathcal{D}$ and $\mathcal{O}_M$-coherent extensions of $\mathcal{H}$ with a logarithmic pole along $\mathcal{D}$.*

Proof: (a) Let $\mathcal{L}$ be an $\mathcal{O}_M$-coherent extension of $\mathcal{H}$ to M with a logarithmic pole along $\mathcal{D}$. The induced extension to $\Delta \times \{0\}$ of the sheaf of sections of $H|_{\Delta^* \times \{0\}}$ is $\mathcal{L}' := \mathcal{L}/(z_2\mathcal{L} + \cdots + z_m\mathcal{L})|_{\Delta\times\{0\}}$. It is a locally free $\mathcal{O}_{\Delta\times\{0\}}$-module of rank μ and has a logarithmic pole in $\{0\} \subset \Delta \times \{0\}$.

The $\mathcal{O}_{(\Delta\times\{0\},0)}$-lattice $\mathcal{L}'_0$ is saturated. By Theorem 7.9, $\mathcal{L}'$ and $\mathcal{L}'_0$ come from an increasing exhaustive filtration of $H|_{\Delta\times\{0\}}$ by flat subbundles. The flat extensions to $M - \mathcal{D}$ of these subbundles are denoted by $F_p H \to M - \mathcal{D}$, $p \in \mathbb{Z}$.

They induce by (8.16) and (8.17) an $\mathcal{O}_M$-locally free extension $\widetilde{\mathcal{L}}$ of $\mathcal{H}$. Choose an $\mathcal{O}_M$-basis $es(A_j, \alpha(j) + p(j))$ of $\widetilde{\mathcal{L}}$ as in (8.16). Their restrictions to $\Delta \times \{0\}$ form an $\mathcal{O}_{\Delta\times\{0\}}$-basis of $\mathcal{L}'_0$. The sections $es(A_j, \alpha(j) + p(j))$ also form an $i_*\mathcal{O}_{M-\mathcal{D}}$-basis of $i_*\mathcal{H}$.

Any germ $\sigma \in \mathcal{L}_0$ is a unique linear combination

$$\sigma = \sum_{j=1}^{\mu} \sigma^{(j)} \cdot es(A_j, \alpha(j) + p(j))$$

with $\sigma^{(j)} \in (i_*\mathcal{O}_{M-\mathcal{D}})_0$. We claim $\sigma^{(j)} \in \mathcal{O}_{M,0}$:
Any derivative of σ by $\frac{\partial}{\partial z_2}, \dots, \frac{\partial}{\partial z_m}$ is also contained in $\mathcal{L}_0$ because it has a logarithmic pole along $\mathcal{D}$; its restriction to $\Delta \times \{0\}$ is contained in $\mathcal{L}'_0$. So the restriction to $\Delta \times \{0\}$ of any derivative of $\sigma^{(j)}$ by $\frac{\partial}{\partial z_2}, \dots, \frac{\partial}{\partial z_m}$ is in $\mathcal{O}_{\Delta\times\{0\}}$.

This shows $\sigma^{(j)} \in \mathcal{O}_{M,0}$ and $\mathcal{L}_0 \subset \widetilde{\mathcal{L}}_0$. Now any germs $\sigma_1, \dots, \sigma_\mu \in \mathcal{L}_0$ with $\sigma_j|_{(\Delta\times\{0\},0)} = es(A_j, \alpha(j) + p(j))|_{(\Delta\times\{0\},0)}$ generate $\widetilde{\mathcal{L}}_0$, because the coefficient matrix expressing $\sigma_1, \dots, \sigma_\mu$ by the elementary sections is invertible with holomorphic entries. Thus $\mathcal{L}_0 = \widetilde{\mathcal{L}}_0$.

(b) This holds for any basis of $\mathcal{L}_0$, because it holds for a basis of elementary sections as in (8.16). For such a basis, the connection matrix is $A_1 \frac{\mathrm{d}z_1}{z_1}$ with A_1 a constant matrix.

(c) A change of coordinates does not change $A_1(0, z_2, \dots, z_m)$ and Res. The definition of the elementary sections depends only on z_1 and they determine ∇^{res,z_1}. One sees easily that a change of z_1 to a coordinate $\widetilde{z}_1$ with $(\widetilde{z}_1/z_1)|_{\mathcal{D}} \equiv$ *constant* does not change $\sum_{i\geq 2} A_i(0, z_2, \dots, z_m)\mathrm{d}z_i$.

(d) The corresponding statements for the restriction to $\Delta \times \{0\}$ hold because of section 7.3. One can use this and (a). But of course, the uniqueness of the extension of $i_*\mathcal{H}_{(\lambda)}$ to an $\mathcal{O}_M$-locally free sheaf with logarithmic pole along $\mathcal{D}$ and α as the only eigenvalue of the residue endomorphism is classical ([**De1**, II §5], [**Man1**], cf. Remark 8.8 (ii)).

(e) The first statement follows from (d), the rest with (a). □

8.3 Logarithmic pole along any divisor

Let $M, \mathcal{D}, g$ be as in section 8.1, M a complex manifold, $\mathcal{D}$ a hypersurface, and $g : M \to \mathbb{C}$ holomorphic with $\mathcal{D} = g^{-1}(0)$ and $(g) \subset \mathcal{O}_M$ reduced everywhere. Let $H \to M - \mathcal{D}$ be a flat vector bundle of rank μ with sheaf of holomorphic sections $\mathcal{H}$.

Section 8.2 gives a correspondence between certain filtrations and coherent extensions to $M - \mathcal{D}_{sing}$ of $\mathcal{H}$ with logarithmic poles along $\mathcal{D} - \mathcal{D}_{sing}$. To make this precise one can choose some Riemannian metric on M and the corresponding distance function $d : M \times M \to \mathbb{R}_{\geq 0}$. For sufficiently small $\varepsilon > 0$ and $\delta > 0$ with $\delta \ll \varepsilon$ the submanifold

$$M_{\delta,\varepsilon} := \{z \in M \mid d(z, \mathcal{D}) < \delta,\ d(z, \mathcal{D}_{sing}) > \varepsilon\} \tag{8.19}$$

is diffeomorphic to a disc bundle over $\mathcal{D} - \{z \in \mathcal{D} \mid d(z, \mathcal{D}_{sing}) \leq \varepsilon\}$, having the same number of components as $\mathcal{D}_{reg}$. Theorem 8.7 implies the following.

Lemma 8.10 *Any $\mathcal{O}_{M-\mathcal{D}_{sing}}$-coherent extension of $\mathcal{H}$ to $M - \mathcal{D}_{sing}$ with a logarithmic pole along $\mathcal{D} - \mathcal{D}_{sing}$ is locally free. There is a one-to-one correspondence between such extensions and increasing exhaustive filtrations by flat subbundles of the restriction of H to $M_{\delta,\varepsilon} - \mathcal{D}$.*

The aim of this section is to study extensions to $\mathcal{D}_{sing}$ of sheaves as in Lemma 8.10. The main result is Theorem 8.11. It will be proved at the end of the section. Let $j : M - \mathcal{D}_{sing} \hookrightarrow M$ be the inclusion.

Theorem 8.11 *Let $\mathcal{L}$ be a locally free extension of $\mathcal{H}$ to $M - \mathcal{D}_{sing}$ with a logarithmic pole along $\mathcal{D} - \mathcal{D}_{sing}$.*

Then $j_*\mathcal{L}$ is $\mathcal{O}_M$-coherent and has a logarithmic pole along $\mathcal{D}$. It is the only reflexive extension of $\mathcal{L}$.

Example 8.12 Consider $M = \mathbb{C}^2$, $g = z_1 z_2$, $\mathcal{D} = g^{-1}(0)$, $H = \mathbb{C} \times (\mathrm{M}{-}\mathrm{D})$ with the trivial flat connection. Each ideal $(z_1^{k_1}, z_2^{k_2})$ for $k_1, k_2 \in \mathbb{Z}_{\geq 0}$ is invariant under $z_1\partial_{z_1}$ and $z_2\partial_{z_2}$. So each such ideal is an $\mathcal{O}_M$-coherent extension of $\mathcal{O}_M|_{M-\mathcal{D}}$ with a logarithmic pole along $\mathcal{D}$. All these extensions coincide on $M - \mathcal{D}_{sing}$. The only reflexive and even locally free extension under them is $\mathcal{O}_M$. The last two statements illustrate Theorem 8.3 (b).

Example 8.13 Consider $M = \mathbb{C}^3$, $g = z_1z_2z_3$, $\mathcal{D} = g^{-1}(0)$, $H = \mathbb{C}^2 \times (M - \mathcal{D})$ with the trivial flat connection and global basis e_1, e_2 of flat sections. The sheaf

$$\mathcal{L} := \mathcal{O}_M \cdot z_2z_3e_1 + \mathcal{O}_M \cdot z_1z_3e_2 + \mathcal{O}_M \cdot z_1z_2(e_1 + e_2)$$

is coherent and has a logarithmic pole along $\mathcal{D}$. The restriction to $M - \{0\}$ is locally free. The sheaf $\mathcal{L}$ satisfies

$$\mathcal{L} = j_*(\mathcal{L}|_{M-\mathcal{D}_{sing}}) \tag{8.20}$$

and is therefore reflexive (Theorem 8.3 (b)), but it is not locally free.

Let us consider now the normal crossing case, $M = \Delta^m$, $g = z_1 \cdot \dots \cdot z_k$, $\mathcal{D} = g^{-1}(0) = \bigcup_{j=1}^k D^{(j)} = \bigcup_{j=1}^k z_j^{-1}(0)$; here $\Delta = \{z \in \mathbb{C} \mid |z| < 1\}$.

The flat bundle $H \to M - \mathcal{D}$ has k commuting monodromies $h^{(j)} = h_s^{(j)} \cdot h_u^{(j)}$, $j = 1, \dots, k$, for the standard loops. A universal covering is

$$\begin{aligned} e : \mathbb{H}^k \times \Delta^{m-k} &\to M - \mathcal{D}, \\ (\zeta_1, \dots, \zeta_k, z_{k+1}, \dots, z_m) &\mapsto (e^{2\pi i \zeta_1}, \dots, e^{2\pi i \zeta_k}, z_{k+1}, \dots, z_m). \end{aligned} \tag{8.21}$$

The space of multivalued flat sections H^∞ is defined as usual,

$$H^\infty = \{pr \circ A : \mathbb{H}^k \times \Delta^{m-k} \to H \mid A \text{ is a global flat section of } e^*H\}. \tag{8.22}$$

All monodromies act on it. The indices of the simultaneous generalized eigenspace decomposition $H^\infty = \bigoplus_\lambda H^\infty_\lambda$ can simply be considered as tuples $\underline{\lambda} = (\lambda^{(1)}, \dots, \lambda^{(k)})$ of eigenvalues for the monodromies $h^{(1)}, \dots, h^{(k)}$.

The notion of elementary sections from sections 7.1 and 8.2 generalizes: if one chooses $A \in H^\infty_{\underline{\lambda}}$ and $\underline{\alpha} = (\alpha^{(1)}, \dots, \alpha^{(k)})$ with $e^{-2\pi i \alpha^{(j)}} = \lambda^{(j)}$ then the natural generalization of the formulas (7.5) and (8.13) yields an elementary section $es(A, \underline{\alpha})$.

Lemma 8.14 will show that any $\mathcal{O}_M$-coherent extension of $\mathcal{H}$ with logarithmic pole along $\mathcal{D}$ is generated by elementary sections. It is essentially due to Deligne, but was first formulated in [**EV**, Appendix C]. We need the following generalization of filtrations on H^∞: a $\mathbb{Z}^k$-*filtration* $(P_{\underline{l}})_{\underline{l} \in \mathbb{Z}^k}$ consists of subspaces

$P_{\underline{l}} \subset H^\infty$ which are invariant with respect to all monodromies $h^{(1)}, \ldots, h^{(k)}$ and which satisfy for a suitable $m > 0$

$$P_{\underline{l}} = 0 \quad \text{if } l_j \le -m \text{ for some } j, \tag{8.23}$$

$$P_{\underline{l}} = H^\infty \quad \text{if } l_j \ge m \text{ for all } j, \tag{8.24}$$

$$P_{\underline{l}} \subset P_{\underline{l}'} \quad \text{if } l_j \le l'_j \text{ for all } j. \tag{8.25}$$

Then $P_{\underline{l}} = \bigoplus_{\underline{\lambda}} P_{\underline{l},\underline{\lambda}}$. A $\mathbb{Z}^k$-filtration $(P_{\underline{l}})$ induces k increasing exhaustive monodromy invariant filtrations $F_\bullet^{(j)}$ on H^∞ by

$$F_p^{(j)} := \bigcup_{l_j = p} P_{\underline{l}}. \tag{8.26}$$

For $\lambda^{(j)}$ an eigenvalue of $h^{(j)}$ let $\alpha^{(j)}$ be defined by $e^{-2\pi i \alpha^{(j)}} = \lambda^{(j)}$ and $-1 < \operatorname{Re} \alpha^{(j)} \le 0$.

Lemma 8.14 [**EV**, Appendix C] *Fix a $\mathbb{Z}^k$-filtration $(P_{\underline{l}})_{\underline{l} \in \mathbb{Z}^k}$.*

(a) For any $\underline{l} \in ([-m, m] \cap \mathbb{Z})^k$ and any $\underline{\lambda}$, choose generators $A^i_{\underline{l},\underline{\lambda}}$ of $P_{\underline{l},\underline{\lambda}}$. The sheaf

$$\mathcal{L} := \sum_{\text{these } \underline{l}} \sum_{\underline{\lambda}} \sum_{i} \mathcal{O}_M \cdot es(A^i_{\underline{l},\underline{\lambda}}, \underline{\alpha} + \underline{l}) \tag{8.27}$$

is an $\mathcal{O}_M$-coherent extension of $\mathcal{H}$ with logarithmic pole along $\mathcal{D}$.

(b) Fix a $\mathbb{Z}^k$-filtration $P_{\underline{l}}$ and $\mathcal{L}$ as in (8.27). The filtrations $F_\bullet^{(j)}$ on H^∞ correspond together to one filtration on $M_{\delta,\varepsilon} - \mathcal{D}$ (formula (8.19)) by flat subbundles. The $\mathcal{O}_{M-\mathcal{D}_{sing}}$-coherent extension of $\mathcal{H}$ which corresponds to this filtration in Lemma 8.10 is $\mathcal{L}|_{M-\mathcal{D}_{sing}}$.

(c) Formula (8.27) yields a one-one correspondence between $\mathbb{Z}^k$-filtrations and $\mathcal{O}_M$-coherent extensions of $\mathcal{H}$ with logarithmic pole along $\mathcal{D}$.

(d) The sheaf $\mathcal{L}$ as in formula (8.27) is reflexive if and only if

$$P_{\underline{l}} = \bigcap_{j=1}^{k} F^{(j)}_{l_j} \quad \text{for all } \underline{l}. \tag{8.28}$$

Then $\mathcal{L} = j_*(\mathcal{L}|_{M-\mathcal{D}_{sing}})$.

(e) The sheaf $\mathcal{L}$ as in formula (8.27) is a locally free $\mathcal{O}_M$-module if and only if (8.28) holds and the filtrations $F_\bullet^{(j)}$ have a common splitting.

(f) If $\mathcal{L}$ as in formula (8.27) is reflexive then it is locally free outside of the intersection of at least three components $D^{(j)}$.

Proof: (a) and (b) follow from the definition of elementary sections.

(c) See [**EV**, Appendix C] for the proof of this.

(d) The sheaf $\mathcal{L}$ is $\mathcal{O}_M$-coherent, and $\mathcal{L}|_{M-\mathcal{D}_{sing}}$ is locally free. By Theorem 8.3 $\mathcal{L}$ is reflexive if and only if $\mathcal{L} = j_*(\mathcal{L}|_{M-\mathcal{D}_{sing}})$. Then it must be generated by the elementary sections which correspond to the $\mathbb{Z}^k$-filtration in (8.28).

(e) The direction '*if*' is clear. For '*only if*' one can compare a generating system of elementary sections as in (8.27) and an $\mathcal{O}_{M,0}$-basis of $\mathcal{L}_0$. It is not hard to see that one can replace the basis elements by certain elements of the generating set. This $\mathcal{O}_{M,0}$-basis of elementary sections yields a common splitting of the filtrations $F_\bullet^{(j)}$.

(f) Any two filtrations admit a common splitting. □

Remarks 8.15 (a) A special case of Lemma 8.14 (e) is the case when the filtrations on the simultaneous generalized eigenspaces H_λ^∞ are all of the type $0 = F^{p(i,\lambda)-1} \subset F^{p(i,\lambda)} = H^\infty$ for some numbers $p(i, \lambda) \in \mathbb{Z}$.

When these numbers $p(i, \lambda)$ depend only on i and $\lambda^{(i)}$ then the residue endomorphisms on the components $D^{(i)}$ satisfy the nonresonance condition (Remark 8.8 (ii)): no two eigenvalues of one residue endomorphism differ by a nonzero integer.

The $\mathcal{O}_M$-locally free extension of $\mathcal{H}$ to M for the case when the real parts of all eigenvalues of all residue endomorphisms are contained in $[0, 1)$ was called the canonical extension by Deligne [**De1**, II 5.5].

(b) The main arguments in the proofs of Theorem 8.4 and Theorem 8.11 are due to Deligne [**De1**, II 5.7].

(c) In both proofs one could use [**KK**, Proposition C.1.2 (ii)] instead of Serre's result Theorem 8.3.

Proof of Theorem 8.11: One chooses a resolution $f : (\widetilde{M}, \widetilde{\mathcal{D}}) \to (M, \mathcal{D})$ as in the proof of Theorem 8.4. The first step is to extend the lift $f^*\mathcal{L}$ to those components of $\widetilde{\mathcal{D}}_{reg}$ which are mapped by f to lower dimensional subsets in M. This can be done by choosing locally around these components for example the filtration $0 = F^{-1} \subset F^0 =$ (whole local flat subbundle) and applying Lemma 8.10.

Then Lemma 8.14 (b) and (d) yields an $\mathcal{O}_{\widetilde{M}}$-coherent extension to $\widetilde{M}$. Just as in the proof of Theorem 8.4, the direct image sheaf on M is an $\mathcal{O}_M$-coherent extension of $\mathcal{L}$ to M by Grauert's proper mapping theorem. Serre's result Theorem 8.3 shows that $j_*\mathcal{L}$ itself is $\mathcal{O}_M$-coherent and reflexive and the unique reflexive extension of $\mathcal{L}$ to M. It has a logarithmic pole along $\mathcal{D}$ because $\mathcal{L}$ has a logarithmic pole along $\mathcal{D} - \mathcal{D}_{sing}$: The sheaf $\mathcal{L}$ and therefore also $j_*\mathcal{L}$ are invariant with respect to logarithmic vector fields. □

8.4 Remarks on regular singular connections

Let $M, \mathcal{D}, g$ be as in section 8.1, namely M a complex manifold, $\mathcal{D} \subset M$ a hypersurface, and $g : M \to \mathbb{C}$ holomorphic with $\mathcal{D} = g^{-1}(0)$ and the ideal sheaf $(g) \subset \mathcal{O}_M$ everywhere reduced. Let $i : M - \mathcal{D} \to M$ denote the inclusion. Let $H \to M - \mathcal{D}$ be a flat vector bundle with sheaf of holomorphic sections $\mathcal{H}$.

Deligne defined the notion *regular singular* not for a single $\mathcal{O}_M$-coherent extension of $\mathcal{H}$, but for equivalence classes of such extensions with respect to a meromorphic equivalence relation ([**De1**, II 2.12]).

This equivalence class determines and is determined by a coherent $\mathcal{O}_M[*]$-module in $i_*\mathcal{H}$ which extends $\mathcal{H}$ (cf. for example [**Mal6**][**Mal7**]). Here $\mathcal{O}_M[*] = \mathcal{O}_M[g^{-1}]$ is the coherent ring sheaf of holomorphic functions on $M - \mathcal{D}$ which are meromorphic along $\mathcal{D}$.

Deligne showed that there is a unique $\mathcal{O}_M[*]$-coherent extension of $\mathcal{H}$ whose sections have moderate growth with respect to the multivalued flat sections (for the result and for the notion of moderate growth see [**De1**, II 5.7, 4.2, 4.1, 2.17, 2.10, ...]. One says that this extension is regular singular along $\mathcal{D}$, or the connection is regular singular with respect to this extension.

Following K. Saito [**SK2**, (5.1)], one can also say that an $\mathcal{O}_M$-coherent extension of $\mathcal{H}$ has a regular singularity along $\mathcal{D}$ if it is contained in this meromorphic regular singular extension.

In the 1-dimensional case $M = \Delta = \{z \in \mathbb{C} \mid |z| < 1\}$, $\mathcal{D} = \{0\}$, the regular singular extension is the $\mathcal{O}_M[*]$-submodule of $i_*\mathcal{H}$ with germ $V^{>-\infty}$ at 0 ([**De1**, II §1]).

In [**De1**, II 4.1] several criteria for regular singular are given. $\mathcal{O}_M$-coherent extensions with a logarithmic pole along $\mathcal{D}$ were used only in the normal crossing case. In view of Theorem 8.11 one has the following.

Theorem 8.16 *Let $\mathcal{L}^{merom}$ be an $\mathcal{O}_M[*]$-coherent subsheaf of $i_*\mathcal{H}$ which extends $\mathcal{H}$. The following conditions are equivalent:*

(i) *The sheaf $\mathcal{L}^{merom}$ is the regular singular extension, that means, its sections have moderate growth with respect to the multivalued flat sections.*

(ii) *The sheaf $\mathcal{L}^{merom}$ contains one (and then all) $\mathcal{O}_M$-coherent extension(s) of $\mathcal{H}$ with a logarithmic pole along $\mathcal{D}$.*

(iii) *The restriction of $\mathcal{L}^{merom}$ to an open set $U \subset M$ which intersects each component of $\mathcal{D}_{reg}$ contains one (and then all) $\mathcal{O}_M$-coherent extension(s) of $\mathcal{H}|_{U-\mathcal{D}}$ with a logarithmic pole along $\mathcal{D}$.*

Of course, this is well known. (e.g. [**Lo2**, 8.10] and references there). It was the starting point for the definition of regular holonomic differential and microdifferential systems ([**KK**][**Bj**]). There the meromorphic regular singular extension is called a Deligne-type module (or D-type module).

Property (iii) says that regular singular is a codimension 2 property: it is sufficient to check it outside a codimension 2 subset.

Chapter 9

Frobenius manifolds and second structure connections

The definition and elementary properties of a Frobenius manifold M are put together in section 9.1. Sections 9.2, 9.4, and 9.5 are devoted to their second structure connections. These are connections over $\mathbb{P}^1 \times M$ on the lifted tangent bundle of M with logarithmic poles along certain hypersurfaces. They come from some twists of the original flat structure by the multiplication and the Euler field. To know them is very instructive for the construction of Frobenius manifolds for singularities, because in that case one of them turns out to be isomorphic to an extension of the Gauß–Manin connection.

Sections 9.2, 9.4, and 9.5 build on the definition and discussion of the second structure connections in [**Man2**] for the case of semisimple Frobenius manifolds, on results in [**Du3**], and on [**SK9**, §5], where they together with many properties had been established much earlier implicitly in the case of singularities.

The second structure connections have some counterparts, the first structure connections, which are better known. The latter are partly Fourier duals. The main purpose of their treatment in section 9.3 (and in section 9.4) is to compare them with the second structure connections.

9.1 Definition of Frobenius manifolds

Frobenius manifolds were defined by Dubrovin [**Du1**][**Du3**]. We follow the notations in Manin's book [**Man2**, chapters I and II].

Here all manifolds will be complex. In the following, M denotes a manifold with $\dim M = m \geq 1$ (in the singularity case $m = \mu$). A (k, l)-tensor is an $\mathcal{O}_M$-linear map $T : \mathcal{T}_M^{\otimes k} \to \mathcal{T}_M^{\otimes l}$. Here $\mathcal{T}_M$ is the holomorphic tangent sheaf, $\mathcal{O}_M$ the structure sheaf.

A metric g is a symmetric nondegenerate $(2, 0)$-tensor, a multiplication on the holomorphic tangent bundle TM is a commutative (i.e. symmetric) and associative $(2, 1)$ tensor.

The Lie derivative $\text{Lie}_X T$ of a (k, l)-tensor along a vector field is again a (k, l)-tensor, as well as the covariant derivative $\nabla_X T$ with respect to a connection ∇ on M. Then ∇T is a $(k+1, l)$-tensor (cf. Remarks 2.7 and 2.13). The Levi–Civita connection ∇ of a metric g is the unique connection which respects the metric, $\nabla g = 0$, and is torsion free, $\nabla_X Y - \nabla_Y X = [X, Y]$ for local vector fields X, Y.

The following definition is a bit more restrictive than in [**Man2**, chapter I] because of the unit field and the Euler field.

Definition 9.1 A *Frobenius manifold* is a tuple $(M, \circ, e, E, g)$ where M is a manifold of dimension $m \geq 1$ with metric g and multiplication $\circ$ on the tangent bundle, e is a global unit field and E is another global vector field, subject to the following conditions:

(1) the metric is multiplication invariant, $g(X \circ Y, Z) = g(X, Y \circ Z)$,
(2) (potentiality) the (3, 1)-tensor $\nabla\circ$ is symmetric (here ∇ is the Levi–Civita connection of the metric),
(3) the metric g is flat,
(4) the unit field e is flat, $\nabla e = 0$,
(5) the Euler field E satisfies $\text{Lie}_E(\circ) = 1 \cdot \circ$ and $\text{Lie}_E(g) = D \cdot g$ for some $D \in \mathbb{C}$.

Remarks 9.2 (a) For $(M, \circ, e, g)$ as in the definition with condition (1) (not necessarily (2)–(5)) the (3, 0)-tensor A defined by $A(X, Y, Z) = g(X \circ Y, Z)$ is symmetric. It arises from the (2, 1)-tensor $\circ$ by contraction with the (2, 0)-tensor g. Because of $\nabla g = 0$ and the symmetry of g, the potentiality (2) is equivalent to the symmetry of the (4, 0)-tensor ∇A.

In Theorem 2.15 it is shown that the potentiality is also equivalent to the closedness of the 1-form $\varepsilon := g(e, .)$ together with the condition

$$\text{Lie}_{X\circ Y}(\circ) = X \circ \text{Lie}_Y(\circ) + Y \circ \text{Lie}_X(\circ) \tag{9.1}$$

for local vector fields $X, Y \in \mathcal{T}_M$.

(b) A manifold $(M, \circ, e)$ with multiplication $\circ$ and unit field e and (9.1) is called F-manifold (Definition 2.8 and [**HM**][**Man2**]). Formula (9.1) for $X = Y = e$ gives $\text{Lie}_e(\circ) = 0 \cdot \circ$. In this context a vector field E is already called Euler field (of the F-manifold) if it satisfies $\text{Lie}_E(\circ) = 1 \cdot \circ$. This implies $[e, E] = e$.

(c) The flatness of e, $\nabla e = 0$, is equivalent to $\text{Lie}_e(g) = 0$ and the closedness of the 1-form $\varepsilon = g(e, .)$ (Lemma 2.16). Because the potentiality implies the closedness of ε (cf. (a)), one could replace (4) in the definition by $\text{Lie}_e(g) = 0$.

(d) The potentiality (2) written out for arbitrary local fields X, Y, Z is

$$\nabla_X(Y\circ Z) - Y\circ\nabla_X(Z) - \nabla_Y(X\circ Z) + X\circ\nabla_Y(Z) - [X,Y]\circ Z = 0. \quad (9.2)$$

This formula can already be found in K. Saito's papers (e.g. [**SK9**, (3.3.2)]). For flat vector fields it is equivalent to the symmetry of $\nabla_X(Y\circ Z)$ in X, Y, Z and to the symmetry of $X\, g(Y\circ Z, W)$ in X, Y, Z, W (cf. (a)). Therefore it is equivalent to the local existence of a *potential* $F\in\mathcal{O}_{M,p}$ with $XYZ(F) = g(X\circ Y, Z)$ for flat fields X, Y, Z.

(e) The property $\mathrm{Lie}_E(g) = D\cdot g$ written out for local vector fields X, Y is

$$E\, g(X,Y) - g([E,X],Y) - g(X,[E,Y]) - D\, g(X,Y) = 0. \quad (9.3)$$

Comparison with $\nabla g = 0$ shows that the (1, 1)-tensor

$$\mathcal{V}: \mathcal{T}_M\to\mathcal{T}_M,\quad X\mapsto\nabla_X E - \frac{D}{2}X,$$

is skewsymmetric,

$$g(\mathcal{V}(X), Y) + g(X, \mathcal{V}(Y)) = 0. \quad (9.4)$$

The property $\mathrm{Lie}_E(g) = D\cdot g$ means that E is a sum of an infinitesimal rotation, a dilation and a constant shift. Therefore $\mathcal{V}$ maps a flat vector field X to a flat vector field $\mathcal{V}(X) = [X,E] - \frac{D}{2}X$. It is an endomorphism of the local system, in other words, it is a flat (1, 1)-tensor, $\nabla\mathcal{V} = 0$. The eigenvalues of $\nabla E = \mathcal{V} + \frac{D}{2}$ are called the *spectrum* of the Frobenius manifold and are denoted by $d_1,\ldots,d_m$. They are symmetric around $\frac{D}{2}$ and one of them is 1 because of $[e,E] = e$. We order them such that $d_i + d_{m+1-i} = D$, $d_1 = 1$.

(f) If $\mathcal{V}$ is semisimple then locally there are flat coordinates $t_1,\ldots,t_m$ such that the flat fields $\delta_i = \frac{\partial}{\partial t_i}$ are eigenvectors of $\mathcal{V}$ with

$$[\delta_i, E] = d_i\cdot\delta_i. \quad (9.5)$$

Then

$$E = \sum_i (d_i t_i + r_i)\cdot\delta_i \quad (9.6)$$

for some $r_i\in\mathbb{C}$.

(g) The multiplication with the Euler field is denoted by

$$\mathcal{U}: \mathcal{T}_M\to\mathcal{T}_M,\quad X\mapsto E\circ X. \quad (9.7)$$

9.2 Second structure connections

Let us fix a Frobenius manifold $(M, \circ, e, E, g)$ and consider the lift

$$\begin{array}{ccc} pr^*TM & \longrightarrow & TM \\ \downarrow & & \downarrow \\ \mathbb{P}^1 \times M & \xrightarrow{pr} & M \end{array} \tag{9.8}$$

of the tangent bundle to $\mathbb{P}^1 \times M$. The canonical lifts to pr^*TM of the tensors $g, \circ, \mathcal{V}, \mathcal{U}$ will be denoted by the same letters. The connection ∇ lifts and extends to a flat connection on pr^*TM such that $\nabla_{\partial_z} Y = 0$ for $Y \in pr^{-1}\mathcal{T}_M$. Here z is a coordinate on $\mathbb{C} \subset \mathbb{P}^1$ and ∂_z the vector field with $\partial_z z = 1$, $\partial_z pr^{-1}\mathcal{O}_M = 0$.

With the multiplication and the Euler field one can twist this connection on pr^*TM in essentially two distinct ways. One obtains two series of flat connections, parameterized by one parameter $s \in \mathbb{C}$. The first structure connections $\hat{\nabla}^{(s)}$, $s \in \mathbb{C}$, are meromorphic along $\{0\} \times M \cup \{\infty\} \times M$. The second structure connections $\check{\nabla}^{(s)}$, $s \in \mathbb{C}$ are meromorphic along $\check{\mathcal{D}} \cup \{\infty\} \times M$, where

$$\check{\mathcal{D}} = \{(z, t) \mid \mathcal{U} - z\mathrm{id} \text{ is not invertible on } \mathrm{T_tM}\}. \tag{9.9}$$

The first structure connections $\hat{\nabla}^{(s)}$ are due to Dubrovin [**Du3**] (Lecture 3). Some of the first and second structure connections are related by some Fourier–Laplace transformations. Here we will concentrate on the second structure connections. We will only give the definition and make some remarks on the first structure connections in section 9.3.

Dubrovin considered also $\check{\nabla}^{(0)}$ and called it the Gauß–Manin connection of the Frobenius manifold [**Du3**, Appendix G]. In fact, we will see in section 11.1 that in the case of a hypersurface singularity $f(x_0, \ldots, x_n)$ the restriction of $\check{\nabla}^{(-\frac{n}{2})}$ to $\mathbb{C} \times M$ is isomorphic to the Gauß–Manin connection on the cohomology bundle of a semiuniversal unfolding of the singularity and that $\check{\nabla}^{(0)}$ and $\check{\nabla}^{(-\frac{n}{2})}$ are isomorphic if n is even and if the intersection form is nondegenerate.

The whole series $\check{\nabla}^{(s)}$, $s \in \mathbb{C}$, was defined by Manin and Merkulov [**MaM**] [**Man2**, II 1.2 and 1.4] for the case of semisimple multiplication. Their definition also works in the general case and is given below in Definition 9.3.

But in the case of hypersurface singularities, K. Saito perceived this series of connections much earlier [**SK9**, §5]: A primitive form for the Gauß–Manin connection (i.e. essentially the connection $\check{\nabla}^{(-\frac{n}{2})}$) gives rise to a period map, which has not so nice properties in the case of a degenerate intersection form.

To ameliorate that he (essentially) defined the whole series $\check{\nabla}^{(s)}$ and proposed to study the period map corresponding to $\check{\nabla}^{(0)}$.

Theorem 9.4 below is a translation and generalization of parts of the results in [**SK9**, §5], [**Du3**, Lecture 3 and Appendix G], [**Man2**, II§1]. Other properties of the connections $\check{\nabla}^{(s)}$ will be discussed in the following sections.

Definition 9.3 Denote $\check{M} := \mathbb{C} \times M - \check{\mathcal{D}}$. Fix $s \in \mathbb{C}$. The second structure connection $\check{\nabla}^{(s)}$ on $pr^*TM|_{\check{M}} \to \check{M}$ is defined by the following formulas for $X, Y \in pr^*\mathcal{T}_M|_{\check{M}}$ (here consider X as a vector field on $\check{M}$ and Y as a section in the bundle)

$$\check{\nabla}^{(s)}_X Y = \nabla_X Y - \left(\mathcal{V} + \frac{1}{2} + s\right)(\mathcal{U} - z)^{-1}(X \circ Y), \tag{9.10}$$

$$\check{\nabla}^{(s)}_{\partial_z} Y = \nabla_{\partial_z} Y + \left(\mathcal{V} + \frac{1}{2} + s\right)(\mathcal{U} - z)^{-1}(Y). \tag{9.11}$$

Theorem 9.4 *(a) The connection $\check{\nabla}^{(s)}$ is a flat connection on $pr^*TM|_{\check{M}}$ and satisfies for $X, Y \in pr^*\mathcal{T}_M|_{\check{M}}$*

$$\check{\nabla}^{(s)}_X((\mathcal{U} - z)Y) = (\mathcal{U} - z)\nabla_X Y - X \circ \left(\mathcal{V} - \frac{1}{2} + s\right)(Y), \tag{9.12}$$

$$\check{\nabla}^{(s)}_{\partial_z}((\mathcal{U} - z)Y) = (\mathcal{U} - z)\nabla_{\partial_z} Y + \left(\mathcal{V} - \frac{1}{2} + s\right)(Y). \tag{9.13}$$

(b) The endomorphism

$$\Delta_s : pr^*TM|_{\check{M}} \to pr^*TM|_{\check{M}}, \quad X \mapsto \left(\mathcal{V} + \frac{1}{2} + s\right)(\mathcal{U} - z)^{-1}X, \tag{9.14}$$

is a homomorphism of flat vector bundles

$$\left(pr^*TM|_{\check{M}}, \check{\nabla}^{(s+1)}\right) \to \left(pr^*TM|_{\check{M}}, \check{\nabla}^{(s)}\right).$$

It satisfies

$$\check{\nabla}^{(s)} \circ \Delta_s = \Delta_s \circ \check{\nabla}^{(s+1)} \tag{9.15}$$

and especially

$$\left(\check{\nabla}^{(s)}_{\partial_z}\right)^k Y = \Delta_s \circ \Delta_{s+1} \circ \ldots \circ \Delta_{s+k-1}(Y) \tag{9.16}$$

for $k \geq 1$ and $Y \in pr^{-1}(\mathcal{T}_M)|_{\check{M}}$. It is an isomorphism if and only if $d_i - \frac{D-1}{2} + s \neq 0$ for all $i = 1, \ldots, m$ (cf. Remark 9.2 (e)).

(c) The $\mathcal{O}_{\check{M}}$-bilinear map

$$\check{g} : pr^*\mathcal{T}_M|_{\check{M}} \times pr^*\mathcal{T}_M|_{\check{M}} \to \mathcal{O}_{\check{M}}$$
$$(X, Y) \mapsto g((\mathcal{U} - z)^{-1}X, Y) \tag{9.17}$$

is a symmetric and nondegenerate and multiplication invariant pairing on the bundle $pr^*TM|_{\check{M}}$ and satisfies

$$\check{g}(\Delta_{-s-1}X, Y) = -\check{g}(X, \Delta_s Y). \tag{9.18}$$

If X is $\check{\nabla}^{(-s)}$-flat and Y is $\check{\nabla}^{(s)}$-flat then $\check{g}(X, Y)$ is constant. Hence the connections $\check{\nabla}^{(-s)}$ and $\check{\nabla}^{(s)}$ are dual and $\check{g}$ induces an isomorphism

$$\left(pr^*TM|_{\check{M}}, \check{\nabla}^{(-s)}\right) \to \left((pr^*TM|_{\check{M}})^*, (\check{\nabla}^{(s)})^*\right), \tag{9.19}$$

here $(\check{\nabla}^{(s)})^*$ is the canonical induced connection.

(d) Consider $s \in \frac{1}{2}\mathbb{Z}_{\geq 0}$. The bilinear form $I^{(s)}$ on $pr^*TM|_{\check{M}}$ which is defined by

$$I^{(s)}(X, Y) = \check{g}(\Delta_{-s} \circ \ldots \circ \Delta_{s-2} \circ \Delta_{s-1}(X), Y) \tag{9.20}$$

is $(-1)^{2s}$-symmetric and $\check{\nabla}^{(s)}$-flat and satisfies

$$I^{(s+1)}(X, Y) = -I^{(s)}(\Delta_s(X), \Delta_s(Y)). \tag{9.21}$$

It is called the intersection form of $\check{\nabla}^{(s)}$. Of course $I^{(0)} = \check{g}$.

(e) The restriction of $\check{\nabla}^{(s)}$ to $\{z\} \times M - \check{\mathcal{D}}$ is torsion free for any $z \in \mathbb{C}$. The restriction of $\check{\nabla}^{(0)}$ to $\{z\} \times M - \check{\mathcal{D}}$ is the Levi–Civita connection for the restriction of $\check{g}$ to $\{z\} \times M - \check{\mathcal{D}}$.

Proof: (a) First we prove (9.12) and (9.13). The potentiality (9.2) and $\mathrm{Lie}_E(\circ) = 1 \cdot \circ$ give for ∇-flat fields X and Y

$$\begin{aligned}
&\check{\nabla}^{(s)}_X((\mathcal{U} - z)Y) \\
&= \check{\nabla}^{(s)}_X(E \circ Y) - z\check{\nabla}^{(s)}_X(Y) \\
&= \nabla_X(E \circ Y) - \left(\mathcal{V} + \frac{1}{2} + s\right)(X \circ Y) \\
&= \nabla_E(X \circ Y) + [X, E] \circ Y - \nabla_{X \circ Y}(E) + \left(\frac{D}{2} - \frac{1}{2} - s\right)(X \circ Y) \\
&= \mathrm{Lie}_E(X \circ Y) - \mathrm{Lie}_E(X) \circ Y + \left(\frac{D}{2} - \frac{1}{2} - s\right)(X \circ Y)
\end{aligned}$$

$$= X \circ Y + X \circ \mathrm{Lie}_E(Y) + X \circ \left(\frac{D}{2} - \frac{1}{2} - s\right) Y$$

$$= -X \circ \left(\mathcal{V} - \frac{1}{2} + s\right)(Y)$$

$$= (\mathcal{U} - z)\nabla_X(Y) - X \circ \left(\mathcal{V} - \frac{1}{2} + s\right)(Y).$$

By $\mathcal{O}_{\check{M}}$-linearity in X and Y one obtains (9.12). Formula (9.13) is a direct consequence of (9.11). For the flatness of $\check{\nabla}^{(s)}$ on $pr^*TM|_{\check{M}}$, it is enough to show

$$[\check{\nabla}^{(s)}_X, \check{\nabla}^{(s)}_Y]((\mathcal{U} - z)Z) = 0 \quad \text{for flat } X, Y, Z \tag{9.22}$$

and

$$[\check{\nabla}^{(s)}_X, \check{\nabla}^{(s)}_{\partial_z}]((\mathcal{U} - z)Y) = 0 \quad \text{for flat } X, Y. \tag{9.23}$$

Formula (9.22) follows with (9.12) and (9.2) for flat X, Y, and (also flat) $(\mathcal{V} - \frac{1}{2} + s)(Z)$ from

$$\check{\nabla}^{(s)}_X \check{\nabla}^{(s)}_Y((\mathcal{U} - z)Z) = \check{\nabla}^{(s)}_X\left(-Y \circ \left(\mathcal{V} - \frac{1}{2} + s\right)(Z)\right)$$

$$= \nabla_X\left(-Y \circ \left(\mathcal{V} - \frac{1}{2} + s\right)(Z)\right)$$

$$+ \left(\mathcal{V} + \frac{1}{2} + s\right)(\mathcal{U} - z)^{-1}\left(X \circ Y \circ \left(\mathcal{V} - \frac{1}{2} + s\right)(Z)\right)$$

$$= \nabla_Y\left(-X \circ \left(\mathcal{V} - \frac{1}{2} + s\right)(Z)\right)$$

$$+ \left(\mathcal{V} + \frac{1}{2} + s\right)(\mathcal{U} - z)^{-1}\left(Y \circ X \circ \left(\mathcal{V} - \frac{1}{2} + s\right)(Z)\right)$$

$$= \check{\nabla}^{(s)}_Y \check{\nabla}^{(s)}_X((\mathcal{U} - z)Z).$$

Formula (9.23) follows from (9.10)–(9.13).

(b) The endomorphism $\mathcal{V} + \frac{1}{2} + s$ maps ∇-flat sections in pr^*TM to ∇-flat sections (cf. Remark 9.2 (e)). So it is an endomorphism of the ∇-flat vector bundle pr^*TM with eigenvalues $d_i - \frac{D-1}{2} + s$, $i = 1, \ldots, m$ on each fibre. The endomorphism $(\mathcal{U} - z)$ is invertible on $pr^*TM|_{\check{M}}$. Therefore Δ_s is an isomorphism if and only if $d_i - \frac{D-1}{2} + s \neq 0$ for all $i = 1, \ldots, m$. (9.16) follows from (9.15) and the definitions of $\check{\nabla}^{(s)}_{\partial_z}$ and Δ_s. It remains to show (9.15).

We obtain with (9.10) and (9.12) for $X \in pr^*TM|_{\check{M}}$ and for ∇-flat Y

$$\begin{aligned}
\check{\nabla}^{(s)}_X(\Delta_s((\mathcal{U}-z)Y) &= \check{\nabla}^{(s)}_X\left(\left(\mathcal{V}+\frac{1}{2}+s\right)Y\right) \\
&= -\left(\mathcal{V}+\frac{1}{2}+s\right)(\mathcal{U}-z)^{-1}\left(X\circ\left(\mathcal{V}+\frac{1}{2}+s\right)Y\right) \\
&= \Delta_s\left(-X\circ\left(\mathcal{V}+\frac{1}{2}+s\right)Y\right) \\
&= \Delta_s(\check{\nabla}^{(s+1)}_X((\mathcal{U}-z)Y)
\end{aligned} \tag{9.24}$$

and in the same manner with (9.11) and (9.13)

$$\check{\nabla}^{(s)}_{\partial_z}(\Delta_s(\mathcal{U}-z)Y) = \Delta_s\big(\check{\nabla}^{(s+1)}_{\partial_z}((\mathcal{U}-z)Y)\big). \tag{9.25}$$

The formulas (9.24) and (9.25) extend to arbitrary $Y \in pr^*\mathcal{T}_M|_{\check{M}}$, formula (9.15) follows.

(c) The pairing $\check{g}$ is symmetric and nondegenerate and multiplication invariant, because g is symmetric and nondegenerate and multiplication invariant. Formula (9.18) holds because $\mathcal{V}$ is an infinitesimal isometry with respect to g. The rest follows from the next two formulas (9.26) and (9.27). Here $X, Y, Z \in pr^*\mathcal{T}_M|_{\check{M}}$ are chosen such that Z is any vector field on $\check{M}$, X is a $\check{\nabla}^{(-s)}$-flat section, Y is a $\check{\nabla}^{(s)}$-flat section.

$$\begin{aligned}
&Z\,\check{g}(X,Y) = Z\,g((\mathcal{U}-z)^{-1}(X),Y) \\
&= g(\nabla_Z(\mathcal{U}-z)^{-1}(X),Y) + g((\mathcal{U}-z)^{-1}(X),\nabla_Z Y) \\
&= g\left((\mathcal{U}-z)^{-1}\left(Z\circ\left(\mathcal{V}-\frac{1}{2}-s\right)(\mathcal{U}-z)^{-1}(X)\right),Y\right) \\
&\quad + g\left((\mathcal{U}-z)^{-1}(X),\left(\mathcal{V}+\frac{1}{2}+s\right)(\mathcal{U}-z)^{-1}(Z\circ Y)\right) \\
&= g\left(\left(\mathcal{V}-\frac{1}{2}-s\right)(\mathcal{U}-z)^{-1}(X),(\mathcal{U}-z)^{-1}(Z\circ Y)+g(\ldots)\right) \\
&= 0,
\end{aligned} \tag{9.26}$$

because $\mathcal{V}$ is an infinitesimal isometry with respect to g. Analogously

$$\partial_z\,\check{g}(X,Y) = 0. \tag{9.27}$$

(d) The endomorphism $\Delta_{-s} \circ \ldots \circ \Delta_{s-2} \circ \Delta_{s-1}$ maps $\check{\nabla}^{(s)}$-flat sections to $\check{\nabla}^{(-s)}$-flat sections. Therefore $I^{(s)}(X, Y)$ is constant for $\check{\nabla}^{(s)}$-flat sections X and Y, and $I^{(s)}$ is $\check{\nabla}^{(s)}$-flat, $\check{\nabla}^{(s)}(I^{(s)}) = 0$. It is $(-1)^{2s}$-symmetric because of (9.18).

(e) The first statement is obvious from (9.10). The second follows from (c).

□

Remarks 9.5 (a) Consider an isolated hypersurface singularity and the base space M of a semiuniversal unfolding. We will see in section 11.1 that some choice (of an opposite filtration to a Hodge filtration and of a generator of a 1-dimensional space) gives the following: a Frobenius manifold structure on M, an isomorphism from the cohomology bundle with its flat structure to $pr^*TM|_{\check{M}}$ with $\check{\nabla}^{(-\frac{n}{2})}$. The form $I^{(\frac{n}{2})}$ corresponds (up to a scalar) to the intersection form on the homology bundle. The map $\Delta_{-\frac{n}{2}} \circ \ldots \circ \Delta_{\frac{n}{2}-2} \circ \Delta_{\frac{n}{2}-1}$ corresponds (up to the same scalar) to the topological map from the homology bundle to the cohomology bundle, which comes from the intersection form.

We have $m = \mu$ and $d_i = 1 + \alpha_1 - \alpha_i$ and $D = 2 - (\alpha_\mu - \alpha_1)$; here $\alpha_1, \ldots, \alpha_\mu$ are the spectral numbers of the singularity (cf. section 10.6. They satisfy $-1 < \alpha_1 \leq \ldots \leq \alpha_\mu < n$ and $\alpha_i + \alpha_{\mu+1-i} = n - 1$.

The map Δ_s is an isomorphism if and only if $\alpha_i \neq \frac{n}{2} + s$ for all i. Therefore $\Delta_{\frac{n}{2}+k}$ and $\Delta_{-\frac{n}{2}-1-k}$ for $k \in \mathbb{Z}_{\geq 0}$ are isomorphisms. All the maps $\Delta_{\frac{n}{2}+k}$ for $k \in \mathbb{Z}$ are isomorphisms if and only if all $\alpha_i \notin \mathbb{Z}$, that means, if and only if the intersection form is nondegenerate. In fact, this also follows from the relation between $\Delta_{-\frac{n}{2}} \circ \ldots \circ \Delta_{\frac{n}{2}-2} \circ \Delta_{\frac{n}{2}-1}$, the form $I^{(\frac{n}{2})}$, and the intersection form.

(b) Consider an isolated hypersurface singularity as in (a) with degenerate intersection form. Then the homology bundle has global flat sections: the monodromy group is generated by the Picard–Lefschetz transformations of a distinguished basis of vanishing cycles (cf. [**AGV2**] for these notions). The elements of the radical of the intersection form in each fibre glue to global flat sections.

The cohomology bundle never has global flat nonzero sections. Such a section would give a flat subbundle of rank $\mu - 1$ in the homology bundle such that the quotient bundle (of rank 1) would have trivial monodromy. Because of the Picard–Lefschetz formulas then this subbundle of rank $\mu - 1$ would contain all the vanishing cycles in the distinguished basis. But they generate the homology, a contradiction.

This implies $\check{\nabla}^{(\frac{n}{2})} \ncong \check{\nabla}^{(-\frac{n}{2})}$ and also $\check{\nabla}^{(\frac{n}{2})} \ncong \check{\nabla}^{(0)} \ncong \check{\nabla}^{(-\frac{n}{2})}$, because $\check{\nabla}^{(\frac{n}{2})}$ and $\check{\nabla}^{(-\frac{n}{2})}$ are dual and $\check{\nabla}^{(0)}$ is selfdual (Theorem 9.4 (c)). Except for K. Saito [**SK9**, §5], nobody in singularity theory considered the intermediate connections $\check{\nabla}^{(s)}$ for $s \in (-\frac{n}{2}, \frac{n}{2}) \cap \frac{1}{2}\mathbb{Z}$. Can one describe their monodromy groups using only topological data? Especially for $\check{\nabla}^{(0)}$?

(c) In the situation of (b), $\check{\nabla}^{(\frac{n}{2})} \not\cong \check{\nabla}^{(0)} \not\cong \check{\nabla}^{(-\frac{n}{2})}$ also follows from another result of K. Saito [**SK5**]: Up to multiplication by scalars, the (degenerate) intersection form is the only flat bilinear form in the homology bundle. This follows essentially from the connectedness of the Coxeter–Dynkin diagram [**Ga**] [**La**]. The connection $\check{\nabla}^{(0)}$ has the nondegenerate flat bilinear form $\check{g} = I^{(0)}$.

(d) One could consider the shift from M to $\mathbb{P}^1 \times M$ as uneconomic because of the following:

(i) Because of $\mathrm{Lie}_e(\circ) = 0$ and $\mathrm{Lie}_e(g) = 0$, the Frobenius manifold is constant along e. One could take locally an $m-1$-dimensional slice U transversal to e and equip $\mathbb{C} \times U$ with the structure of a Frobenius manifold locally isomorphic to M. Because of $[e, E] = e$ the Euler field also extends to $\mathbb{C} \times U$.

(ii) Suppose $M = \mathbb{C} \times U$. One can recover $\check{\nabla}^{(s)}$ and $\check{g}$ on $pr^*TM|_{\check{M}}$ from the restrictions to $\{0\} \times M - \check{\mathcal{D}}$. They are canonically extended from $\{0\} \times M - \check{\mathcal{D}}$ with the flow of $\partial_z + e$: With respect to this flow on pr^*TM one has Lie derivatives $\mathrm{Lie}_{\partial_z + e}(\mathcal{U} - z) = 0$, $\mathrm{Lie}_{\partial_z + e}(\check{g}) = 0$, and also $\check{\nabla}^{(s)}_{\partial_z + e} Y = 0$ for Y with $\mathrm{Lie}_e Y = \mathrm{Lie}_{\partial_z} Y = 0$.

(iii) In singularity theory, usually the Gauß–Manin connection is considered on the base M of a semiuniversal unfolding, and not on a space $\Delta \times M$ for some disc $\Delta \subset \mathbb{C}$.

But it is more convenient to have the whole Frobenius manifold as $\{\infty\} \times M$ in the base space $\mathbb{P}^1 \times M$ of $\check{\nabla}^{(s)}$. One can treat all vector fields in $\mathcal{T}_M$ in the same way and does not have to distinguish those invariant under Lie_e. Also, the second and first structure connections are closely related, and for the first structure connections one does not have this possibility to reduce the dimension.

9.3 First structure connections

Also the first structure connections are defined on the lift pr^*TM (cf. (9.8)) of the tangent bundle of a Frobenius manifold $(M, \circ, e, E, g)$. The following definition and statements are known and can be found in different versions in [**Du3**][**Man2**][**Sab4**].

Definition 9.6 Denote $\hat{M} = \mathbb{C}^* \times M$. Fix $s \in \mathbb{C}$. The first structure connection $\hat{\nabla}^{(s)}$ on $pr^*TM|_{\hat{M}} \to \hat{M}$ is defined by the following formulas for $X, Y \in pr^*(\mathcal{T}_M)|_{\hat{M}}$ (again consider X as a vector field on $\hat{M}$ and Y as a section in the bundle)

$$\hat{\nabla}^{(s)}_X Y = \nabla_X Y + z \cdot X \circ Y, \tag{9.28}$$

$$\hat{\nabla}^{(s)}_{\partial_z} Y = \nabla_{\partial_z} Y + \frac{1}{z}\left(\mathcal{V} + \frac{1}{2} + s\right)(Y) + E \circ Y. \tag{9.29}$$

Remark 9.7 If one pulls back the connection $\hat{\nabla}^{(s)}$ with the involution $i : \mathbb{P}^1 \times M \to \mathbb{P}^1 \times M, (z,t) \mapsto (-z,t)$, one obtains a connection $i^*\hat{\nabla}^{(s)}$ on $pr^*TM|_{\hat{M}} \to \hat{M}$ which is given by the formulas

$$i^*\hat{\nabla}^{(s)}_X Y = \nabla_X Y - z \cdot X \circ Y, \tag{9.30}$$

$$i^*\hat{\nabla}^{(s)}_{\partial_z} Y = \nabla_{\partial_z} Y + \frac{1}{z}\left(\mathcal{V} + \frac{1}{2} + s\right)(Y) - E \circ Y. \tag{9.31}$$

Theorem 9.8 *(a) The connection $\hat{\nabla}^{(s)}$ is flat.*

(b) The multiplication by z is an isomorphism of flat vector bundles

$$\left(pr^*TM|_{\hat{M}}, \hat{\nabla}^{(s+1)}\right) \to \left(pr^*TM|_{\hat{M}}, \hat{\nabla}^{(s)}\right).$$

(c) If X is $\hat{\nabla}^{(s)}$-flat and Y is $i^*\hat{\nabla}^{(-1-s)}$-flat then $g(X,Y)$ is constant. Hence the connections $\hat{\nabla}^{(s)}$ and $i^*\hat{\nabla}^{(-1-s)}$ are dual.

(d) For each $z \in \mathbb{C}^*$ the restrictions to $\{z\} \times M$ of all the connections $\hat{\nabla}^{(s)}$, $s \in \mathbb{C}^*$ coincide. These restrictions are torsion free.

Proof: (a) For flat $X, Y, Z \in pr^*\mathcal{T}_M$ one has

$$\begin{aligned}\hat{\nabla}^{(s)}_X \hat{\nabla}^{(s)}_Y Z &= z\,\hat{\nabla}^{(s)}_X (Y \circ Z) = z\,\nabla_X (Y \circ Z) + z^2\, X \circ Y \circ Z \\ &= z\,\nabla_Y (X \circ Z) + z^2\, Y \circ X \circ Z = \hat{\nabla}^{(s)}_Y \hat{\nabla}^{(s)}_X Z,\end{aligned}$$

here the symmetry $\nabla_X(Y \circ Z) = \nabla_Y(X \circ Z)$ follows from the potentiality.

For flat $X, Y \in pr^*\mathcal{T}_M$ one has $\nabla_{\partial_z} Y = 0$, $\nabla_{\partial_z}(X \circ Y) = 0$, and $\mathcal{V}(Y)$ is flat, so

$$\hat{\nabla}^{(s)}_X \hat{\nabla}^{(s)}_{\partial_z} Y = X \circ \left(\mathcal{V} + \frac{1}{2} + s\right)(Y) + \nabla_X(E \circ Y) + z \cdot X \circ E \circ Y,$$

$$\hat{\nabla}^{(s)}_{\partial_z} \hat{\nabla}^{(s)}_X Y = X \circ Y + \left(\mathcal{V} + \frac{1}{2} + s\right)(X \circ Y) + z \cdot E \circ X \circ Y.$$

These two terms are equal because of the following calculation. It uses $\mathrm{Lie}_E(\circ) = 1 \cdot \circ$ and the potentiality (cf. (9.2))

$$\begin{aligned}X \circ Y + \nabla_{X \circ Y} E &= X \circ Y - \mathrm{Lie}_E(X \circ Y) + \nabla_E(X \circ Y) \\ &= -\mathrm{Lie}_E(X) \circ Y - X \circ \mathrm{Lie}_E(Y) + \nabla_X(E \circ Y) + [E, X] \circ Y \\ &= X \circ \nabla_Y E + \nabla_X(E \circ Y).\end{aligned}$$

(b) Formula (9.29) shows that $z\,Y$ is $\hat{\nabla}^{(s)}$-flat if Y is $\hat{\nabla}^{(s+1)}$-flat.

(c) The endomorphism $\mathcal{V}$ is an infinitesimal isometry with respect to g and g is multiplication invariant.

(d) This is obvious from (9.28). □

In this paper we will not make use of the first structure connections. Concerning them we restrict ourselves to the following remarks.

Remarks 9.9 (a) In Dubrovin's papers the first structure connection $\hat{\nabla}^{(-\frac{1}{2})}$ is a main tool for studying the Frobenius manifolds. Sabbah [**Sab2**, 4.1][**Sab4**, VII 1.1] approaches and characterizes Frobenius manifolds by the first structure connection $\hat{\nabla}^{(\frac{D-1}{2})}$. More precisely, he considers $j^*\hat{\nabla}^{(\frac{D-1}{2})}$ where $j : \mathbb{P}^1 \times M \to \mathbb{P}^1 \times M$ is the map $(z, t) \mapsto (-\frac{1}{z}, t)$. Also Manin [**Man2**] considers the first structure connection $\hat{\nabla}^{(\frac{D-1}{2})}$.

(b) All the connections $\hat{\nabla}^{(s)}$ give the same family of flat connections on the submanifolds $\{z\} \times M$ for $z \in \mathbb{C}^*$. The Euler field is not necessary for their definition. One can regard this family of flat connections as the primary datum and the extensions along ∂_z via an Euler field as mere refinements.

But the definition of the second structure connections requires the Euler field right from the beginning.

(c) Some of the first and second structure connections are related by Fourier–Laplace transformations. One can obtain precise informations following Sabbah [**Sab4**, V 2]. One has to consider the whole vector bundle pr^*TM and global meromorphic sections, in order to make everything algebraic with respect to the variable z. Then one can check that the inverse Fourier–Laplace transformation in the precise sense of [**Sab4**, V 2.10] gives a correspondence between $i^*\hat{\nabla}^{(s-1)}$ and $\check{\nabla}^{(s)}$ for those $s \in \mathbb{C}$ such that $\mathcal{V} + \frac{1}{2} + s - k$ is invertible for all $k \in \mathbb{Z}_{\geq 1}$.

This is not satisfied precisely for s in $\bigcup_{i=1}^m (-d_i + \frac{D-1}{2} + \mathbb{Z}_{\geq 1})$. In the case of a hypersurface singularity with degenerate intersection form (cf. Remarks 9.5) the correspondence is valid for $s = -\frac{n}{2}$, but not for $s = \frac{n}{2}$.

(d) One can write the defining equations for $\check{\nabla}^{(s)}$ and $\hat{\nabla}^{(s)}$ in a different way, emphasizing their similarity. For $X, Y \in pr^{-1}(\mathcal{T}_M) \subset pr^*\mathcal{T}_M$ one has

$$\hat{\nabla}^{(s)}_X Y = \nabla_X Y + z\, X \circ Y, \tag{9.32}$$

$$\hat{\nabla}^{(s)}_{E - z\partial_z} Y = [E, Y] - \left(\frac{1}{2} + s - \frac{D}{2}\right) Y, \tag{9.33}$$

$$\check{\nabla}^{(s)}_X Y = \nabla_X Y - \check{\nabla}^{(s)}_{\partial_z}(X \circ Y), \tag{9.34}$$

$$\check{\nabla}^{(s)}_{E + z\partial_z} Y = [E, Y] - \left(\frac{1}{2} + s - \frac{D}{2}\right) Y. \tag{9.35}$$

Using (9.13), one can check that (9.34) and (9.35) define $\check{\nabla}^{(s)}$.

(e) In the case of a hypersurface singularity $f(x_0, \ldots, x_n)$ the second structure connection $\check{\nabla}^{(-\frac{n}{2})}$ corresponds to the Gauß–Manin connection on the cohomology bundle. In fact, a main point of the construction of a Frobenius manifold for a singularity f is to enrich the Gauß–Manin connection to a connection isomorphic to $\check{\nabla}^{(-\frac{n}{2})}$.

Analogously the first structure connection $\hat{\nabla}^{(-\frac{n}{2}-1)}$ corresponds to a connection coming from oscillating integrals. One can also construct a Frobenius manifold for a singularity f via oscillating integrals and the first structure connection $\hat{\nabla}^{(-\frac{n}{2}-1)}$. Sabbah generalized this to the case of certain global functions $f : Y \to \mathbb{C}$ with isolated singularities on affine manifolds Y. We discuss this informally in section 11.4.

9.4 From the structure connections to metric and multiplication

The first and second structure connections of a Frobenius manifold $(M, \circ, e, E, g)$ are defined on the restrictions of pr^*TM to the submanifolds $\hat{M}$ and $\check{M}$ of $\mathbb{P}^1 \times M$. But $pr^*\mathcal{T}_M$ is a canonical extension to $\mathbb{P}^1 \times M$ of their sheaves of sections.

One can apply the notions from chapter 8. We will see that the pairs $(pr^*\mathcal{T}_M, \check{\nabla}^{(s)})$ and $(pr^*\mathcal{T}_M, \hat{\nabla}^{(s)})$ carry most of the structure of the Frobenius manifold. The contents of Lemmas 9.10, 9.13, 9.14, and 9.15 are summarized in figure 9.1 and figure 9.2.

The lower half of the first diagram is only proved for a massive Frobenius manifold.

First we discuss the metric. The following canonical isomorphisms are important in order to shift information from pr^*TM to TM. The sheaf $pr_*pr^*\mathcal{T}_M$ of fibrewise global sections of pr^*TM is a free $\mathcal{O}_M$-module of rank m, because pr^*TM can be seen as a family of trivial bundles on $\mathbb{P}^1$. One has canonical isomorphisms

$$\mathcal{T}_M \cong \mathcal{T}_{\{0\}\times M} \cong \mathcal{T}_{\{\infty\}\times M} \cong pr_*pr^*\mathcal{T}_M = pr_*pr^{-1}\mathcal{T}_M. \qquad (9.36)$$

Second structure connection $\check{\nabla}^{(s)}$ on $\mathbb{P}^1 \times M$

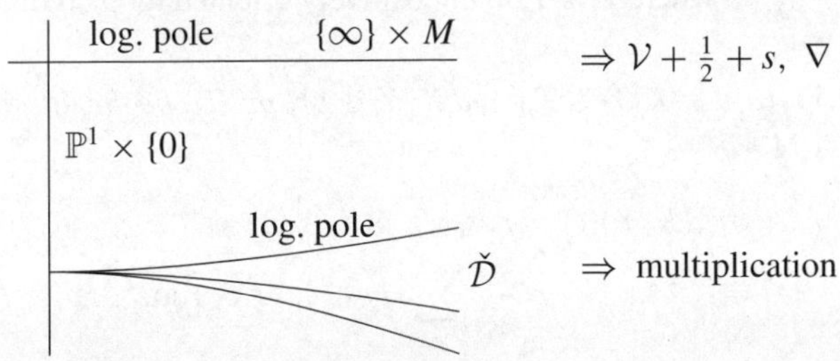

Figure 9.1

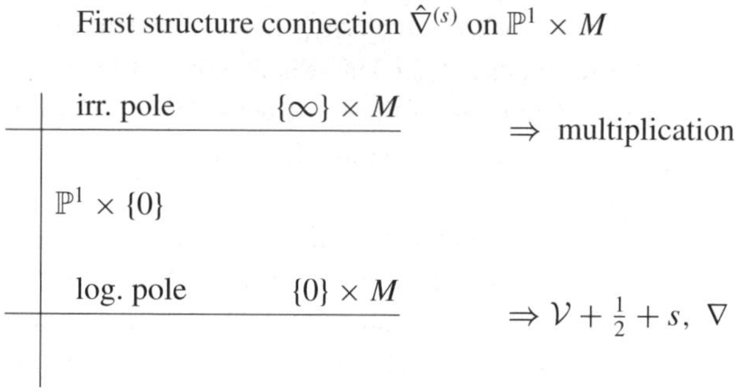

Figure 9.2

Lemma 9.10 [**Man2**, II 2.1.1] *(a) The pair $(pr^*\mathcal{T}_M, \check{\nabla}^{(s)})$ has a logarithmic pole along $\{\infty\} \times M$. The residue endomorphism and the residual connection with respect to $\frac{1}{z}$ are defined on $\mathcal{T}_{\{\infty\}\times M}$. Under the identification $\mathcal{T}_{\{\infty\}\times M} \cong \mathcal{T}_M$ the residue endomorphism is $\mathcal{V} + \frac{1}{2} + s$, and the residual connection is the flat Levi–Civita connection ∇ of g on M.*

(b) The pair $(pr^*\mathcal{T}_M, \hat{\nabla}^{(s)})$ has a logarithmic pole along $\{0\} \times M$. Under the identification $\mathcal{T}_{\{0\}\times M} \cong \mathcal{T}_M$ the residue endomorphism is $\mathcal{V} + \frac{1}{2} + s$, and the residual connection with respect to z is the flat connection ∇ on M.

Proof: (a) One can use Lemma 8.9 (b). One has to write out (9.10) and (9.11) for a basis of global ∇-flat sections of pr^*TM, using the coordinate $\widetilde{z} = \frac{1}{z}$ and $\widetilde{z}\partial_{\widetilde{z}} = -z\partial_z$.

(b) Similarly without $\widetilde{z}$. □

Using $\check{g}$ one can read off not only the flat connection ∇, but also the metric g from the logarithmic poles along $\{\infty\} \times M$ of the pairs $(pr^*\mathcal{T}_M, \check{\nabla}^{(s)})$.

In fact, Lemma 9.11 (a) shows that one can read it off also from the poles along $\check{\mathcal{D}}$. Lemma 9.11 (a) and (b) are not profound, but they will be very informative when we come to our version (Theorem 10.13) of K. Saito's higher residue pairing in the singularity case.

As usual, the residue of a meromorphic 1-form ω on $\mathbb{P}^1$ at a point $q \in \mathbb{P}^1$ is $res_q\omega = \frac{1}{2\pi i}\int_\gamma \omega$ where γ is a small positively oriented loop around q.

Lemma 9.11 *(a) Let $X, Y \in \mathcal{T}_M$ and $\widetilde{X}, \widetilde{Y} \in pr_*pr^*\mathcal{T}_M$ be the lifts to pr^*TM. For any $t \in M$*

$$
\begin{aligned}
g(X, Y)(t) &= res_{(\infty,t)}\, \check{g}(\widetilde{X}, \widetilde{Y})dz \qquad (9.37)\\
&= -\sum_{(\zeta,t)\in\check{\mathcal{D}}} res_{(\zeta,t)}\, \check{g}(\widetilde{X}, \widetilde{Y})dz.
\end{aligned}
$$

(b) Consider an $s \in \frac{1}{2}\mathbb{Z}_{\geq 0}$ *and suppose that* $\widetilde{\Delta} := \Delta_{-s} \circ \ldots \circ \Delta_{s-1}$ *is an isomorphism (compare Theorem 9.4 (b) and (d)). Then* $I^{(s)}$ *is nondegenerate and induces another nondegenerate bilinear form* $I^{(-s)}$ *on* $pr^*TM|_{\check{M}}$ *by* $I^{(-s)} = I^{(s)} \circ (\widetilde{\Delta}^{-1} \times \widetilde{\Delta}^{-1})$. *The form* $I^{(-s)}$ *is* $\check{\nabla}^{(-s)}$*-flat. It satisfies for any* $\widetilde{X}, \widetilde{Y}$ *as in (a)*

$$\check{g}(\widetilde{X}, \widetilde{Y}) = (-1)^{2s} I^{(-s)}((\check{\nabla}^{(-s)}_{\partial_z})^{2s} \widetilde{X}, \widetilde{Y}). \tag{9.38}$$

Proof: (a) The first equality follows from the definition of $\check{g}$. The sum of the residues of a meromorphic 1-form on a compact Riemann surface is 0.

(b) The form $I^{(-s)}$ is $\check{\nabla}^{(-s)}$-flat because of Theorem 9.4 (b) and (d). Formula (9.38) follows with (9.16), (9.20), and the $(-1)^{2s}$-symmetry of $I^{(s)}$. □

For the first structure connections there is no analogon of $\check{g}$. Theorem 9.8 (c) leads to another interplay between metric and the first structure connections. The following version for $\hat{\nabla}^{(-\frac{1}{2})}$ is close to [**Sab4**, VI 2.b]. A quite different version is in Dubrovin's papers (e.g. [**Du4**, chapter 3]). We will not subsequently use it. But again, it will be informative to compare K. Saito's higher residue pairings in the singularity case with it.

As in Remark , $i : \mathbb{P}^1 \times M \to \mathbb{P}^1 \times M$ denotes the involution $(z, t) \mapsto (-z, t)$; the induced involutions on $pr^*\mathcal{T}_M$ and $\mathcal{O}_{\mathbb{P}^1 \times M}$ and other sheaves on $\mathbb{P}^1 \times M$ are all denoted by i^*.

Lemma 9.12 *Fix a point* $q \in \{0\} \times M \subset \mathbb{P}^1 \times M$. *Let* $\hat{g}_q$ *be defined by*

$$\begin{aligned} \hat{g}_q : (pr^*\mathcal{T}_M)_q \times (pr^*\mathcal{T}_M)_q &\to \mathcal{O}_{\mathbb{P}^1 \times M, q} \\ (X, Y) &\mapsto g(X, i^*Y). \end{aligned} \tag{9.39}$$

It is a nondegenerate and $\mathcal{O}_{\mathbb{P}^1 \times M}$*-sesquilinear pairing, that is*

$$\hat{g}_q(f \cdot X, Y) = f \cdot \hat{g}_q(X, Y) = \hat{g}_q(X, i^* f \cdot Y) \tag{9.40}$$

for $f \in \mathcal{O}_{\mathbb{P}^1 \times M, q}$. *It is i-hermitian, that means,*

$$\hat{g}_q(Y, X) = i^*(\hat{g}_q(X, Y)), \tag{9.41}$$

and satisfies for $Z \in (pr^{-1}\mathcal{T}_M)_q$, $X, Y \in (pr^*\mathcal{T}_M)_q$

$$Z\, \hat{g}_q(X, Y) = \hat{g}_q\big(\hat{\nabla}^{(-\frac{1}{2})}_Z X, Y\big) + \hat{g}_q\big(X, \hat{\nabla}^{(-\frac{1}{2})}_Z Y\big), \tag{9.42}$$

$$\partial_z\, \hat{g}_q(X, Y) = \hat{g}_q\big(\hat{\nabla}^{(-\frac{1}{2})}_{\partial_z} X, Y\big) - \hat{g}_q\big(X, \hat{\nabla}^{(-\frac{1}{2})}_{\partial_z} Y\big). \tag{9.43}$$

Proof: Everything follows easily from the definitions. □

Now we come to the multiplication. The multiplication of a Frobenius manifold $(M, \circ, e, E, g)$ is encoded in that pole of the first and second structure connections which does not encode the flat structure. The following is an explicit description for the first structure connection. A more general discussion is given in [**Sab4**, 0 14.c].

Lemma 9.13 *The pair $(pr^*\mathcal{T}_M, \hat{\nabla}^{(s)})$ has a pole of Poincaré rank 1 (in the sense of Definition 8.5) along $\{\infty\} \times M$. Fix coordinates $t_1, \ldots, t_m$ on M around a point q, the coordinate $\widetilde{z} = \frac{1}{z}$ on $(\mathbb{P}^1, \infty)$, and fix a basis of pr^*TM in a neighbourhood of $(\infty, q) \in \mathbb{P}^1 \times M$. The connection matrix of $\hat{\nabla}^{(s)}$ with respect to this basis takes the form*

$$\Omega_0 \frac{d\widetilde{z}}{\widetilde{z}^2} + \sum_{i=1}^{m} \Omega_i \frac{dt_i}{\widetilde{z}}, \tag{9.44}$$

where all coefficients are in $\mathcal{O}_{\mathbb{P}^1 \times M, (\infty, q)}$.

Under the identification $\mathcal{T}_M \cong \mathcal{T}_{\{\infty\}\times M}$, the matrix $-\Omega_0(0, t)$ encodes $\mathcal{U}$ and the matrix $\Omega_i(0, t)$ encodes the multiplication by $\frac{\partial}{\partial t_i}$.

Proof: All statements follow directly from Definition 9.6. □

For the second structure connections, the multiplication is encoded in the geometry of $\check{\mathcal{D}}$ and the poles of $(pr^*\mathcal{T}_M, \check{\nabla}^{(s)})$ along it. This will be made precise in the next section in the case of a massive Frobenius manifold.

Here we give only a weak general statement: If $X, Y \in pr_* pr^* \mathcal{T}_M$ then also $\nabla_X Y \in pr_* pr^* \mathcal{T}_M$ and $\check{\nabla}^{(s)}_X Y|_{\{\infty\}\times M} = \nabla_X Y|_{\{\infty\}\times M}$. Therefore one can recover from $\check{\nabla}^{(s)}_X Y$ the fibrewise global section $\nabla_X Y$ and the difference $\check{\nabla}^{(s)}_X Y - \nabla_X Y$ and then with (9.10) the product $X \circ Y$, if $\mathcal{V} + \frac{1}{2} + s$ is invertible.

9.5 Massive Frobenius manifolds

A Frobenius manifold $(M, \circ, e, E, g)$ is massive if it is generically semisimple. It is semisimple if locally a basis of vector fields $e_1, \ldots, e_m$ exists with $e_i \circ e_j = \delta_{ij} e_i$. Then these vector fields are unique up to renumbering. They are called the idempotent vector fields. Semisimple Frobenius manifolds have been studied thoroughly ([**Du3**], [**Man2**], [**Hi**]).

The potentiality, or more precisely the condition (9.1) for an F-manifold implies $[e_i, e_j] = 0$ (Theorem 2.11). Coordinates $u_1, \ldots, u_m$ with $e_i = \frac{\partial}{\partial u_i}$ are unique up to shifts and are called *canonical coordinates*, following Dubrovin. An Euler field E with $\mathrm{Lie}_E(\circ) = 1 \cdot \circ$ takes the form $E = \sum_{i=1}^{m}(u_i + r_i)e_i$

for some $r_i \in \mathbb{C}$ (Theorem 2.11). Therefore the eigenvalues of $\mathcal{U}(= E\circ)$ are locally canonical coordinates.

Let us fix a massive Frobenius manifold $(M, \circ, e, E, g)$. The *bifurcation diagram* $\mathcal{B} \subset M$ is the set of points where some of the eigenvalues of $\mathcal{U}$ coincide. It is empty or a hypersurface and it is e-invariant, because of $\text{Lie}_e \mathcal{U} = \text{id}$. It contains the *caustic* $\mathcal{K} \subset M$, the set of points where M is not semisimple. A function whose zero set is the caustic will be considered in section 14.1. Another important hypersurface is the *discriminant*

$$\mathcal{D} = \{t \in M \mid \mathcal{U} \text{ is not invertible on } T_t M\}. \tag{9.45}$$

The geometry of the discriminant was discussed in sections 4.1 and 4.3. It is very rich. It is a free divisor with sheaf $\text{Der}_M(\log \mathcal{D}) = E \circ \mathcal{T}_M$ (Theorem 4.9). The tangent hyperplanes to $\mathcal{D}$ at smooth points have only a finite number of limit hyperplanes at the singular points. All tangent hyperplanes and the limit hyperplanes are transversal to e (Remark 4.2 (i)).

Consider for a moment a Frobenius manifold of the form $M = \mathbb{C} \times M'$ with $\mathbb{C} \times \{t'\}$ the orbits of e. Then one has (Corollaries 4.5 and 4.6):

(i) The projection $\mathcal{D} \to M'$ is a branched covering of degree m.
(ii) The bifurcation diagram is the set of e-orbits through $\mathcal{D}_{sing}$, the caustic $\mathcal{K}$ is the subset of e-orbits through those points of $\mathcal{D}_{sing}$ where the singularities are more complicated than the transversal intersection of local smooth components of $\mathcal{D}$.
(iii) The discriminant determines the multiplication.

For any massive Frobenius manifold, without assuming $M = \mathbb{C} \times M'$, one can recover the multiplication from $\check{\mathcal{D}} \subset \mathbb{P}^1 \times M$, because $\check{\mathcal{D}}$ and $\mathbb{P}^1 \times M$ are automatically big enough.

Lemma 9.14 *(a) The discriminants $\check{\mathcal{D}}$ and $\mathcal{D}$ are related by $\check{\mathcal{D}} \cap \{0\} \times M = \{0\} \times \mathcal{D}$. The canonical projection $\check{\mathcal{D}} \to \{\infty\} \times M$ (or $\{z\} \times M$ for any $z \in \mathbb{C}$) is a branched covering of degree m. $\check{\mathcal{D}}$ is a free divisor with*

$$\text{Der}_{\mathbb{C}\times M}(\log \check{\mathcal{D}}) = (\mathcal{U} - z)(pr^* \mathcal{T}_M|_{\mathbb{C}\times M}) \oplus \mathcal{O}_{\mathbb{C}\times M}(\partial_z + e). \tag{9.46}$$

(b) Let $U \subset M - \mathcal{B}$ be a sufficiently small open subset with canonical coordinates $u_1, \ldots, u_m$ with $E = \sum_i u_i e_i$. Then

$$\check{\mathcal{D}} \cap \mathbb{P}^1 \times U = \bigcup_{i=1}^{m} \{(z, u) \in \mathbb{C} \times U \mid z = u_i\}$$

consists of m smooth components which do not intersect and which project isomorphically to $\{\infty\} \times U$. The standard lifts $\widetilde{e}_1, \ldots, \widetilde{e}_m$ to $\mathbb{P}^1 \times U$ of

$e_1, \dots, e_m$ *are uniquely determined by the following conditions:*

(i) Their sum is $\sum \widetilde{e}_i = \widetilde{e}$ *(the lift of e),*
(ii) Each $\widetilde{e}_i$ *is tangent to all components* $\{(z,u) \in \mathbb{C} \times U \mid z = u_j\}$ *of* $\check{\mathcal{D}} \cap \mathbb{P}^1 \times U$ *with* $j \neq i$.

This determines the e_i *and the multiplication on M.*

Proof: (a) One has to regard the definitions of $\mathcal{D}$ and $\check{\mathcal{D}}$ and needs Theorem 4.9. Compare also Remark 9.5 (d).

(b) The intersection $\check{\mathcal{D}} \cap \mathbb{P}^1 \times U$ decomposes as described because $U \subset M - \mathcal{B}$. The conditions (i) and (ii) are obviously satisfied, (ii) because of $\widetilde{e}_i(z-u_j) = 0$ for $j \neq i$.

On the other hand, $\widetilde{e}_i$ is not tangential to $\{(z,u) \in \mathbb{C} \times U \mid z = u_i\}$ because of $\widetilde{e}_i(z-u_i) = -1$. Therefore $\widetilde{e}_1, \dots, \widetilde{e}_m$ are uniquely determined by (i) and (ii) and by the fact that they are in $pr_* pr^* \mathcal{T}_M$. □

Lemma 9.14 (b) characterizes the lifts to $pr^*\mathcal{T}_M$ of $e_1, \dots, e_m$ as vector fields on $\mathbb{P}^1 \times M$ by their relation to $\check{\mathcal{D}}$. Remark 9.16 (d) will characterize them as sections of pr^*TM by their relation to $\check{\nabla}^{(s)}$. In the case of a massive Frobenius manifold Lemma 9.10 (a) is supplemented by the following.

Lemma 9.15 [**Man2**, II 2.1.1] *Let* $(M, \circ, e, E, g)$ *be a massive Frobenius manifold.*

(a) The pair $(pr^*\mathcal{T}_M, \check{\nabla}^{(s)})$ *has a logarithmic pole along* $\check{\mathcal{D}}$. *The residue endomorphism along any smooth piece of* $\check{\mathcal{D}}$ *has eigenvalues* $(-(\frac{1}{2}+s), 0, \dots, 0)$. *For* $s \neq -\frac{1}{2}$ *it is semisimple. For* $s = -\frac{1}{2}$ *it is 0 or nilpotent with one* 2×2 *Jordan block.*

(b) The monodromy of $\check{\nabla}^{(s)}$ *around a smooth piece of* $\check{\mathcal{D}}$ *is semisimple with eigenvalues* $(-e^{2\pi i s}, 1, \dots, 1)$ *for* $s \notin \frac{1}{2} + \mathbb{Z}$. *For* $s \in \frac{1}{2} + \mathbb{Z}$ *it is the identity or unipotent with one* 2×2 *Jordan block.*

Proof: Equation (9.46) together with (9.10) and (9.11) show that $pr^*\mathcal{T}_M$ is invariant with respect to $\check{\nabla}^{(s)}$ for any logarithmic vector field. By Definition 8.5 and (8.9), then $(pr^*\mathcal{T}_M, \check{\nabla}^{(s)})$ has a logarithmic pole along $\check{\mathcal{D}}$.

For the residue endomorphism we have to be more explicit. Let $U \subset M - \mathcal{B}$, $u_1, \dots, u_m$, $E = \sum u_i e_i$, $\check{\mathcal{D}} \cap \mathbb{P}^1 \times U$, and $\widetilde{e}_1, \dots, \widetilde{e}_m$ be as in Lemma 9.14 (b). In a neighbourhood of one component $\{(z,u) \mid z - u_i = 0\}$ of $\check{\mathcal{D}} \cap \mathbb{P}^1 \times U$

$$(z-u_i)\check{\nabla}^{(s)}_{\partial_z}\widetilde{e}_j = -\left(\mathcal{V} + \frac{1}{2} + s\right)\frac{z-u_i}{z-u_j}\widetilde{e}_j. \tag{9.47}$$

For $j \neq i$ locally this is contained in $(z-u_i)pr^*\mathcal{T}_M$. Because $\mathcal{V}$ is skewsymmetric, $\mathcal{V}(\widetilde{e}_i) \in \sum_{j\neq i} \mathcal{O}_{\mathbb{P}^1\times M}\widetilde{e}_j$. Therefore the eigenvalues of the residue endomorphism are $(-(\frac{1}{2}+s), 0, \ldots, 0)$. The remainder of (a) and (b) follows from Example 7.11 and section 8.2. □

Remarks 9.16 (a) In the case of $s \in \frac{1}{2}+\mathbb{Z}$, it is not easy to say when the local monodromy of $\check{\nabla}^{(s)}$ around a smooth piece of $\check{\mathcal{D}}$ has a 2×2 Jordan block and when it is the identitiy.

In the situation in the proof it has a 2×2 Jordan block for $s = -\frac{1}{2}$ if and only if $\mathcal{V}(e_i) \neq 0$. This is often satisfied. For example, if the discriminant is irreducible one has essentially only one local monodromy and it has a 2×2 Jordan block if and only if $\mathcal{V} \neq 0$. But in the case of the trivial Frobenius manifold A_1^m, that is,

$$M=\mathbb{C}^m,\quad e_i=\frac{\partial}{\partial u_i},\quad e_i\circ e_j=\delta_{ij},\quad E=\sum u_ie_i,\quad g(e_i,e_j)=\delta_{ij},\quad (9.48)$$

one has $\mathcal{V}=0$, and the discriminant is $\{u \mid u_1\cdot\ldots\cdot u_m = 0\}$. For $s\in -\frac{1}{2}+\mathbb{Z}_{\neq 0}$ it is even more difficult to give precise conditions.

(b) For $s \notin \frac{1}{2}+\mathbb{Z}$ the locally free $\mathcal{O}_{\mathbb{C}\times M}$-module $pr^*\mathcal{T}_M|_{\mathbb{C}\times M}$ is the unique coherent and reflexive extension of $pr^*\mathcal{T}_M|_{\check{M}}$ to $\mathbb{C}\times M$ with logarithmic poles along $\check{\mathcal{D}}$ such that all residue endomorphisms along smooth pieces of $\check{\mathcal{D}}$ have eigenvalues $(-(\frac{1}{2}+s), 0, \ldots, 0)$. This follows from chapter 8.

(c) The situation in the proof has been studied extensively in [**Man2**, II §3] for $\check{\nabla}^{(0)}$. One needs to know only the restriction of $(pr^*\mathcal{T}_M, \check{\nabla}^{(0)}, \check{g})$ to one slice $\mathbb{P}^1\times\{u^{(0)}\}$. Then one can recover the whole structure because $(pr^*\mathcal{T}_M, \check{\nabla}^{(0)})$ is the unique extension to a flat connection with logarithmic poles along $(\check{\mathcal{D}}\cap \mathbb{P}^1\times U)\cup\{\infty\}\times M$. The metric $\check{g}$ is the $\check{\nabla}^{(0)}$-flat extension of its restriction to $\mathbb{P}^1\times\{u^{(0)}\}-\check{\mathcal{D}}$. The data for the slice $\mathbb{P}^1\times\{u^{(0)}\}$ are called special initial conditions. They completely determine the Frobenius manifold locally. In order to recover the metric g one needs Lemma 9.11 (a).

(d) In the situation in the proof, the coordinate $z-u_i$ for the component $\{(z,u)\mid z-u_i=0\}$ of $\check{\mathcal{D}}\cap\mathbb{P}^1\times U$ induces spaces $C^\alpha_{(i)}$ of elementary sections of order α for $\check{\nabla}^{(s)}$ near this component, as in (8.13). Here $\alpha\in\mathbb{Z}\cup -(\frac{1}{2}+s)+\mathbb{Z}$.

The sections $\widetilde{e}_1,\ldots,\widetilde{e}_m$ decompose locally uniquely into a sum (in general infinite) of elementary sections, their elementary parts. For $s\notin\frac{1}{2}+\mathbb{Z}$ the fibrewise global sections $\widetilde{e}_1,\ldots,\widetilde{e}_m$ satisfy the following and are uniquely determined by it:

> they do not have elementary parts in $C^\alpha_{(i)}$ for $\alpha\in -(\frac{1}{2}+s)+\mathbb{Z}_{<0}$ and any i. The section $\widetilde{e}_i$ is the only one of them with a nonvanishing elementary part in $C^{-(\frac{1}{2}+s)}_{(i)}$.

This follows from the residue endomorphism. There is a similar statement for

$s \in \frac{1}{2} + \mathbb{Z}$, but it is more complicated and less satisfactory because of Remark 9.16 (a).

In all examples of massive Frobenius manifolds which I know, the eigenvalues of $\mathcal{V}$ are rational numbers. A partial explanation is given by the following application of the second structure connections and of a result of Kashiwara [**Kas1**].

Theorem 9.17 *Let $(M, \circ, e, E, g)$ be a massive Frobenius manifold with a point $t \in M$ such that all eigenvalues of $\mathcal{U} : T_t M \to T_t M$ coincide. Then the eigenvalues of $\mathcal{V}$ and thus also the numbers $d_1, \ldots, d_m$ and D are rational.*

Proof: Kashiwara calls a constructible sheaf quasiunipotent if for any map from a sufficiently small disc to the base manifold the induced monodromy around $0 \in$ *disc* is quasiunipotent.

He showed that a constructible sheaf is quasiunipotent if its restriction to the complement of a codimension 2 subset is quasiunipotent [**Kas1**, Theorem 3.1]. We can apply this to $\check{\nabla}^{(0)}$ on $\mathbb{C} \times M$, taking $\check{\mathcal{D}}_{sing}$ as the codimension 2 subset. The monodromies around smooth pieces of $\check{\mathcal{D}}$ are quasiunipotent, so all local monodromies on discs embedded in $\mathbb{C} \times M$ are quasiunipotent.

If $z^{(0)}$ denotes the single eigenvalue of $\mathcal{U} : T_t M \to T_t M$, then $\check{\mathcal{D}} \cap \mathbb{P}^1 \times \{t\} = \{(z^{(0)}, t)\}$. The monodromy on $\mathbb{C} - \{(z^{(0)}, t)\}$ is quasiunipotent. Its inverse is the monodromy around $\{\infty\} \times M$ with residue endomorphism $\mathcal{V}$ (Lemma 9.10 (a)). Its eigenvalues are the numbers $d_i - \frac{D}{2}, i = 1, \ldots, m$ with $d_1 = 1$. □

The assumption for t and $\mathcal{U}$ is satisfied for example, if the algebra $T_t M$ is irreducible, that means, local (Lemma 2.1). Then $E|_t$ is a sum of a multiple of $e|_t$ and a vector in the maximal ideal of $T_t M$.

Chapter 10

Gauß–Manin connections for hypersurface singularities

Section 10.1 resumes the definition of a semiuniversal unfolding of a singularity and the resulting structure of an F manifold on its base space M. This was treated in greater detail in section 5.1. The cohomology bundle is discussed from the point of view of chapter 8 in section 10.2. Sections 10.3, 10.4, and 10.6 put together most of the known results on the Gauß–Manin connection of (the semiuniversal unfolding of) a singularity. The presentation is as plain as possible. The detailed discussion of the extensions $\mathcal{H}^{(k)}$ for $k > 0$ (Lemma 10.2, Theorem 10.7) is new and leads to a very explicit approach to the microlocal Gauß–Manin system (Theorem 10.10).

Also instructive and not well known are two alternative descriptions for K. Saito's higher residue pairings, both under some restrictions (Theorem 10.13, Theorem 10.28). The second one gives the link to a polarizing form on the cohomology of the Milnor fibre. The definition and general facts from **[He4]** on polarized mixed Hodge structures are provided in section 10.5.

10.1 Semiuniversal unfoldings and F-manifolds

Let $f : (\mathbb{C}^{n+1}, 0) \to (\mathbb{C}, 0)$ be a holomorphic function germ with an isolated singularity at 0. Its Milnor number $\mu \in \mathbb{N}$ is the dimension of the Jacobi algebra $\mathcal{O}/J_f := \mathcal{O}_{\mathbb{C}^{n+1},0}/(\frac{\partial f}{\partial x_0}, \ldots \frac{\partial f}{\partial x_n})$.

Unfoldings of f, morphisms between them, semiuniversal unfoldings, and the germ of an F-manifold which belongs to f have been discussed in section 5.1. A semiuniversal unfolding of f is a holomorphic function germ $F : (\mathbb{C}^{n+1} \times \mathbb{C}^\mu, 0) \to (\mathbb{C}, 0)$ with $F|(\mathbb{C}^{n+1} \times \{0\}, 0) = f$ and coordinates $t_1, \ldots, t_\mu$ on $(\mathbb{C}^\mu, 0) =: (M, 0)$ such that the reduced Kodaira–Spencer map

$$\mathbf{a}|_0 : T_0M \to \mathcal{O}/J_f, \quad \frac{\partial}{\partial t_j} \mapsto \left[\frac{\partial F}{\partial t_j}|(\mathbb{C}^{n+1} \times \{0\}, 0)\right] \tag{10.1}$$

is an isomorphism.

The germ $(C,0)\subset(\mathbb{C}^{n+1}\times M,0)$ of the critical space is defined by the ideal $J_F=(\frac{\partial F}{\partial x_0},\dots,\frac{\partial F}{\partial x_n})$. It is smooth. It is the normalization of the discriminant $(\check{\mathcal{D}},0)=\varphi((C,0))\subset(\mathbb{C}\times M,0)$ where

$$\varphi:(\mathbb{C}^{n+1}\times M,0)\to(\mathbb{C}\times M,0),\quad (x,t)\mapsto(F(x,t),t). \tag{10.2}$$

One can choose representatives of all these germs with good properties in the following way ([**Lo2**, 2.D], [**AGV2**, 10.3.1]). First, $\varepsilon>0$ is chosen such that $f^{-1}(0)$ intersects the boundaries $\partial B^{n+1}_{\varepsilon'}$ of the balls $B^{n+1}_{\varepsilon'}=\{x\in\mathbb{C}^{n+1}\mid |x|<\varepsilon'\}$ transversally for all $0<\varepsilon'\le\varepsilon$. Then $\delta>0$ is chosen such that all the fibres $f^{-1}(z)$ for $z\in\Delta:=B^1_\delta\subset\mathbb{C}$ intersect $\partial B^{n+1}_\varepsilon$ transversally. Finally, we choose $\theta>0$ and define $M:=B^\mu_\theta$ such that $\check{\mathcal{D}}$ does not intersect $(\partial\Delta)\times M$ and all the fibres $\varphi^{-1}(z,t)$ for $(z,t)\in\Delta\times M$ intersect $\partial B^{n+1}_\varepsilon\times\{t\}$ transversally.

Then the space $\mathcal{X}:=F^{-1}(\Delta)\cap(B^{n+1}_\varepsilon\times M)$ and the maps $F:\mathcal{X}\to\Delta$ and $\varphi:\mathcal{X}\to\Delta\times M$ are good representatives with critical space $C\subset\mathcal{X}$ of φ and discriminant $\check{\mathcal{D}}=\varphi(C)\subset\Delta\times M$.

The projections $pr_{C,M}:C\to M$ and $\check{\mathcal{D}}\to M$ are finite and flat of degree μ. The Kodaira–Spencer map

$$\mathbf{a}:\mathcal{T}_M\to(pr_{C,M})_*\mathcal{O}_C,\quad X\mapsto\widetilde{X}(F)|_C, \tag{10.3}$$

where $\widetilde{X}$ is any lift of $X\in\mathcal{T}_M$ to $\mathcal{X}$, is welldefined and an isomorphism of free $\mathcal{O}_M$-modules of rank μ. It induces a multiplication $\circ$ on $\mathcal{T}_M$. Then M becomes a massive F-manifold (Theorem 5.3). Generic semisimplicity follows from the smoothness of C and corresponds to the fact that for generic $t\in M$ the function

$$F_t:\mathcal{X}\cap(B^{n+1}_\varepsilon\times\{t\})\to\Delta \tag{10.4}$$

has μ A_1-singularities. The caustic $\mathcal{K}\subset M$ is the hypersurface of parameters t for which F_t has less than μ singularities.

The field $e=\mathbf{a}^{-1}(1|_C)$ is the unit field. One could choose $F=t_1+\widetilde{F}(x,t_2,\dots,t_\mu)$; then one would have $e=\frac{\partial}{\partial t_1}$; but we will not need this in the following.

The field $E:=\mathbf{a}^{-1}(F|_C)$ is an Euler field of the F-manifold $(M,\circ,e)$ (Theorem 5.3). The Kodaira–Spencer map in (10.3) gives canonical isomorphisms for all $t\in M$,

$$(T_tM,\circ,E|_t)\cong\left(\bigoplus_{x\in Sing\,F_t}\text{Jacobi algebra of }(F_t,x),\text{ mult., }[F_t]\right). \tag{10.5}$$

The eigenvalues of $\mathcal{U}:T_tM\to T_tM,\ X\mapsto E\circ X$, are the critical values of F_t for each $t\in M$. They form on $M-\mathcal{K}$ locally canonical coordinates $u_1,\dots,u_m$. Locally on $M-\mathcal{K}$ the fields $e_i=\frac{\partial}{\partial u_i}$ are defined and satisfy $e_i\circ e_j=\delta_{ij}e_i$.

There are two discriminants,

$$\mathcal{D} = (\det \mathcal{U})^{-1}(0) \subset M \quad \text{with} \tag{10.6}$$

$$\mathcal{D} = \varphi(C \cap F^{-1}(0)) \subset \{0\} \times M$$

and

$$\check{\mathcal{D}} = \varphi(C) = (\det(\mathcal{U} - z\,id))^{-1}(0) \subset \Delta \times M. \tag{10.7}$$

The critical space C and the restriction $C \cap F^{-1}(0)$ are smooth. The projections $C \to \check{\mathcal{D}}$ and $C \cap F^{-1}(0) \to \mathcal{D}$ are generically one-to-one. Therefore they are the normalizations of $\check{\mathcal{D}}$ and $\mathcal{D}$, and $\check{\mathcal{D}} \subset \Delta \times M$ and $\mathcal{D} \subset M$ are irreducible hypersurfaces.

The function $\check{h} := \det(\mathcal{U} - z\,id) : \Delta \times M \to \mathbb{C}$ gives an everywhere reduced equation for $\check{\mathcal{D}}$, similarly $\det \mathcal{U}$ for $\mathcal{D}$. Both discriminants $\mathcal{D}$ and $\check{\mathcal{D}}$ are free divisors (Theorem 4.9),

$$\mathrm{Der_M}(\log \mathcal{D}) = E \circ \mathcal{T}_M = \mathcal{U}(\mathcal{T}_M), \tag{10.8}$$

$$\mathrm{Der_{\Delta\times M}}(\log \check{\mathcal{D}}) = (\mathcal{U} - z)(\pi^* \mathcal{T}_M|_{\Delta\times M}) \oplus \mathcal{O}_{\Delta\times M}(\partial_z + e). \tag{10.9}$$

Here we use notations similar to those in section 9.2: the tangent bundle TM is lifted to $\Delta \times M$,

$$\begin{array}{ccc} \pi^* TM & \longrightarrow & TM \\ \downarrow & & \downarrow \\ \Delta \times M & \stackrel{\pi}{\longrightarrow} & M, \end{array} \tag{10.10}$$

the canonical lifts of the tensors $\circ, \mathcal{U}, e$ to $\pi^* TM$ are denoted by the same letters. Then (10.9) follows from (10.8) and the definition of $\check{\mathcal{D}}$, (cf. Lemma 9.14, Remark 9.5 (d)(ii)).

A careful discussion of the isomorphisms between semiuniversal unfoldings shows the following (Theorem 5.4).

Theorem 10.1 *The germ $((M, 0), \circ, e, E)$ of the F-manifold and the germs $(\mathcal{D}, 0) \subset (M, 0)$, $(\check{\mathcal{D}}, 0) \subset (\Delta \times M, 0)$ depend only on f, that means, they are unique up to canonical isomorphism and independent of the choices of $\varepsilon, \delta, \theta$ and F.*

10.2 Cohomology bundle

Let $f : (\mathbb{C}^{n+1}, 0) \to (\mathbb{C}, 0)$, $F : \mathcal{X} \to \Delta$, $\varphi : \mathcal{X} \to \Delta \times M$ and $\check{\mathcal{D}} \subset \Delta \times M$ be as in section 10.1. We make the additional assumption that $n \geq 1$. That excludes only the A_μ-singularities in one variable (but not their suspensions in several variables). Their Gauß–Manin connections and Brieskorn lattices are exceptional. One can find a treatment of some Fourier duals of their Gauß–Manin connections and also their Frobenius manifolds in [**Sab4**, VII 4.b and 5.c].

In this section the cohomology bundle of φ is discussed using chapter 8 and an argument of Varchenko. In the next section things will be compared with the Gauß–Manin connection, that is, with sections coming from holomorphic differential forms.

Good references for the following facts are [**Lo2**] and [**AGV2**] (but, of course, many of the facts are much older, e.g. [**Mi**]).

A fibre $\varphi^{-1}(z,t)$ of $\varphi : \mathcal{X} \to \Delta \times M,\ (x,t) \mapsto (F(x,t),t)$ is singular if and only if $(z,t) \in \check{\mathcal{D}}$. Each regular fibre is homotopy equivalent to a bouquet of μ n-spheres. The restriction

$$\varphi : \varphi^{-1}(\Delta \times M - \check{\mathcal{D}}) \to \Delta \times M - \check{\mathcal{D}} \tag{10.11}$$

is a locally trivial C^∞-bundle. One may call it a generalized Milnor fibration. The restriction to $(\Delta - \{0\}) \times \{0\}$ is a Milnor fibration.

The cohomology bundle

$$H^n := \bigcup_{(z,t)\in\Delta\times M-\check{\mathcal{D}}} H^n(\varphi^{-1}(z,t),\mathbb{C}) \to \Delta \times M - \check{\mathcal{D}} \tag{10.12}$$

has rank μ and a canonical flat structure. Its connection is called ∇.

The cohomology bundle and its monodromy group are essentially independent of the choice of $\varepsilon, \delta, \theta$ in section 10.1: first, the fibres of a representative for some choice of smaller $\varepsilon', \delta', \theta'$ are deformation retracts of the corresponding larger fibres of φ by [**LeR**, Lemma 2.2]; second, the representative $\varphi : \mathcal{X} \to \Delta \times M$ is *excellent* in the sense of [**Lo2**, 2.D], and therefore the monodromy group does not change when one restricts the cohomology bundle H^n to a smaller base $B^1_{\varepsilon'} \times B^\mu_{\theta'} \subset \Delta \times M$.

The subbundle of H^n of the $\mathbb{Z}$-lattices $H^n(\varphi^{-1}(z,t),\mathbb{Z})$ is invariant under the monodromy group. The local monodromy around $\check{\mathcal{D}}_{reg}$ is given by a Picard–Lefschetz transformation: for even n it is semisimple with eigenvalues $(-1, 1, \ldots, 1)$, for odd n it is unipotent with one 2×2 Jordan block; in both cases the invariant subspace has dimension $\mu - 1$.

Let $\mathcal{H}$ be the sheaf of holomorphic sections of H^n and $i : \Delta \times M - \check{\mathcal{D}} \to \Delta \times M$ the inclusion. With chapter 8 in mind, there are distinguished extensions of $\mathcal{H}$ to $\Delta \times M$.

Lemma 10.2 *For any $k \in \mathbb{Z}$ there is a unique extension $\mathcal{H}^{(k)} \subset i_*\mathcal{H}$ of $\mathcal{H}$ to $\Delta \times M$ with the properties: it is an $\mathcal{O}_{\Delta\times M}$-coherent and reflexive subsheaf* *of $i_*\mathcal{H}$, and $(\mathcal{H}^{(k)}, \nabla)$ has a logarithmic pole along $\check{\mathcal{D}}$ whose residue endomorphism along $\check{\mathcal{D}}_{reg}$ is semisimple with eigenvalues $(\frac{n-1}{2} - k, 0, \dots, 0)$ for $\frac{n-1}{2} - k \neq 0$ and nilpotent with a 2×2 Jordan block for $\frac{n-1}{2} - k = 0$.*

Proof: By Example 7.11 and Theorem 8.7 there is a unique extension of $\mathcal{H}$ to a coherent and, in fact, locally free sheaf on $\Delta \times M - \check{\mathcal{D}}_{sing}$ with a logarithmic pole along $\check{\mathcal{D}}_{reg}$ and residue endomorphism as above. By Theorem 8.11, the direct image under j_* for $j : \Delta \times M - \check{\mathcal{D}}_{sing} \to \Delta \times M$ is the sheaf $\mathcal{H}^{(k)}$. □

Which of these extensions $\mathcal{H}^{(k)}$ are free $\mathcal{O}_{\Delta\times M}$-modules? Following an argument of Varchenko one can read this off from the restrictions to $\Delta \times \{0\}$. Let $(t_1, \dots, t_\mu) \subset \mathcal{O}_{\Delta\times M}$ be the ideal which defines $\Delta \times \{0\}$ and

$$\mathcal{B}^{(k)} := \left(\mathcal{H}^{(k)} / (t_1, \dots, t_\mu)\mathcal{H}^{(k)}\right)|_{\Delta\times\{0\}}. \tag{10.13}$$

The sheaf $\mathcal{B}^{(k)}$ is an extension of the sheaf of holomorphic sections $\mathcal{B}$ of $H^n|\Delta^* \times \{0\}$ to $\Delta \times \{0\}$. It is a free $\mathcal{O}_\Delta$-module of rank μ. Its sections have moderate growth with respect to the multivalued flat sections of $H^n|\Delta^* \times \{0\}$, because they are restrictions of the sections of $\mathcal{H}^{(k)}$, and those have moderate growth (cf. section 8.4, [**De1**]).

We can use the notions of sections 7.1 and 7.2. The germ $\mathcal{B}_0^{(k)}$ is a $\mathbb{C}\{z\}$-lattice in $V^{>-\infty}$ and has μ spectral numbers $Sp(\mathcal{B}_0^{(k)}) = (\alpha_1^{(k)}, \dots, \alpha_\mu^{(k)})$, defined by (7.21) and (7.23).

The following is essentially due to Varchenko ([**Va1**, ch. 2], [**AGV2**, ch. 12]). It is a relative statement and will be complemented by the next section 10.3.

Theorem 10.3 *The space $\mathcal{H}_0^{(k)}$ is a free $\mathcal{O}_{\Delta\times M,0}$-module if and only if $\sum_{i=1}^{\mu} \alpha_i^{(k)} = \mu(\frac{n-1}{2} - k)$.*

Proof: Consider μ sections $\omega_1, \dots, \omega_\mu$ of $\mathcal{H}^{(k)}$ in a neighbourhood of $0 \in \Delta \times M$. Choose a basis of $H_n(\varphi^{-1}(z, t), \mathbb{Z})$ for some $(z, t) \in \Delta \times M - \check{\mathcal{D}}$ and extend the vectors to flat multivalued sections $\delta_1, \dots, \delta_\mu$ of the homology bundle over $\Delta \times M - \check{\mathcal{D}}$. Then $\det^2((\langle\omega_i, \delta_j\rangle)_{ij})$ is a univalued holomorphic function on $\Delta \times M - \check{\mathcal{D}}$, because $\det^2 A = 1$ for any monodromy transformation matrix $A \in \mathrm{GL}(\mu, \mathbb{Z})$.

The function $\check{h} = \det(\mathcal{U} - z\mathrm{id})$ gives an everywhere reduced equation for $\check{\mathcal{D}}$. The eigenvalues of the residue endomorphism of $(\mathcal{H}^{(k)}, \nabla)$ along $\check{\mathcal{D}}_{reg}$ imply that

$$\det{}^2((\langle\omega_i, \delta_j\rangle)_{ij}) = g \cdot \check{h}^{n-1-2k} \tag{10.14}$$

for some function $g \in \mathcal{O}_{\Delta\times M-\check{\mathcal{D}}_{sing},0} = \mathcal{O}_{\Delta\times M,0}$. The following statements are equivalent:

(i) The μ sections $\omega_1, \dots, \omega_\mu$ form a $\mathcal{O}_{\Delta\times M,0}$-basis of $\mathcal{H}_0^{(k)}$ (especially, then $\mathcal{H}_0^{(k)}$ is a free $\mathcal{O}_{\Delta\times M,0}$-module).
(ii) They form a basis of $\mathcal{H}^{(k)}$ in (a neighbourhood of 0 in) $\Delta \times M - \check{\mathcal{D}}_{sing}$.
(iii) The function g does not vanish in (a neighbourhood of 0 in) $\Delta \times M - \check{\mathcal{D}}_{sing}$.
(iv) It has value $g(0) \neq 0$.

On the other hand, the definition of the spectral numbers $\alpha_1^{(k)}, \dots, \alpha_\mu^{(k)}$ in (7.21) and (7.23) shows that

$$\det{}^2((\langle\omega_i, \delta_j\rangle)_{ij})|_{(\Delta\times\{0\},0)} = \widetilde{g}\cdot z^{2\sum\alpha_i^{(k)}}$$

for some function $\widetilde{g} \in \mathcal{O}_{\Delta,0}$ and that there exist sections $\omega_1, \dots, \omega_\mu$ with $\widetilde{g}(0) \neq 0$.

Finally, $\check{h}|_\Delta = (-z)^\mu$ because $\mathcal{U}$ is nilpotent on T_0M. Therefore there exist sections $\omega_1, \dots, \omega_\mu$ with $g(0) \neq 0$ if and only if $2\sum\alpha_i^{(k)} = \mu(n-1-2k)$. □

10.3 Gauß–Manin connection

Let $f,\ F,\ \varphi : \mathcal{X} \to \Delta \times M,\ \check{\mathcal{D}}$, and H^n be as in sections 10.1 and 10.2 with $n \geq 1$. Holomorphic differential forms yield holomorphic sections in H^n. The investigation of these sections and their relation to the flat connection on H^n is summarized in the notion of the Gauß–Manin connection for φ.

The Gauß–Manin connection for f was introduced by Brieskorn [**Bri2**]. It has been generalized to complete intersections (as φ) by Greuel [**Gre**] and K. Saito [**SK2**]. The Gauß–Manin connection for singularities has been studied and applied further by many people. [**Mal1**], [**Ph1**], [**Va1**], [**SK6**][**SK9**], [**Lo2**], [**Od2**], [**SM3**], [**AGV2**], [**He1**], and [**Ku**] are some references with the character of a partial survey.

The languages are very different, from an explicit use of integrals to a sophisticated use of $\mathcal{D}$-modules and $\mathcal{E}$-modules. Many of the results which we have to cite have been proved several times and in different styles. We will try to present them in the most explicit way and point to more technical ways in remarks.

Remark 10.4 Usually the Gauß–Manin connection for the semiuniversal unfolding of f is developed for the fibration $F^{-1}(0) \to M$. One can choose $F = t_1 + F_1(x, t_2, \dots, t_\mu)$. Then t_1 takes the role of z.

In fact, the cohomology bundle H^n on $\Delta \times M - \check{\mathcal{D}}$ with its flat structure is the trivial extension along $\partial_z + e$ of its restriction to $\{0\} \times M - \mathcal{D}$.

We take the uneconomic version with base $\Delta \times M$ because then one can treat all coordinates in M equally without an uncanonical choice of t_1. Also it fits better to the second structure connections of Frobenius manifolds (cf. Remark 9.5 (d)).

Let $\eta \in \Omega^n_{\mathcal{X},0}$ and $\omega \in \Omega^{n+1}_{\mathcal{X},0}$. By diminishing $\varepsilon, \delta, \theta$ from section 10.1, one can arrange that η and ω are defined on $\mathcal{X}$. The restriction of η to a smooth fibre $\varphi^{-1}(z,t)$ of $\varphi : \mathcal{X} \to \Delta \times M$ is closed and can be integrated over cycles in the homology. One obtains a holomorphic section in H^n. The germ in $(i_*\mathcal{H})_0$ is independent of any choice of $\varepsilon, \delta, \theta$. So there is a welldefined map

$$\Omega^n_{\mathcal{X},0} \to (i_*\mathcal{H})_0. \tag{10.15}$$

The form ω induces a relative n-form, the Gelfand–Leray form, which gives on each smooth fibre $\varphi^{-1}(z,t)$ a holomorphic n-form $\frac{\omega}{\mathrm{d}F}|\varphi^{-1}(z,t)$. It is the Poincaré residue of the form $\frac{\omega}{F-z}|B^{n+1}_\varepsilon \times \{t\}$. Again one obtains a holomorphic section in H^n. Again this gives a welldefined map

$$\Omega^{n+1}_{\mathcal{X},0} \to (i_*\mathcal{H})_0. \tag{10.16}$$

This also all works for the restriction to $\Delta \times \{0\}$, that is, for the Milnor fibration of f instead of φ. The sheaf of holomorphic sections of the restriction of H^n to $\Delta^* \times \{0\}$ is called $\mathcal{B}$ as in section 10.2. Let $i_0 : \Delta^* \to \Delta$ be the inclusion. Then analogously to (10.15) and (10.16) one has welldefined maps

$$\Omega^n_{\mathbb{C}^{n+1},0} \to (i_{0*}\mathcal{B})_0, \tag{10.17}$$

$$\Omega^{n+1}_{\mathbb{C}^{n+1},0} \to (i_{0*}\mathcal{B})_0, \tag{10.18}$$

and the images are called H_0' and H_0''. They are the restrictions of the images of (10.15) and (10.16) to $(\Delta \times \{0\}, 0)$.

An elementary and classical, but crucial task is the analysis of (10.17) and (10.18) for the case of the A_1-singularity $f = x_0^2 + \ldots + x_n^2$. Let $\delta(z)$, $z \in \Delta^*$ denote a (1 or 2)-valued family of representatives of the vanishing cycles in the fibres $f^{-1}(z)$. Then one finds (e.g. [**AGV2**, p. 294])

$$\int_{\delta(z)} x_0 \mathrm{d}x_1 \ldots \mathrm{d}x_n = c_1 \cdot z^{\frac{n+1}{2}}, \quad \int_{\delta(z)} \frac{\mathrm{d}x_0 \ldots \mathrm{d}x_n}{\mathrm{d}f}\bigg|_{f^{-1}(z)} = c_2 \cdot z^{\frac{n-1}{2}} \tag{10.19}$$

for some $c_1, c_2 \in \mathbb{C}^*$. Therefore the images in (10.17) and (10.18) are $V^{\frac{n+1}{2}}$ and $V^{\frac{n-1}{2}}$ where V^α is defined as in section 7.1.

If f is any singularity with F and φ as above then the fibres $\varphi^{-1}(z,t)$ for $(z,t) \in \check{\mathcal{D}}_{reg}$ have only A_1-singularities. The discussion of the A_1-singularity and the definition of $\mathcal{H}^{(k)}$ in Lemma 10.2 imply that the images of the maps (10.15) and (10.16) are contained in $\mathcal{H}_0^{(-1)}$ and $\mathcal{H}_0^{(0)}$ and that $H_0' \subset \mathcal{B}_0^{(-1)}$ and $H_0'' \subset \mathcal{B}_0^{(0)}$. We will not prove the following theorem, but comment upon the difficult and simple parts of it and several ways to prove it.

Theorem 10.5 *(i) There are equalities $H_0'' = \mathcal{B}_0^{(0)}$ and $H_0' = \mathcal{B}_0^{(-1)}$. The spectral numbers $(\alpha_1^{(0)}, \dots, \alpha_\mu^{(0)}) =: (\alpha_1, \dots, \alpha_\mu) =: Sp(f)$ of H_0'' are contained in $(-1, n) \cap \mathbb{Q}$ and satisfy $\alpha_i + \alpha_{\mu+1-i} = n-1$, the spectral numbers of H_0' are $(\alpha_1^{(-1)}, \dots, \alpha_\mu^{(-1)}) = Sp(f) + 1$.*

(ii) The spaces $\mathcal{H}_0^{(0)}$ and $\mathcal{H}_0^{(-1)}$ are the images of the maps (10.15) and (10.16). They are free $\mathcal{O}_{\Delta\times M,0}$-modules. The sheaves $\mathcal{H}^{(0)}$ and $\mathcal{H}^{(-1)}$ are free $\mathcal{O}_{\Delta\times M}$-modules.

(iii) The inclusion $\mathcal{H}_0^{(-1)} \subset \mathcal{H}^{(0)}$ is represented by $[\eta] \mapsto [\mathrm{d}F \wedge \eta]$ for $\eta \in \Omega^n_{\mathcal{X},0}$.

(iv) The covariant derivatives of sections in $\mathcal{H}^{(-1)}$ and $\mathcal{H}^{(0)}$ can be described in terms of n-forms η and $n+1$-forms ω by

$$\nabla_{\partial_z}[\mathrm{d}F \wedge \eta] = [\mathrm{d}\eta], \tag{10.20}$$

$$\nabla_{\partial/\partial t_i}[\mathrm{d}F \wedge \eta] = [\mathrm{d}F \wedge \mathrm{Lie}_{\partial/\partial t_i}\eta] - \left[\frac{\partial F}{\partial t_i}\mathrm{d}\eta\right], \tag{10.21}$$

$$\nabla_{\partial/\partial t_i}[\omega] = [\mathrm{Lie}_{\partial/\partial t_i}\omega] - \nabla_{\partial_z}\left[\frac{\partial F}{\partial t_i}\omega\right]. \tag{10.22}$$

(v) The operator ∇_{∂_z} yields isomorphisms $\nabla_{\partial_z} : H_0' \to H_0''$ and $\nabla_{\partial_z} : \mathcal{H}_0^{(-1)} \to \mathcal{H}_0^{(0)}$. The space H_0'' is a free $\mathbb{C}\{\{\partial_z^{-1}\}\}$-module.

Remarks 10.6 (a) The $\mathbb{C}\{z\}$-lattices H_0'' and H_0' were considered first by Brieskorn [**Bri2**]. The lattice H_0'' is called the *Brieskorn lattice*. Its spectral numbers $(\alpha_1, \dots, \alpha_\mu)$ form the *spectrum* $Sp(f)$ of the singularity f [**AGV2**].

(b) On the one hand, H_0'' is defined via $n+1$-forms. On the other hand, it is determined by the discriminant $\check{\mathcal{D}} \subset \Delta \times M$ and the monodromy group of the cohomology bundle, because of $H_0'' = \mathcal{B}_0^{(0)}$ and the definition of $\mathcal{B}_0^{(0)}$ and $\mathcal{H}^{(0)}$. The discriminant determines the singularity up to right equivalence (cf. Corollary 4.6, Remarks 5.5 (iv) and (v), [**Sche2**], and [**Wir**]). One may ask how much information H_0'' contains. This leads to Torelli type questions (cf. [**He1**]–[**He4**] and section 12.2).

(c) Parts (iii) and (iv) are not deep. Part (iii) follows from the definition of the Gelfand–Leray form, (iv) can be proved with the residue theorem of Leray (cf. for example [**Bri2**]).

(d) Formula (10.20) shows that $\nabla_{\partial_z} : H_0' \to H_0''$ and $\nabla_{\partial_z} : \mathcal{H}_0^{(-1)} \to \mathcal{H}_0^{(0)}$ are surjective. The injectivity follows from the algebraic descriptions in (10.23)–(10.26) below.

The injectivity of $\nabla_{\partial_z} : H_0' \to H_0''$ also follows from $H_0' \subset V^{>0}$, which is part of (i) and which was proved by Malgrange [**Mal1**]. Together with Lemma 7.4 this shows that H_0'' is a $\mathbb{C}\{\{\partial_z^{-1}\}\}$-lattice.

The injectivity of $\nabla_{\partial_z} : \mathcal{H}_0^{(-1)} \to \mathcal{H}_0^{(0)}$ also follows from the fact that $H^n|_{\Delta\times\{t\}}$ for t generic does not have global flat sections, see Remark 9.5 (b). We will come back to it in Theorem 10.7.

(e) The relation $(\alpha_1^{(-1)}, \ldots, \alpha_\mu^{(-1)}) = Sp(f) + 1$ in (i) follows simply from the isomorphism $\nabla_{\partial_z} : H_0' \to H_0''$. But the symmetry $\alpha_i + \alpha_{\mu+1-i} = n - 1$ and also $-1 < \alpha_1 \leq \ldots \leq \alpha_\mu < n$ are profound. There are two ways to prove them. One way uses Varchenko's mixed Hodge structure, which comes from H_0'', see Remark 10.31. The other way uses K. Saito's higher residue pairings, see Theorem 10.28 (v).

(f) Part (i) gives the identity $\sum \alpha_i = \mu \frac{n-1}{2}$. In view of Theorem 10.3, it implies (ii) and is equivalent to the freeness of $\mathcal{H}_0^{(0)}$ as an $\mathcal{O}_{\Delta\times M,0}$-module. Varchenko [**Va1**][**AGV2**] gave a proof of the freeness of $\mathcal{H}_0^{(0)}$ and thus also of this identity, which is simpler than either of the two ways to prove (i).

First, one has to somehow see it for the special singularities $x_0^N + \cdots + x_n^N$. Then he shows that any singularity turns up in a semiuniversal unfolding of such a special singularity. Using determinants as in (10.14) he proves that the freeness of $\mathcal{H}_0^{(0)}$ for $x_0^N + \cdots + x_n^N$ implies the freeness of $\mathcal{H}_0^{(0)}$ for the other singularity.

(g) Greuel [**Gre**] gave the first proof that the images of the maps (10.15) and (10.16) are free $\mathcal{O}_{\Delta\times M,0}$-modules. In view of the discussion before Theorem 10.5 this implies that the images are $\mathcal{H}_0^{(-1)}$ and $\mathcal{H}_0^{(0)}$. He also determined the kernels (see also [**SK6**] [**SK9**] [**SM3**]).

We set $S := \Delta \times M,\ S' := \Delta \times M - \check{\mathcal{D}},\ \mathcal{X}' := \varphi^{-1}(S')$. Then

$$\mathcal{H}_0^{(0)} \cong \Omega^{n+1}_{\mathcal{X}/M,0}/\mathrm{d}F \wedge \mathrm{d}\Omega^{n-1}_{\mathcal{X}/M,0}, \tag{10.23}$$

$$\mathcal{H}_0^{(-1)} \cong \Omega^{n}_{\mathcal{X}/M,0}/\mathrm{d}F \wedge \Omega^{n-1}_{\mathcal{X}/M,0} + \mathrm{d}\Omega^{n-1}_{\mathcal{X}/M,0} \tag{10.24}$$

$$\cong \Omega^{n}_{\mathcal{X}/S,0}/\mathrm{d}\Omega^{n-1}_{\mathcal{X}/S,0},$$

$$H_0'' \cong \Omega^{n+1}_{\mathbb{C}^{n+1},0}/\mathrm{d}f \wedge \mathrm{d}\Omega^{n-1}_{\mathbb{C}^{n+1},0}, \tag{10.25}$$

$$H_0' \cong \Omega^{n}_{\mathbb{C}^{n+1},0}/\mathrm{d}f \wedge \Omega^{n-1}_{\mathbb{C}^{n+1},0} + \mathrm{d}\Omega^{n-1}_{\mathbb{C}^{n+1},0}. \tag{10.26}$$

The morphism $\varphi : \mathcal{X} \to \Delta \times M$ is Stein. This allows cohomology classes to be represented in smooth fibres $\varphi^{-1}(z, t)$ by fibrewise global holomorphic n-forms.

Greuel and K. Saito obtained

$$\mathcal{H}^{(0)} \cong \varphi_*\Omega^{n+1}_{\mathcal{X}/M}/\mathrm{d}F \wedge \mathrm{d}\varphi_*\Omega^{n-1}_{\mathcal{X}/M}, \tag{10.27}$$

$$\begin{aligned}\mathcal{H}^{(-1)} &\cong \varphi_*\Omega^{n}_{\mathcal{X}/M}/\mathrm{d}F \wedge \varphi_*\Omega^{n-1}_{\mathcal{X}/M} + \mathrm{d}\varphi_*\Omega^{n-1}_{\mathcal{X}/M} \\ &\cong \varphi_*\Omega^{n}_{\mathcal{X}/S}/\mathrm{d}\varphi_*\Omega^{n-1}_{\mathcal{X}/S}.\end{aligned} \tag{10.28}$$

Coherence and freeness of these extensions of the relative de Rham cohomology for $\varphi : \mathcal{X}' \to S'$ are profound results. But we will not explicitly make use of the precise form of the denominators in (10.23)–(10.26).

The next result, Theorem 10.7, shows which of the ($\mathcal{O}_{\Delta\times M}$-coherent) sheaves $\mathcal{H}^{(k)}$ from Lemma 10.2 are free $\mathcal{O}_{\Delta\times M}$-modules and extends part (v) of Theorem 10.5 to all of the sheaves $\mathcal{H}^{(k)}$. For $k \leq 0$ this can be found essentially in **[SK6][SK9]**, but not for $k > 0$. We give a proof. But for the fact in part (c), that $\mathcal{H}^{(0)}_0$ is a free $\mathcal{O}_{M,0}\{\{\partial_z^{-1}\}\}$-module of rank μ, we have to refer to **[SM3]**.

$$\mathcal{O}_{M,0}\{\{\partial_z^{-1}\}\} := \left\{ \sum_{i=0}^{\infty} a_i(t)\partial_z^{-i} \;\middle|\; \sum_{i=0}^{\infty} a_i(t)\frac{1}{i!}z^i \in \mathcal{O}_{\Delta\times M,0} \right\} \tag{10.29}$$

is the ring of microdifferential operators of order ≤ 0 at $(\mathrm{d}z, 0) \in T^*(\Delta \times M)$ **[Ph1][SM3]**.

As in (10.10), $\pi : \Delta \times M \to M$ is the projection. The sheaves $\pi_*\mathcal{H}^{(k)}$ of fibrewise global sections are $\mathcal{O}_M$-modules, but not coherent $\mathcal{O}_M$-modules, they are too big.

Theorem 10.7 *(a) The covariant derivative ∇_{∂_z} maps $\mathcal{H}^{(k)}$ surjectively to $\mathcal{H}^{(k+1)}$ for any $k \in \mathbb{Z}$. For any $k \in \mathbb{Z}$ it yields an isomorphism*

$$\nabla_{\partial_z} : \pi_*\mathcal{H}^{(k)} \to \pi_*\mathcal{H}^{(k+1)} \tag{10.30}$$

of $\mathcal{O}_M$-modules and an isomorphism

$$\nabla_{\partial_z} : \mathcal{H}^{(k)}_0 \to \mathcal{H}^{(k+1)}_0. \tag{10.31}$$

(b) The sheaf $\mathcal{H}^{(k)}$ is a free $\mathcal{O}_{\Delta\times M}$-module if and only if it is a locally free $\mathcal{O}_{\Delta\times M}$-module and if and only if the germ $\mathcal{H}^{(k)}_0$ is a free $\mathcal{O}_{\Delta\times M,0}$ module.

If $\alpha_i \notin \mathbb{Z}$ for all i then all $\mathcal{H}^{(k)}_0$ are free $\mathcal{O}_{\Delta\times M,0}$-modules. Else, $\mathcal{H}^{(k)}_0$ is a free $\mathcal{O}_{\Delta\times M,0}$-module if and only if $k \leq \min(\alpha_i \mid \alpha_i \in \mathbb{Z})$.

(c) Each germ $\mathcal{H}^{(k)}_0$ is a free $\mathcal{O}_{M,0}\{\{\partial_z^{-1}\}\}$-module of rank μ.

Proof: (a) From the definition of $\mathcal{H}^{(k)}$ it follows that ∇_{∂_z} maps $\mathcal{H}^{(k)}|(\Delta \times M - \check{\mathcal{D}}_{sing})$ surjectively to $\mathcal{H}^{(k+1)}|(\Delta \times M - \check{\mathcal{D}}_{sing})$. Therefore ∇_{∂_z} maps $\mathcal{H}^{(k)}$ to $\mathcal{H}^{(k+1)}$, but the surjectivity near $\check{\mathcal{D}}_{sing}$ still has to be proved.

For (10.30) and (10.31), first consider $H^n|(\Delta \times \{t\} - \check{\mathcal{D}})$ for some generic t. Then F_t has μ A_1-singularities and μ critical values. We claim:

(i) Any global holomorphic section σ in $H^n|(\Delta \times \{t\} - \check{\mathcal{D}})$ is the image under ∇_{∂_z} of a unique holomorphic section in $H^n|(\Delta \times \{t\} - \check{\mathcal{D}})$.

This claim shows that (10.30) and (10.31) are injective. It extends Remark 9.5 (b) that $H^n|(\Delta \times \{t\} - \check{\mathcal{D}})$ has no nonzero global flat sections and follows in the same way.

Proof of the claim: One chooses a system of nonintersecting paths (e.g. also ordered anticlockwise) in $\Delta \times \{t\} - \check{\mathcal{D}}$ from a regular value to the μ points in $\Delta \times \{t\} \cap \check{\mathcal{D}}$. One obtains a (distinguished) basis of vanishing cycles. The monodromy around a point $(z, t) \in \Delta \times \{t\} \cap \check{\mathcal{D}}$ is given by a Picard–Lefschetz transformation.

Therefore the space of preimages of σ under ∇_{∂_z} in a neighbourhood of (z, t) in $\Delta \times \{t\} \cap \check{\mathcal{D}}$ forms a $\mu - 1$ dimensional affine space. The difference of two preimages is contained in the $\mu - 1$ dimensional vector space of flat sections which vanish on the vanishing cycle that corresponds to (z, t). One extends the μ affine spaces of dimension $\mu - 1$ of local preimages of σ under ∇_{∂_z} along the paths to the regular value. There they intersect in one point because the vanishing cycles form a basis. This shows the claim. ◊

Now suppose that $\sigma \in \pi_* \mathcal{H}^{(k+1)}$, that is, σ is a section of $\mathcal{H}^{(k+1)}$ defined in $\pi^{-1}(U)$ for some open subset $U \subset M$. Let $\mathcal{B}_M \subset M$ be the bifurcation diagram, the set of $t \in M$ such that F_t has less than μ different critical values. It contains the caustic $\mathcal{K}$, the set of parameters $t \in M$ such that F_t has less than μ critical points. We claim:

(ii) The unique preimages under ∇_{∂_z} in generic slices $\Delta \times \{t\} - \check{\mathcal{D}}$ as above glue to a holomorphic section in $\pi^{-1}(U - \mathcal{B}_M) - \check{\mathcal{D}}$.
(iii) The holomorphic section is in $\pi_* \mathcal{H}^{(k)}$.
(iv) It extends to $\pi^{-1}(M - \mathcal{K})$.
(v) It extends to $\pi^{-1}(M)$.

(ii)–(iv) are easy to see. But unfortunately, (v) is not so clear a priori. We claim that it follows from

(vi) The map in (10.31) is an isomorphism in the case of an A_2-singularity $f = x_0^3 + x_1^2 + \cdots + x_n^2$.

Proof of (vi) ⇒ (v): It is sufficient to show that the preimage under ∇_{∂_z} of the above section σ extends to slices $\Delta \times \{t\} - \check{\mathcal{D}}$ for generic $t \in \mathcal{K}$. Then it is defined outside of a codimension 2 subset of $\pi^{-1}(U)$ and extends globally. For generic $t^{(0)} \in \mathcal{K}$ the function $F_{t^{(0)}}$ has $\mu - 2$ A_1-singularities and 1 A_2-singularity and $\mu - 1$ critical values. One can extend the proof of claim (i) if one can show that σ has a preimage under ∇_{∂_z} in a neighbourhood in $\Delta \times M$ of the critical value of the A_2-singularity.

In such a neighbourhood one can split the cohomology bundle into a subbundle of rank 2 for the versal unfolding of the A_2-singularity and into a subbundle of rank $\mu - 2$ which is invariant with respect to the local monodromies. The subbundle of rank 2 with its flat connection is induced by the cohomology bundle of a semiuniversal unfolding of the A_2-singularity. Therefore (vi) would give locally a preimage of σ under ∇_{∂_z} and would allow the proof of claim (i) to be applied. ◊

The eigenvalues of the monodromy of an A_2-singularity are $\pm i$ for any number of variables. Therefore (vi) follows from the claim

(vii) The map $\nabla_{\partial_z} : \mathcal{H}_0^{(k)} \to \mathcal{H}_0^{(k+1)}$ is an isomorphism in the case of a singularity with $\alpha_i \notin \mathbb{Z}$ for all i.

Proof of claim (vii): When $k \leq -1$ it follows from Theorem 10.5 (v) and from claim (iii) (in fact, without any condition on the spectral numbers). When $k \geq 0$ one can argue as follows. Let $\sigma_1, \ldots, \sigma_\mu$ be an $\mathcal{O}_{\Delta\times M,0}$-basis of $\mathcal{H}_0^{(0)}$. Their restrictions to $(\Delta \times \{0\}, 0)$ generate the $\mathbb{C}\{z\}$-lattice H_0'' with spectral numbers $\alpha_1, \ldots, \alpha_\mu$. The restrictions to $(\Delta \times \{0\}, 0)$ of the derivatives $\nabla_{\partial_z}^{k+1}\sigma_1, \ldots, \nabla_{\partial_z}^{k+1}\sigma_\mu$ generate a $\mathbb{C}\{z\}$-lattice with spectral numbers $\alpha_1 - k - 1, \ldots, \alpha_\mu - k - 1$. Because of claim (iii) and Theorem 10.3 this $\mathbb{C}\{z\}$-lattice is $\mathcal{B}_0^{(k+1)}$, and $\mathcal{H}_0^{(k+1)}$ is a free $\mathcal{O}_{\Delta\times M,0}$-module generated by $\nabla_{\partial_z}^{k+1}\sigma_1, \ldots, \nabla_{\partial_z}^{k+1}\sigma_\mu$. By induction on k, starting with $k = -1$, one can show that $\nabla_{\partial_z}\mathcal{H}_0^{(k)}$ is an $\mathcal{O}_{\Delta\times M,0}$-module. Then $\nabla_{\partial_z}^{k+1}\mathcal{H}_0^{(0)} = \nabla_{\partial_z}\mathcal{H}_0^{(k)} = \mathcal{H}_0^{(k+1)}$. ◊

(b) The first statement is clear because $\Delta \times M$ is Stein. The second statement has been shown in the proof of claim (vii).

Suppose that there exist $\alpha_i \in \mathbb{Z}$. Because of (a) we have $\mathcal{B}^{(k)} = \nabla_{\partial_z}^k H_0''$ for any $k \in \mathbb{Z}$ (with $\mathcal{B}^{(k)} \subset V^{>0}$ for $k \leq -1$). Looking at the derivatives of principal parts, one sees easily that the spectral numbers $\alpha_1^{(k)}, \ldots, \alpha_\mu^{(k)}$ of $\mathcal{B}^{(k)}$ satisfy

$$\left(\alpha_1^{(k)}, \ldots, \alpha_\mu^{(k)}\right) = Sp(f) - k \quad \text{for } k \le \min(\alpha_i \mid \alpha_i \in \mathbb{Z}), \tag{10.32}$$

$$\sum_i \alpha_i^{(k)} > \mu\left(\frac{n-1}{2} - k\right) \quad \text{for } k > \min(\alpha_i \mid \alpha_i \in \mathbb{Z}). \tag{10.33}$$

One can apply Theorem 10.3. ◇

(c) In the notation of [**SM3**, §2], $\mathcal{H}_0^{(0)}$ is the germ $(F_{-n} \int_\varphi^0 \mathcal{O}_\mathcal{X})_0$ of the filtered Gauß–Manin system for $\varphi : \mathcal{X} \to \Delta \times M$ which is constructed there. By [**SM3**, §2 (5.17) and (5.18)] it is a free $\mathcal{O}_{M,0}\{\{\partial_z^{-1}\}\}$-module of rank μ. Because of the isomorphism (10.31), then all germs $\mathcal{H}_0^{(k)}$ are free $\mathcal{O}_{M,0}\{\{\partial_z^{-1}\}\}$-modules of rank μ. This finishes the proof of Theorem 10.7. □

The following considerations on the quotients $\mathcal{H}^{(k)}/\mathcal{H}^{(k-1)}$ are due to K. Saito [**SK6**][**SK9**] (for $k \le 0$). Because of Theorem 10.5 (iii), (10.27), and (10.28) there is an exact sequence

$$0 \longrightarrow \mathcal{H}^{(-1)} \longrightarrow \mathcal{H}^{(0)} \xrightarrow{r^{(0)}} \varphi_* \Omega^{n+1}_{\mathcal{X}/\Delta\times M} \longrightarrow 0. \tag{10.34}$$

Here $r^{(0)}$ is the projection.

$$\Omega^{n+1}_{\mathcal{X}/\Delta\times M} = \Omega^{n+1}_\mathcal{X} / \mathrm{d}F \wedge \Omega^n_\mathcal{X} + \sum_{i=1}^{\mu} \mathrm{d}t_i \wedge \Omega^n_\mathcal{X}$$

is the sheaf of relative $(n+1)$-forms with respect to $\varphi : \mathcal{X} \to \Delta \times M$. Its support is the critical space $C \subset \mathcal{X}$ of φ. It is a free $\mathcal{O}_C$-module of rank 1. The sheaf

$$\Omega_F := (pr_M)_* \Omega^{n+1}_{\mathcal{X}/\Delta\times M} \tag{10.35}$$

is a free $\mathcal{O}_M$-module of rank μ. Via the Kodaira–Spencer isomorphism $\mathbf{a}$ in (10.3) it is a free $\mathcal{T}_M$-module of rank 1. We write the action as

$$(X, [\omega]) \mapsto \mathbf{a}(X) \cdot [\omega] \quad \text{for } X \in \mathcal{T}_M,\ [\omega] \in \Omega_F. \tag{10.36}$$

Vector fields $X \in \mathcal{T}_M$ will be identified with their canonical lifts to $\Delta \times M$, that is, with the fibrewise global lifts $\widetilde{X}$ which satisfy $\widetilde{X}(z) = 0$.

Corollary 10.8 *For any $k \in \mathbb{Z}$ one has the following.*

(i)

$$\nabla : \mathcal{H}^{(k)} \to \Omega^1_{\Delta\times M} \otimes \mathcal{H}^{(k+1)}. \tag{10.37}$$

(ii) There are exact sequences

$$0 \longrightarrow \pi_* \mathcal{H}^{(k)} \longrightarrow \pi_* \mathcal{H}^{(k+1)} \overset{r^{(k+1)}}{\longrightarrow} \Omega_F \longrightarrow 0,$$

$$0 \longrightarrow \mathcal{H}_0^{(k)} \longrightarrow \mathcal{H}_0^{(k+1)} \overset{r^{(k+1)}}{\longrightarrow} \Omega_{F,0} \longrightarrow 0, \tag{10.38}$$

where $r^{(k+1)} := r^{(0)} \circ \nabla_{\partial_z}^{-k-1}$.

(iii) If $X \in \mathcal{T}_M$ *and* $\sigma \in \pi_* \mathcal{H}^{(k)}$ *then*

$$r^{(k)}\big(\nabla_X \nabla_{\partial_z}^{-1} \sigma\big) = -\mathbf{a}(X) \cdot r^{(k)}(\sigma). \tag{10.39}$$

(iv) Suppose that $v_1 \in \mathcal{H}_0^{(0)}$ *is represented by a volume form* $u(x,t)\mathrm{d}x_0 \ldots \mathrm{d}x_n \in \Omega^{n+1}_{\mathcal{X},0}$ *with* $u(0) \neq 0$. *Then the period map*

$$v : \mathcal{T}_{M,0} \to \mathcal{H}_0^{(0)}, \quad X \mapsto -\nabla_X \nabla_{\partial_z}^{-1} v_1 \tag{10.40}$$

is injective and yields a splitting of the sequence (10.38) for $k = -1$

$$\mathcal{H}_0^{(0)} = v(\mathcal{T}_{M,0}) \oplus \mathcal{H}_0^{(-1)}. \tag{10.41}$$

Proof: (ii) follows from Theorem 10.7 (a) and (10.34), (i) and (iii) follow from the formulas (10.20) and (10.21) for the covariant derivatives of sections in $\mathcal{H}^{(-1)}$. (iv) follows from (10.39) and the fact that a volume form represents a generator of $\Omega_{F,0}$ as $\mathcal{T}_{M,0}$-module. □

Remarks 10.9 (a) By definition of $\mathcal{H}^{(k)}$, the pair $(\mathcal{H}^{(k)}, \nabla)$ has a logarithmic pole along $\check{\mathcal{D}}$, thus besides (10.37) one has

$$\nabla : \mathcal{H}^{(k)} \to \Omega^1_{\Delta\times M}(\log \check{\mathcal{D}}) \otimes \mathcal{H}^{(k)}. \tag{10.42}$$

One can also see this algebraically [**SK6**][**SK9**], using (10.39) and the fact that precisely the vector fields in $\mathrm{Der}_{\Delta\times M}(\log \check{\mathcal{D}})$ are liftable along φ to $\mathcal{X}$ (cf. for example [**Lo2**, (6.14)]).

(b) A primitive form [**SK6**][**SK9**] is a section v_1 in $\mathcal{H}_0^{(0)}$ (or in $\pi_*\mathcal{H}^{(0)}$) which is represented by a volume form and induces a splitting (10.41) with good properties with respect to K. Saito's higher residue pairings (cf. sections 10.4 and 11.1).

Our quite elementary approach to the sheaves $\mathcal{H}^{(k)}$ and their properties should be compared with the constructions in [**Ph1**] [**Od2**] [**SM3**]. The following theorem comprises the relations. A detailed proof would require precise definitions of all the notions and objects, from which we refrain. We give only a rough sketch. We refer to [**Ph1**] [**KK**] [**Bj**] and to the survey articles

[**Od1**] [**No**] [**Kas2**] for the definition of regular holonomic $\mathcal{D}$-modules, characteristic varieties, and good filtrations. We will not subsequently make use of Theorem 10.10.

Theorem 10.10 *The union* $\bigcup_{k\in\mathbb{Z}} \mathcal{H}^{(k)}$ *is a regular holonomic* $\mathcal{D}_{\Delta\times M}$*-module with characteristic variety* $T^*_{\Delta\times M}(\Delta\times M)\cup T^*_{\check{\mathcal{D}}}(\Delta\times M)$ *and good filtration* $\mathcal{H}^{(\bullet)}$*. As a* $\mathcal{D}_{\Delta\times M}$*-module it is isomorphic to the microlocal Gauß–Manin system constructed in* [**Ph1**] [**Od2**] [**SM3**].

Sketch of a proof: The $\mathcal{D}$-module $\bigcup_{k\in\mathbb{Z}} \mathcal{H}^{(k)}$ is a local system on $\Delta\times M-\check{\mathcal{D}}$ and singular along $\check{\mathcal{D}}$. Therefore its characteristic variety contains the zero section $T^*_{\Delta\times M}(\Delta\times M)\subset T^*(\Delta\times M)$ and the conormal bundle $T^*_{\check{\mathcal{D}}}(\Delta\times M)$ of $\check{\mathcal{D}}$, that is, the closure of $T^*_{\check{\mathcal{D}}_{reg}}(\Delta\times M)$.

In order to see that this is all, first one has to check that $\mathcal{H}^{(\bullet)}$ is a good filtration. This follows from Theorem 10.7 (a). Then one has to study the ideal in $\mathcal{O}_{T^*M}$ which is defined by that filtration and which determines the characteristic variety. It has homogeneous (in the fibre variables) generators of degree 2 and degree 1.

The generators of degree 2 come from the defining relations $\frac{\partial}{\partial t_i}\circ\frac{\partial}{\partial t_j}=\sum_k a_{ij}^k\frac{\partial}{\partial t_k}$ for the multiplication in $\mathcal{T}_M$ and from (10.39). The zero set of the ideal of these generators is the zero section and the $\mu+2$ dimensional union of the conormal bundles of all shifts of $\check{\mathcal{D}}$ along ∂_z. The generators of degree one come from the logarithmic vector fields. They cut out the zero section and $T^*_{\check{\mathcal{D}}}(\Delta\times M)$ (see also [**SK9**, p. 1254], [**Od2**, (2.3)]).

Therefore $\bigcup_{k\in\mathbb{Z}} \mathcal{H}^{(k)}$ is a holonomic $\mathcal{D}_{\Delta\times M}$-module with characteristic variety as claimed. It is regular holonomic because the subsheaves $\mathcal{H}^{(k)}$ have a logarithmic pole along $\check{\mathcal{D}}$.

The microlocal Gauß–Manin system [**Ph1**][**Od2**][**SM3**] is the unique extension of $\mathcal{H}^{(0)}$ to a $\mathcal{D}_{\Delta\times M}$ module $\bigcup_{k\geq 0}\partial_z^k\mathcal{H}^{(0)}$ such that ∂_z is invertible on the germ at 0. Because of Theorem 10.7 (a), the union $\bigcup_{k\geq 0}\mathcal{H}^{(k)}$ is isomorphic to this $\mathcal{D}_{\Delta\times M}$-module. □

10.4 Higher residue pairings

Let $f,\ F,\ \varphi:\mathcal{X}\to\Delta\times M,\ H^n,\ \nabla$, and $\mathcal{H}^{(k)}$ be as in sections 10.1 and 10.2 with $n\geq 1$. In this section we formulate the properties of K. Saito's higher residue pairings on $\pi_*\mathcal{H}^{(0)}$. For the algebraic construction and proof we refer to [**SK7**], for sketches to [**SK6**] [**Nam**] [**SK9**]. Constructions in the framework of microlocal differential systems can be found in [**Od2**] [**SM3**]. There it is emphasized that Kashiwara observed that the higher residue pairings can be identified with a microlocal Poincaré duality.

We will offer two completely different descriptions of the higher residue pairings. But both are valid only under some restrictions. The first in Theorem 10.13 holds only for singularities with nondegenerate intersection form, the second in section 10.6, Theorem 10.28, holds only for the restriction to $(\Delta \times \{0\}, 0)$. Nevertheless both are instructive and explain the properties of the higher residue pairings.

For the family of functions F_t, $t \in M$, and the fixed coordinates $x_0, \ldots, x_n$ on the fibres of the projection $pr_M : \mathcal{X} \to M$, $(x, t) \mapsto t$, the Grothendieck residue associates to a relative $n+1$-form $\omega \in (pr_M)_* \Omega^{n+1}_{\mathcal{X}/M}$ a function

$$\operatorname{Res}_{\mathcal{X}/M} \left[\frac{\omega}{\frac{\partial F}{\partial x_0}, \ldots, \frac{\partial F}{\partial x_n}} \right] \in \mathcal{O}_M. \tag{10.43}$$

It can be defined algebraically or analytically [**GrH**]. The value at $t \in M$ is $1/(2\pi(i)^{n+1}$ times the integral of the form $\omega / \frac{\partial F}{\partial x_0} \cdot \ldots \cdot \frac{\partial F}{\partial x_n}$ over the cycle $\{(x,t) \mid |\frac{\partial F}{\partial x_i}| = \gamma\} \subset B^{n+1}_\varepsilon \times \{t\}$ for a sufficiently small $\gamma > 0$. It induces a welldefined pairing

$$J_F : \Omega_F \times \Omega_F \to \mathcal{O}_M,$$

$$([g_1 \mathrm{d}x_0 \ldots \mathrm{d}x_n], [g_2 \mathrm{d}x_0 \ldots \mathrm{d}x_n]) \mapsto \operatorname{Res}_{\mathcal{X}/M} \left[\frac{g_1 g_2 \mathrm{d}x_0 \ldots \mathrm{d}x_n}{\frac{\partial F}{\partial x_0}, \ldots, \frac{\partial F}{\partial x_n}} \right] \tag{10.44}$$

on $\Omega_F = (pr_M)_* \Omega^{n+1}_{\mathcal{X}/\Delta \times M}$ (cf. (10.35)). K. Saito observed that J_F is independent of the coordinates $x_0, \ldots, x_n$. By Grothendieck it is nondegenerate.

The sheaf Ω_F is not only a free $\mathcal{O}_M$-module of rank μ, but also a free $(pr_{C,M})_* \mathcal{O}_C$-module of rank 1. By definition, J_F satisfies

$$J_F(g[\omega_1], [\omega_2]) = J_F([\omega_1], g[\omega_2]) \tag{10.45}$$

for $g \in (pr_{C,M})_* \mathcal{O}_C$, $[\omega_1], [\omega_2] \in \Omega_F$.

So, J_F is a canonical nondegenerate pairing on $\Omega_F \cong \pi_* \mathcal{H}^{(0)} / \pi_* \mathcal{H}^{(-1)}$ (cf. (10.38)). K. Saito extended it to a series of pairings on $\pi_* \mathcal{H}^{(0)}$, his higher residue pairings $K_F^{(-k)}$.

Theorem 10.11 (**[SK7][SK6][SK9]**) *There exists an $\mathcal{O}_M$-bilinear pairing*

$$K_F : \pi_* \mathcal{H}^{(0)} \times \pi_* \mathcal{H}^{(0)} \to \mathcal{O}_M[[\partial_z^{-1}]] \cdot \partial_z^{-n-1} \tag{10.46}$$

$$K_F(\omega_1, \omega_2) = \sum_{k \geq 0} K_F^{(-k)}(\omega_1, \omega_2) \cdot \partial_z^{-n-1-k}$$

with $K_F^{(-k)}(\omega_1, \omega_2) \in \mathcal{O}_M$ and the following properties.

(i) *The pairing* $K_F^{(-k)}$ *is* $(-1)^k$*-symmetric.*

(ii) *For* $\omega_1, \omega_2 \in \pi_*\mathcal{H}^{(0)}$

$$K_F(\omega_1, \omega_2) \cdot \partial_z^{-1} = K_F\big(\nabla_{\partial_z}^{-1}\omega_1, \omega_2\big) = K_F\big(\omega_1, -\nabla_{\partial_z}^{-1}\omega_2\big).$$

(iii) *For* $a, b \le 0,\ a + b < -k$

$$K_F^{(-k)}\big(\pi_*\mathcal{H}^{(a)}, \pi_*\mathcal{H}^{(b)}\big) = 0$$

and for $\omega_1, \omega_2 \in \pi_*\mathcal{H}^{(0)}$

$$K_F^{(0)}(\omega_1, \omega_2) = J_F([\omega_1], [\omega_2]).$$

(iv) *For* $\omega_1, \omega_2 \in \pi_*\mathcal{H}^{(0)}$

$$K_F(z \cdot \omega_1, \omega_2) - K_F(\omega_1, z \cdot \omega_2) = [z, K_F(\omega_1, \omega_2)],$$

where $[z, \partial_z^{-k}] := k\, \partial_z^{-k-1}$.

(v) *For* $\omega_1, \omega_2 \in \pi_*\mathcal{H}^{(-1)}$, $X \in \mathcal{T}_M$ *(*$X \in \mathcal{T}_M$ *is lifted to* $\Delta \times M$ *such that* $X(z) = 0$*)*

$$X\, K_F(\omega_1, \omega_2) = K_F(\nabla_X\omega_1, \omega_2) + K_F(\omega_1, \nabla_X\omega_2).$$

Remarks 10.12 (a) In view of Theorem 10.7 (a) and the properties of K_F, one can extend K_F to a pairing on $\bigcup_{k\in\mathbb{Z}} \pi_*\mathcal{H}^{(k)}$ with values in $\mathcal{O}_M[[\partial_z^{-1}]][\partial_z]$ and similar properties.

(b) This extension is studied in [**SM3**, 2.7]. From the microlocal properties of the Gauß–Manin system one obtains

$$K_F(\omega_1, \omega_2) \in \mathcal{O}_{M,0}\big\{\{\partial_z^{-1}\}\big\} \cdot \partial_z^{-n-1} \quad \text{for } \omega_1, \omega_2 \in \mathcal{H}_0^{(0)} \cong \big(\pi_*\mathcal{H}^{(0)}\big)_0$$

(c) It is also proved in [**SM3**, 2.7] that K_F is uniquely determined by its properties. This follows essentially from the fact that the germ $(\int_\varphi^0 \mathcal{O}_\mathcal{X})_0$ of the Gauß–Manin system is simple holonomic as a microdifferential system, so that the only endomorphisms of it are multiplications by scalars ([**SM3**, 2.7], [**Kas2**]).

The intersection form I on $H_n(\varphi^{-1}(z,t), \mathbb{Z})$ for a regular fibre $\varphi^{-1}(z,t)$ induces a map $H_n(\varphi^{-1}(z,t), \mathbb{Z}) \to H^n(\varphi^{-1}(z,t), \mathbb{Z})$, $\delta \mapsto I(\delta, .)$ [or put the other way, the intersection form is induced by the canonical map

$$H_n(\varphi^{-1}(z,t), \mathbb{Z}) \to H_n(\varphi^{-1}(z,t), \partial\varphi^{-1}(z,t), \mathbb{Z}) \cong H^n(\varphi^{-1}(z,t), \mathbb{Z})\ ].$$

The intersection form I is flat on the homology bundle

$$H_n = \bigcup_{(z,t)\in\Delta\times M - \check{\mathcal{D}}} H_n(\varphi^{-1}(z,t), \mathbb{C})$$

and induces a homomorphism $H_n \to H^n$ of flat bundles. If the intersection form is nondegenerate then this map is an isomorphism and induces a flat bilinear form I^* on the cohomology bundle H^n.

The residue pairing J_F was defined by residues in the fibres of $pr_M : \mathcal{X} \to M$. We can present the pairings $K_F^{(-k)}$ by residues in the fibres of $\pi : \Delta \times M \to M$ in the case of a nondegenerate intersection form. The following result is informative, but we will not subsequently use it. The residue at (z_0, t) of a meromorphic function $g_t(z)$ in $\Delta \times \{t\}$ is by definition $res_{(z_0,t)}(g_t) = \frac{1}{2\pi i}\int_{|z-z_0|=\gamma} g_t(z)dz$ for some small $\gamma > 0$.

Theorem 10.13 *Suppose that the intersection form I of the singularity f is nondegenerate. Let I^* be the induced bilinear form on the cohomology bundle H^n. Then for $\omega_1, \omega_2 \in \pi_*\mathcal{H}^{(0)}$, fixed $t \in M$, and any $k \geq 0$*

$$K_F^{(-k)}(\omega_1, \omega_2)(t) = (-1)^{\frac{n(n-1)}{2}} \frac{1}{(2\pi i)^n} \sum_{(z,t)\in\check{\mathcal{D}}} res_{(z,t)}\, I^*\big(\nabla_{\partial_z}^{n+k}\omega_1, \omega_2\big). \quad (10.47)$$

Sketch of two proofs: I can offer two quite indirect proofs, but not a direct proof which would relate (10.47) to K. Saito's definition of K_F.

The first proof uses a result of Varchenko and the result of M. Saito in Remark 10.12 (c). By [**Va6**, §3.2, Theorem], the right hand side of (10.47) is equal to $J_F([\omega_1], [\omega_2])$ for $k = 0$. One can and has to check that the series of bilinear forms defined by the right hand side of (10.47) satisfies also the other properties in Theorem 10.11. Property (i) holds because I is $(-1)^n$-symmetric. The properties (ii), (iv), and (v) are simple. The first line of (iii) follows from the definition of $\mathcal{H}^{(k)}$ in Lemma 10.2: for generic t (and thus for all t) the function $I^*(\nabla_{\partial_z}^{n+k}\omega_1, \omega_2)$ is holomorphic on $\Delta \times \{t\}$ if $\omega_1 \in \mathcal{H}^{(a)}$, $\omega_2 \in \mathcal{H}^{(b)}$ with $a + b < -k$ (cf. [**Va6**, §3.2]). Because of Remark 10.12 (c), K_F must now coincide with the series of bilinear forms defined by the right hand side of (10.47).

The second proof will follow from the construction of a Frobenius manifold in section 11.1, from the induced isomorphism between (H^n, ∇, I^*) and $(\pi^*TM, \check{\nabla}^{(-\frac{n}{2})}, I^{(-\frac{n}{2})}$ up to a scalar) (cf. chapter 9 for the notations), and from Lemma 9.11. In order to determine the scalar between $I^{(-\frac{n}{2})}$ and I^* one again needs [**Va6**, §3.2, Theorem] □

Remarks 10.14 (a) Formula (10.47) explains the properties of the pairing K_F, as one could see in the sketch of the first proof.

(b) One can obtain an analogous formula for the general case of a singularity with any intersection form if one embeds the Milnor fibration into a fibration with projective fibres as in [**Va6**, §3.3]. But then one has to check that things are welldefined and independent of choices. We refrain from carrying it out here. Theorem 10.13 is informative, but we will not subsequently use it in any case.

10.5 Polarized mixed Hodge structures and opposite filtrations

There are several possible definitions of a *polarized* mixed Hodge structure. The definition presented in [**CaK**] [**He4**] is motivated by Schmid's limit mixed Hodge structure. It is the correct one in the case of isolated hypersurface singularities. Steenbrink's mixed Hodge structure [**Stn**] is polarized in this sense. One can recover it from the Brieskorn lattice. Varchenko [**Va2**] first saw that the Brieskorn lattice induces mixed Hodge structures. In section 10.6 we will regard this as a feature of the Brieskorn lattice.

Here we want to resume the definition of polarized mixed Hodge structures and the construction of classifying spaces for them from [**He4**, ch. 2]. Opposite filtrations and classifying spaces for them will also be discussed. The choice of an opposite filtration is necessary for the construction of a Frobenius manifold in section 11.1.

The weight filtration of a polarized mixed Hodge structure in the sense of Definition 10.16 comes from a nilpotent endomorphism. Its properties are given in the following lemma from [**Schm**, Lemma 6.4], (cf. also [**Gri**, pp. 255–256]).

Lemma 10.15 *Let $m \in \mathbb{N}$, $H_\mathbb{Q}$ a finite dimensional $\mathbb{Q}$-vector space, S a nondegenerate bilinear form on $H_\mathbb{Q}$, $S : H_\mathbb{Q} \times H_\mathbb{Q} \to \mathbb{Q}$, which is $(-1)^m$-symmetric, and $N : H_\mathbb{Q} \to H_\mathbb{Q}$ a nilpotent endomorphism with $N^{m+1} = 0$, which is an infinitesimal isometry, i.e. $S(Na, b) + S(a, Nb) = 0$ for $a, b \in H_\mathbb{Q}$.*

(a) There exists a unique increasing filtration $0 = W_{-1} \subset W_0 \subset \ldots \subset W_{2m} = H_\mathbb{Q}$ such that $N(W_l) \subset W_{l-2}$ and such that $N^l : \mathrm{Gr}^W_{m+l} \to \mathrm{Gr}^W_{m-l}$ is an isomorphism.

(b) If $l + l' < 2m$ then $S(W_l, W_{l'}) = 0$.

(c) A nondegenerate $(-1)^{m+l}$-symmetric bilinear form S_l is welldefined on Gr^W_{m+l} for $l \geq 0$ by the requirement: $S_l(a, b) = S(\widetilde{a}, N^l\widetilde{b})$ if $\widetilde{a}, \widetilde{b} \in W_{m+l}$ represent $a, b \in \mathrm{Gr}^W_{m+l}$.

(d) The primitive subspace $P_{m+l}(H_\mathbb{Q})$ of Gr^W_{m+l} is defined by

$$P_{m+l} = \ker\left(N^{l+1} : \mathrm{Gr}^W_{m+l} \to \mathrm{Gr}^W_{m-l-2}\right) \tag{10.48}$$

if $l \geq 0$ and $P_{m+l} = 0$ if $l < 0$. Then

$$\mathrm{Gr}^W_{m+l} = \bigoplus_{i \geq 0} N^i P_{m+l+2i}, \tag{10.49}$$

and this decomposition is orthogonal with respect to S_l if $l \geq 0$.

Definition 10.16 [CaK][He4] A *polarized mixed Hodge structure* (abbreviation: PMHS) of weight m is given by the following data: a lattice $H_\mathbb{Z}$ with $H_\mathbb{Z} \subset H_\mathbb{Q} \subset H_\mathbb{R} \subset H_\mathbb{C} = H_\mathbb{Z} \otimes \mathbb{C}$, a bilinear form S on $H_\mathbb{Q}$ and an endomorphism N of $H_\mathbb{Q}$ such that $m,\ H_\mathbb{Q},\ S,\ N,\ W_\bullet,\ S_l,\ P_{m+l}$ satisfy all properties in Lemma 10.15, and a decreasing Hodge filtration $F^\bullet$ on $H_\mathbb{C}$ with the properties:

(i) The induced filtration $F^\bullet \mathrm{Gr}_k^W$ gives a pure Hodge structure of weight k on Gr_k^W, i.e. $\mathrm{Gr}_k^W = F^p\mathrm{Gr}_k^W \oplus \overline{F^{k+1-p}\mathrm{Gr}_k^W}$,

(ii) The endomorphism N satisfies $N(F^p) \subset F^{p-1}$, i.e. N is a $(-1,-1)$-morphism of mixed Hodge structures,

(iii) The Hodge filtration and the pairing satisfy $S(F^p, F^{m+1-p}) = 0$,

(iv) the pure Hodge structure $F^\bullet P_{m+l}$ of weight $m+l$ on P_{m+l} is polarized by S_l, i.e. (α) $S_l(F^p P_{m+l}, F^{m+l+1-p} P_{m+l}) = 0$, and ($\beta$) $i^{2p-m-l} S_l(u, \bar{u}) > 0$ if $u \in F^p P_{m+l} \cap \overline{F^{m+l-p} P_{m+l}},\ u \neq 0$.

Remarks 10.17 (a) The primitive subspace P_{m+l} carries a pure Hodge structure, because it is the kernel of the morphism $N^{l+1} : \mathrm{Gr}_{m+l}^W \to \mathrm{Gr}_{m-l-2}^W$ of pure Hodge structures. The strictness of the $(-1,-1)$-morphism N also implies

$$F^p\mathrm{Gr}_k^W = \bigoplus_{j\geq 0} F^p N^j P_{k+2j} \quad \text{and} \quad F^p N^j P_{k+2j} = N^j F^{p+j} P_{k+2j}.$$

(b) In Lemma 10.15 the number m could be replaced by some bigger number, but in Definition 10.16 the weight m is essential for condition (iii). Also the assumption that S is $(-1)^m$-symmetric is not important in Lemma 10.15, but essential for (iv) (β).

(c) Condition (iii) implies (iv) (α), but in general it is not equivalent to (iv) (α). One can easily see the following. Under the assumption of all conditions except for (iii) and (iv), condition (iv) (α) for all p and l is equivalent to

$$S(F^p \cap W_{m+l}, F^{m+1-p} \cap W_{m-l}) = 0 \quad \text{for all } p \text{ and } l.$$

(d) The form S is called a *polarizing form*.

Deligne [**De2**, Lemma 1.2.8] defined subspaces $I^{p,q}$ for a mixed Hodge structure, which give a simultaneous splitting of the Hodge filtration and the weight filtration. They also behave nicely with respect to a polarizing form ([**SM3**, Lemma 2.8], [**He4**, Lemma 2.3]).

Lemma 10.18 *For a PMHS as in Definition 10.16 let*

$$I^{p,q} := (F^p \cap W_{p+q}) \cap \left(\overline{F^q} \cap W_{p+q} + \sum_{j>0} \overline{F^{q-j}} \cap W_{p+q-j-1} \right). \tag{10.50}$$

Then

$$\bigoplus_q I^{p,q} = F^p \cap \left(\sum_q \overline{F}^q \cap W_{p+q} \right), \tag{10.51}$$

$$F^p = \bigoplus_{i,q:i\geq p} I^{i,q}, \tag{10.52}$$

$$W_l = \bigoplus_{p+q\leq l} I^{p,q}, \tag{10.53}$$

$$N(I^{p,q}) \subset I^{p-1,q-1}, \tag{10.54}$$

$$S(I^{p,q}, I^{r,s}) = 0 \quad \textit{for } (r,s) \neq (m-p, m-q). \tag{10.55}$$

For $p+q \geq m$ let $I_0^{p,q}$ be the primitive subspace of $I^{p,q}$,

$$I_0^{p,q} = \ker(N^{p+q-m+1} : I^{p,q} \to I^{m-q-1,m-p-1}). \tag{10.56}$$

Then

$$I^{p,q} = \bigoplus\nolimits_{j\geq 0} N^j I_0^{p+j,q+j}, \tag{10.57}$$

$$S\big(N^i I_0^{p,q}, N^j I_0^{r,s}\big) = 0 \quad \textit{for } (r,s,i+j) \neq (q,p,p+q-m). \tag{10.58}$$

Definition 10.19 An opposite filtration for a PMHS as in Definition 10.16 is an increasing filtration $U_\bullet$ on $H_\mathbb{C}$ with

$$H_\mathbb{C} = \bigoplus_p F^p \cap U_p, \tag{10.59}$$

$$N(U_p) \subset U_{p-1}, \tag{10.60}$$

$$S(U_p, U_{m-1-p}) = 0. \tag{10.61}$$

Remarks 10.20 (a) The decomposition (10.59) is equivalent to (10.62) and also to (10.63), see Definition 7.15,

$$F^p = \bigoplus\nolimits_{q\geq p} F^q \cap U_q \quad \text{and} \quad U_p = \bigoplus\nolimits_{q\leq p} F^q \cap U_q, \tag{10.62}$$

$$\mathrm{Gr}_F^p \mathrm{Gr}_q^U = 0 \quad \text{for } p \neq q. \tag{10.63}$$

(b) Condition (10.60) and condition (ii) in Definition 10.16 imply

$$N(F^p \cap U_p) \subset F^{p-1} \cap U_{p-1}. \tag{10.64}$$

Therefore N and all powers N^j of N are strict morphisms for $U_\bullet$, that is, $N^j(U_p) = \mathrm{Im} N^j \cap U_{p-j}$. The powers N^j are strict morphisms with respect to $F^\bullet$ and $W_\bullet$ because of Deligne's $I^{p,q}$ (Lemma 10.18).

(c) Condition (10.61) and condition (iii) in Definition 10.16 imply

$$S(F^p \cap U_p, F^q \cap U_q) = 0 \quad \text{for } p+q \neq m. \tag{10.65}$$

The pairing $S : F^p \cap U_p \times F^{m-p} \cap U_{m-p} \to \mathbb{C}$ is nondegenerate because S is nondegenerate.

(d) If there is a holomorphic family of Hodge filtrations, each giving a PMHS, then an opposite filtration for one PMHS is an opposite filtration for all PMHSs nearby. Being an opposite filtration is an open condition. This follows from codim $F^p \cap U_p = \text{codim} F^p + \text{codim} U_p$.

(e) There exist opposite filtrations. Lemma 10.18 induces a canonical opposite filtration $U_\bullet^{(0)}$ by

$$U_p^{(0)} := \bigoplus_{i,q:\, i \leq p} I^{i,q}. \tag{10.66}$$

It will be useful in the discussion of the symmetries of a single singularity in section 13.2. But it will not be useful in the discussion of μ-constant families of singularities in chapter 12. In general the Hodge filtration varies there holomorphically. One can check that the opposite filtration $U_\bullet^{(0)}$ varies antiholomorphically. But we need a constant opposite filtration.

Lemma 10.21 *The space of all opposite filtrations as in Definition 10.19 for a PMHS as in Definition 10.16 is an algebraic manifold isomorphic to $\mathbb{C}^{N_{opp}}$ for some $N_{opp} \in \mathbb{N}$.*

Proof: The algebraic group

$$\{g \in \text{Aut}(H_\mathbb{C}, S, N, F^\bullet) \mid g = id \text{ on } \text{Gr}_p^F \text{ for all } p\} \tag{10.67}$$

is unipotent. We will see that it acts transitively on the space of all opposite filtrations. Then this space is isomorphic to $\mathbb{C}^{N_{opp}}$ for some $N_{opp} \in \mathbb{N}$ (cf. for example [**Bore**, 11.13]).

One starts with a basis $a_1^{(0)}, \ldots, a_\mu^{(0)}$ of $H_\mathbb{C}$ (here $\mu := \dim H_\mathbb{C}$) which fits to Deligne's I^{pq}, that means, $a_i^{(0)} \in I^{p(i)q(i)}$ for each i and some $p(i), q(i) \in \mathbb{Z}$. It fits to the decomposition $H_\mathbb{C} = \bigoplus_p F^p \cap U_p^{(0)}$, that is, $a_i^{(0)} \in F^{p(i)} \cap U_{p(i)}^{(0)}$. One has

$$N a_i^{(0)} = \sum_j \nu_{ij} a_j^{(0)} \quad \text{and} \quad S\big(a_i^{(0)}, a_j^{(0)}\big) = \sigma_{ij} \tag{10.68}$$

for some $\nu_{ij}, \sigma_{ij} \in \mathbb{C}$. If $U_\bullet$ is any opposite filtration, then there are canonical isomorphisms

$$F^p \cap U_p^{(0)} \to \text{Gr}_F^p \leftarrow F^p \cap U_p. \tag{10.69}$$

There exists a unique basis $a_1, \dots, a_\mu$ of $H_\mathbb{C}$ which satisfies

$$a_i \in F^{p(i)} \cap U_{p(i)} \quad \text{and} \tag{10.70}$$

$$a_i - a_i^{(0)} \in F^{p(i)+1}. \tag{10.71}$$

We claim that $a_1, \dots, a_\mu$ satisfy precisely the same relations with respect to N and S as $a_1^{(0)}, \dots, a_\mu^{(0)}$,

$$Na_i = \sum_j \nu_{ij} a_j \quad \text{and} \quad S(a_i, a_j) = \sigma_{ij}. \tag{10.72}$$

The relation for N follows from $p(j) = p(i) - 1$ for $\nu_{ij} \neq 0$, from

$$N\, a_i - \sum_j \nu_{ij} a_j \in U_{p(i)-1} = N\left(a_i - a_i^{(0)}\right) - \sum_j \nu_{ij}\left(a_j - a_j^{(0)}\right) \in F^{p(i)}, \tag{10.73}$$

and from $F^{p(i)} \cap U_{p(i)-1} = \{0\}$. The relation for S follows from (10.61) and condition (iii) in Definition 10.16.

The automorphism of $H_\mathbb{C}$ which maps $a_i^{(0)}$ to a_i is an element of the group in (10.67) and maps $U_\bullet^{(0)}$ to $U_\bullet$. Therefore the group in (10.67) acts transitively on the space of all opposite filtrations. □

Often the situation is more complicated. In the singularity case one has the following setting:

(i) a lattice $H_\mathbb{Z} \subset H_\mathbb{Q} \subset H_\mathbb{R} \subset H = H_\mathbb{C} = H_\mathbb{Z} \otimes_\mathbb{Z} \mathbb{C}$;

(ii) a quasiunipotent monodromy $h = h_s \cdot h_u$ on $H_\mathbb{Z}$ with semisimple part h_s, unipotent part h_u, and nilpotent part $N = \log h_u$ on $H_\mathbb{Q}$; the eigenspaces $H_\lambda = \ker(h_s - \lambda : H \to H)$; a weight filtration $W_\bullet$ on $H_\mathbb{Q}$ whose restrictions to H_1 and $H_{\neq 1} := \bigoplus_{\lambda \neq 1} H_\lambda$ come from N as in Lemma 10.15 and are centred at $n+1$ and n;

(iii) a nondegenerate bilinear form S on $H_\mathbb{Q}$ (respectively on $H_\mathbb{Z}$ with values in $\mathbb{Q}$) whose restrictions to H_1 and $H_{\neq 1}$ are $(-1)^{n+1}$- and $(-1)^n$-symmetric; S is monodromy invariant; then N is an infinitesimal isometry;

(iv) an h_s-invariant Hodge filtration $F_0^\bullet$ whose restrictions to H_1 and $H_{\neq 1}$ form PMHSs of weight $n+1$ and n together with the other data.

We also call this sum of PMHSs on H_1 and $H_{\neq 1}$ a PMHS. Deligne's $I^{p,q}$ and the $I_0^{p,q}$ are h_s-invariant and decompose into eigenspaces $(I^{p,q})_\lambda$ and $(I_0^{p,q})_\lambda$ of h_s. One should consider only monodromy invariant opposite filtrations. Condition (10.61) has to be taken with $m = n+1$ on H_1 and $m = n$ on $H_{\neq 1}$. Lemma 10.21 also holds for such opposite filtrations.

In [**He4**] the space

$$D_{PMHS} := \{\text{filtrations } F^\bullet \text{ on } H_\mathbb{C} \mid \dim F^p P_{l,\lambda} = \dim F_0^p P_{l,\lambda},$$
$$F^\bullet \text{ is } h_s\text{-invariant and induces PMHSsof weight } n+1$$
$$\text{and } n \text{ on } H_1 \text{ and } H_{\neq 1}\} \quad (10.74)$$

is defined. It is a classifying space for Hodge filtrations $F^\bullet$ giving PMHSs with the same dimensions as $F_0^\bullet$. It is studied together with the group

$$G_\mathbb{Z} := \mathrm{Aut}(H_\mathbb{Z}, h, S) \quad (10.75)$$

and other spaces and groups.

Theorem 10.22 [**He4**] *In the above setting, the classifying space D_{PMHS} is a real homogeneous space and a complex manifold. The group $G_\mathbb{Z}$ acts properly discontinuously on it. The moduli space $D_{PMHS}/G_\mathbb{Z}$ for isomorphism classes of polarized mixed Hodge structures as above is a normal complex space and has only quotient singularities.*

Remarks 10.23 In [**He4**] a bigger space $\check{D}_{PMHS}$ is also defined. It is the space of the filtrations $F^\bullet$ which are similar to $F_0^\bullet$ with respect to all conditions except those involving complex conjugation. $\check{D}_{PMHS}$ is an algebraic manifold and a complex homogeneous space. The group $G_\mathbb{C} = \mathrm{Aut}(H_\mathbb{C}, h, S)$ acts transitively on it. It is a bundle over a product $\check{D}_{prim}$ of projective algebraic manifolds with fibres isomorphic to $\mathbb{C}^{N_{fibre}}$ for some number $N_{fibre} \in \mathbb{N}$. The space D_{PMHS} is the restriction of this bundle to an open submanifold $D_{prim} \subset \check{D}_{prim}$, which is a product of classifying spaces for polarized pure Hodge structures.

10.6 Brieskorn lattice

We use the notations from sections 10.1–10.3; the function $f(x_0, \ldots, , x_n)$ is an isolated hypersurface singularity with $n \geq 1$. Its Brieskorn lattice H_0'' is the germ at 0 of a distinguished extension to 0 of the sheaf of holomorphic sections of the cohomology bundle $H^n|_{\Delta^* \times \{0\}}$ of a Milnor fibration for f. It was defined as the set of germs of sections of Gelfand–Leray forms of $(n+1)$-forms, $H_0'' \cong \Omega^{n+1}_{\mathbb{C}^{n+1},0}/\mathrm{d}f \wedge \mathrm{d}\Omega^{n-1}_{\mathbb{C}^{n+1},0}$ (cf. (10.18) and (10.25)). Its relation to the Gauß–Manin connection for a semiuniversal unfolding and some of its properties were already formulated in Theorem 10.5.

The Brieskorn lattice is a key to the singularity. It has a rich structure and a long history, which started with [**Bri2**]. Here we give a concise report on its structure, but not on its history. We follow [**He4**].

The notions of sections 7.1 and 7.2 (e.g. the spaces C^α of elementary sections and the V-filtration) will be used freely. The space H^∞ is the μ-dimensional space of multivalued global flat sections on the cohomology bundle $H^n|_{\Delta^*\times\{0\}}$ over the punctured disc Δ^* (cf. (7.2) and (7.3)). Now it contains the lattice $H^\infty_\mathbb{Z} \cong \mathbb{Z}^\mu$ of multivalued sections of cohomology classes in $H^n(f^{-1}(z), \mathbb{Z})$. Here $f^{-1}(z)$ is a Milnor fibre of the representative $f : B^{n+1}_\varepsilon \cap f^{-1}(\Delta) \to \Delta$ of f.

In addition to the monodromy h on H^∞, there is a nondegenerate bilinear form $S : H^\infty_\mathbb{Z} \times H^\infty_\mathbb{Z} \to \mathbb{Q}$ [**He4**]. It is less well known than it deserves. It makes Steenbrink's mixed Hodge structure [**Stn**] on H^∞ into a polarized mixed Hodge structure. It induces a series of bilinear forms on H''_0 which coincide with the restriction to H''_0 of K. Saito's higher residue pairings. It is defined in (10.78).

The variation operator $\mathrm{Var} : H^n(f^{-1}(z), \mathbb{Z}) \to H_n(f^{-1}(z), \mathbb{Z})$ for a Milnor fibre $f^{-1}(z)$ uses the identification $H^n(f^{-1}(z), \mathbb{Z}) \cong H_n(f^{-1}(z), \partial f^{-1}(z), \mathbb{Z})$ and maps a relative cycle γ to the absolute cycle $\widetilde{h}(\gamma) - \gamma$, where $\widetilde{h}$ is a geometric monodromy on the Milnor fibre which fixes the boundary of it. Var is an isomorphism and determines the monodromy h and the intersection form I on $H_n(f^{-1}(z), \mathbb{Z})$ (cf. for example [**Lo2**][**AGV2**]).

The monodromy has a semisimple part h_s and a unipotent part h_u with $N := \log h_u$. It acts on $H_n(f^{-1}(z), \mathbb{Z})$, $H^n(f^{-1}(z), \mathbb{Z})$, and $H^\infty_\mathbb{Z}$. We use the notations $H^\infty_\lambda = \ker(h_s - \lambda)$ and $H^\infty_{\neq 1} = \bigoplus_{\lambda\neq 1} H^\infty_\lambda$, and the same for $H^n(f^{-1}(z), \mathbb{C})$ and $H_n(f^{-1}(z), \mathbb{C})$.

The intersection form I on $H_n(f^{-1}(z), \mathbb{C})_{\neq 1}$ is nondegenerate. It induces an isomorphism to $H^n(f^{-1}(z), \mathbb{C})_{\neq 1}$ and a form I^* on it.

An isomorphism $\nu : H^n(f^{-1}(z), \mathbb{Q}) \to H^n(f^{-1}(z), \mathbb{Q})$ is defined by

$$\nu = (h - id)^{-1} \quad \text{on } H^n(f^{-1}(z), \mathbb{C})_{\neq 1}, \tag{10.76}$$

$$\nu = -\sum_{l\geq 1} \tfrac{1}{l}(-1)^{l-1}(h - id)^{l-1} \quad \text{on } H^n(f^{-1}(z), \mathbb{C})_1 \left(' = \tfrac{-N}{h-id}{}' \right). \tag{10.77}$$

The polarizing form S on $H^n(f^{-1}(z), \mathbb{Q})$ is defined by

$$S(a, b) = (-1)^{n(n-1)/2} \langle a, \mathrm{Var} \circ \nu(b)\rangle. \tag{10.78}$$

Lemma 10.24 [**He4**] *The form S is nondegenerate and monodromy invariant. The restriction to $H^n(f^{-1}(z), \mathbb{C})_{\neq 1}$ is $(-1)^n$-symmetric and is equal to $(-1)^{n(n-1)/2} I^*$. The restriction to $H^n(f^{-1}(z), \mathbb{C})_1$ is $(-1)^{n+1}$-symmetric.*

Remarks 10.25 Unfortunately, in [**He4**] it was not noticed that the monodromy there is the inverse of that in [**Schm**]. Therefore here ν and S on $H^n(f^{-1}(z), \mathbb{C})_1$

differ by a sign from those in [**He4**]. Also, in (10.83) a sign turns up, and in Theorem 10.30 (i) one needs $-N$.

The form S on $H^n(f^{-1}(z),\mathbb{Q})$ induces a form S with the same properties on $H^\infty_\mathbb{Q}$. This form S can be used to define a $\mathbb{C}\{\{\partial_z^{-1}\}\}$-sesquilinear pairing

$$K_f : V^{>-1}\times V^{>-1}\to \mathbb{C}\{\{\partial_z^{-1}\}\}\cdot\partial_z^{-1} \tag{10.79}$$

on the space $V^{>-1}=\bigoplus_{-1<\alpha\le 0}\mathbb{C}\{z\}C^\alpha=\bigoplus_{-1<\alpha\le 0}\mathbb{C}\{\{\partial_z^{-1}\}\}C^\alpha$ (cf. section 7.1). One needs the homomorphisms ψ_α in (7.6) from multivalued flat sections to elementary sections,

$$\psi_\alpha : H^\infty_{e^{-2\pi i\alpha}}\to C^\alpha,\quad A\mapsto es(A,\alpha), \tag{10.80}$$

in order to go from H^∞ to $V^{>-1}$. The pairing K_f and its parts $K_f^{(-k)}$ with values in $\mathbb{C}$ are defined by the formulas (10.81)–(10.85), here $\alpha,\beta\in(-1,0]$, $a\in C^\alpha$, $b\in C^\beta$, $g_1,g_2\in\mathbb{C}\{\{\partial_z^{-1}\}\}$,

$$K_f(a,b)=0\quad\text{for }\alpha+\beta\notin\mathbb{Z}, \tag{10.81}$$

$$K_f(a,b)=\frac{1}{(2\pi i)^n}S\big(\psi_\alpha^{-1}(a),\psi_\beta^{-1}(b)\big)\cdot\partial_z^{-1} \tag{10.82}$$

for $\alpha+\beta=-1$,

$$K_f(a,b)=\frac{-1}{(2\pi i)^{n+1}}S\big(\psi_\alpha^{-1}(a),\psi_\beta^{-1}(b)\big)\cdot\partial_z^{-2} \tag{10.83}$$

for $\alpha=\beta=0$,

$$K_f(g_1a,g_2b)=g_1\big(\partial_z^{-1}\big)g_2\big(-\partial_z^{-1}\big)K_f(a,b), \tag{10.84}$$

$$K_f(g_1a,g_2b)=\sum_{k\ge -n}K_f^{(-k)}(g_1a,g_2b)\cdot\partial_z^{-n-1-k}. \tag{10.85}$$

Lemma 10.26 [**He4**] *(i) The pairing $K_f^{(-k)}$ is $(-1)^k$-symmetric.*

(ii) For the restrictions of K_f to the eigenspaces one has $K_f : C^\alpha\times C^\beta\to 0$ for $\alpha+\beta\notin\mathbb{Z}$, $\alpha,\beta>-1$, and $K_f : C^\alpha\times C^\beta\to\mathbb{C}\cdot\partial_z^{-\alpha-\beta-2}$ is nondegenerate for $\alpha+\beta\in\mathbb{Z}$, $\alpha,\beta>-1$.

(iii) The pairing K_f and the multiplication by z are related by $K_f(z\,a,b)-K_f(a,z\,b)=[z,K_f(a,b)]$, where $[z,\partial_z^{-k}]=k\partial_z^{-k-1}$.

The pairing K_f shares the properties (i), (iii), and (10.84) with K. Saito's higher residue pairings (Theorem 10.11). Theorem 10.28 will show that it coincides on H_0'' with the restriction of K_F to H_0''.

Now we come to the properties of H_0'' (cf. (10.18), Theorem 10.5, (10.25)). Brieskorn [**Bri2**] showed that its germs of sections have moderate growth (that means $H_0'' \subset V^{>-\infty}$), that they generate the cohomologies of the Milnor fibres and that the map $\nabla_{\partial_z} : H_0' \to H_0''$ from the sublattice $H_0' \subset H_0''$ defined by n-forms is an isomorphism (cf. (10.17), Theorem 10.5 (v)). Malgrange [**Mal1**] showed $H_0' \subset V^{>0}$. Together with Lemma 7.4 this gives the following.

Theorem 10.27 *(i) The Brieskorn lattice H_0'' is a free $\mathbb{C}\{z\}$-module of rank μ, equivalently $\mathbb{C}\{z\}[z^{-1}]H_0'' = V^{>-\infty}$.*

(ii) The Brieskorn lattice and H_0' satisfy $H_0' \subset V^{>0}$, $H_0'' \subset V^{>-1}$, $H_0' = \nabla_{\partial_z}^{-1} H_0'' \subset H_0''$.

(iii) The Brieskorn lattice H_0'' is a free $\mathbb{C}\{\{\partial_z^{-1}\}\}$-module of rank μ.

The nondegenerate Grothendieck residue pairing $J_F : \Omega_F \times \Omega_F \to \mathcal{O}_M$ from (10.44) for the semiuniversal unfolding F of f induces a nondegenerate pairing

$$J_f : \Omega_f \times \Omega_f \to \mathbb{C} \tag{10.86}$$

on the μ-dimensional space

$$\Omega_f := \Omega^{n+1}_{\mathbb{C}^{n+1},0}/\mathrm{d}f \wedge \Omega^{n}_{\mathbb{C}^{n+1},0} = H_0''/\nabla_{\partial_z}^{-1} H_0''. \tag{10.87}$$

The following theorem from [**He4**] gives the relation between H_0'', K_f, J_f, and K_F. Remarks on the proof will be given afterwards.

Theorem 10.28 *(i) For $k \geq 0$, the pairing $K_f^{(-k)}$ on H_0'' is the restriction of K. Saito's higher residue pairing $K_F^{(-k)}$ on $\mathcal{H}_0^{(0)}$ (cf. Theorem 10.11) to H_0''.*

(ii) The pairing and the Brieskorn lattice satisfy $K_f^{(-k)}(H_0'', H_0'') = 0$ for $-n \leq k \leq -1$, i.e. $K_f(H_0'', H_0'') = \mathbb{C}\{\{\partial_z^{-1}\}\} \cdot \partial_z^{-n-1}$.

(iii) The pairings $K_f^{(0)}$ and J_f are related by $K_f^{(0)}(\omega_1, \omega_2) = J_f([\omega_1], [\omega_2])$ for $\omega_1, \omega_2 \in H_0''$.

(iv) The Brieskorn lattice H_0'' is isotropic of maximal size with respect to the antisymmetric bilinear form $K_f^{(1)}$, that means, a section $\omega \in V^{>-1}$ satisfies $K_f^{(1)}(\omega, H_0'') = 0$ if and only if $\omega \in H_0''$.

(v) Therefore $H_0'' \supset V^{n-1}$ and $\dim H_0''/V^{n-1} = \frac{1}{2} \dim V^{>-1}/V^{n-1}$, and the spectral numbers $\alpha_1, \ldots, \alpha_\mu$ are contained in $(-1, n)$ and satisfy the symmetry $\alpha_i + \alpha_{\mu+1-i} = n - 1$.

Remarks 10.29 (a) Part (i) can be found essentially in [**SM2**, Appendix], [**SM3**, 2.7]. But he is not very explicit about the definition of K_f. Another way to check it is with the results in [**Va6**, §3.3]. Varchenko embeds the Milnor fibration (and its extension to the semiuniversal unfolding) into a fibration with projective fibres. He relates the Grothendieck pairing J_f (and J_F) to the primitive cohomology bundle of this extended fibration and to the nondegenerate intersection form on it.

He considers formulas of the same type as (10.47), but for the primitive cohomology bundle. I checked that these formulas induce a bilinear form on $\mathcal{H}_0^{(0)}$ with the same properties as K_F, that they induce K_f on $V^{>-1}$, and that the properties (ii)–(v) hold.

Now one needs M. Saito's result ([**SM3**, 2.7], see Remark 10.12 (c)) that the form K_F on $\mathcal{H}_0^{(0)}$ is uniquely determined by its properties. One obtains (i) and a generalization of Theorem 10.13 (see Remark 10.14 (b)).

(b) The existence of a relative compactification of the Milnor fibration to a fibration with projective fibres and other good properties is due to [**Bri2**][**Sche1**]. It is not only essential for the proof of Theorem 10.28, but also for that of Theorem 10.30.

(c) Part (iv) follows from the facts that J_f is nondegenerate and that K_f is $\mathbb{C}\{\{\partial_z^{-1}\}\}$-sesquilinear (cf. (10.84)). The form $K_f^{(1)}$ satisfies $K_f^{(1)}(V^{>-1}, V^{n-1}) = 0$ and induces a nondegenerate form on the quotient $V^{>-1}/V^{n-1}$. This shows the inclusions $V^{n-1} \subset H_0''$ and $\alpha_1, \ldots, \alpha_\mu \in (-1, n)$ and the dimension formula in (v). A refinement of this argument yields the symmetry of the spectral numbers.

Varchenko [**Va2**][**Va1**][**AGV2**] showed that the principal parts of the sections in H_0'' induce a holomorphic family of Hodge filtrations on the cohomology bundle, giving a mixed Hodge structure on each fibre. There is a limit Hodge filtration $F_{Va}^\bullet$ on H^∞. He used a relative compactification of the Milnor fibration and results of Griffiths, Schmid, and Scherk. His construction was modified [**SchSt**] [**SM1**] [**Ph3**] to obtain Steenbrink's [**Stn**] mixed Hodge structure $F^\bullet$ on H^∞. It is polarized by S because of an exact sequence of Steenbrink, connecting it with the limit mixed Hodge structure of Schmid for a relative compactification of the Milnor fibration. This was emphasized in [**He4**] (for the sign in $-N$ in part (i) see Remark 10.25).

Theorem 10.30 *(i) The subspaces*

$$F^p H_\lambda^\infty := \psi_\alpha^{-1} \nabla_{\partial_z}^{n-p} \mathrm{Gr}_V^{\alpha+n-p} H_0'' \big(= F_{n-p}^{alg} \text{ from (7.27)}\big) \qquad (10.88)$$

for $\alpha \in (-1, 0]$, $e^{-2\pi i\alpha} = \lambda$, define an h_s-invariant decreasing filtration on H^∞ with $0 = F^{n+1} \subset F^n \subset \ldots \subset F^0 = H^\infty$. It is Steenbrink's Hodge filtration.

Together with S and $-N$ it gives a PMHS of weight n on $H^\infty_{\neq 1}$ and a PMHS of weight $n+1$ on H^∞_1 (in the sense of Definition 10.16).

(ii) The subspaces

$$F^p_{Va}H^\infty_\lambda := \psi_\alpha^{-1} z^{-n+p}\mathrm{Gr}_V^{\alpha+n-p} H_0'' \ (= F_{n-p} \textit{ from } (7.24)) \tag{10.89}$$

define an h_s-invariant decreasing filtration on H^∞. Together with the weight filtration $W_\bullet$ from N this filtration also gives a mixed Hodge structure. The filtrations $F^\bullet_{Va}$ and $F^\bullet$ coincide on the quotients Gr^W_l (cf. Lemma 7.4 (b)).

Remark 10.31 In **[AGV2][Stn]** μ spectral pairs in $\mathbb{Q}\times\mathbb{Z}$ are defined. The multiplicity of $(\alpha, l) \in \mathbb{Q}\times\mathbb{Z}$ as spectral pair is

$$d(\alpha, l) := \dim \mathrm{Gr}_F^{[n-\alpha]}\mathrm{Gr}^W_{l+[\alpha+1]} H^\infty_\lambda \tag{10.90}$$

for $e^{-2\pi i\alpha} = \lambda$. Because of the PMHS and the strict morphism N, they satisfy the symmetries (any two of the symmetries determine the third)

$$d(\alpha, l) = d(n-1-\alpha, 2n-l), \tag{10.91}$$

$$d(\alpha, l) = d(2n-1-l-\alpha, l), \tag{10.92}$$

$$d(\alpha, l) = d(\alpha-n+l, 2n-l). \tag{10.93}$$

The first entries of the spectral pairs are the spectral numbers. It is unlikely that one can recover all the symmetries only with Theorem 10.28, the strictness of the morphism N is probably too profound.

The following result of Varchenko will be used in chapter 12.

Theorem 10.32 **[Va3]** *The spectral pairs are constant within a μ-constant family of singularities.*

The next result of M. Saito is essential for the construction of a (good) primitive form in section 11.1. The μ-dimensional space Ω_f from (10.87) is a free module of rank 1 over the Jacobi algebra $\mathcal{O}/J_f$. Generators for this module are represented by volume forms in $\Omega^{n+1}_{\mathbb{C}^{n+1},0}$, that is, forms $u(x)\mathrm{d}x_0 \ldots \mathrm{d}x_n$ with $u(0) \neq 0$. The space Ω_f inherits a finite decreasing filtration $V^\bullet\Omega_f$ from the V-filtration on H_0''. The dimension $\dim \mathrm{Gr}_V^\alpha \Omega_f = d(\alpha)$ is the multiplicity of α as spectral number. Parts (ii) and (iii) in Theorem 10.33 are corollaries of part (i).

Theorem 10.33 **[SM4**, 3.11 Remark]

(i) The maximal ideal $\mathbf{m}/J_f$ in the Jacobi algebra maps $V^\alpha\Omega_f$ to $V^{>\alpha}\Omega_f$.

(ii) $\alpha_1 < \alpha_2$, that means, $d(\alpha_1) = 1$ and $\dim \mathrm{Gr}_F^{[n-\alpha_1]} H^\infty_{e^{-2\pi i\alpha_1}} = 1$.

(*iii*) *A form* $u(x)\mathrm{d}x_0 \dots \mathrm{d}x_n \in \Omega^{n+1}_{\mathbb{C}^{n+1},0}$ *represents a section in* H_0'' *with principal part in* C^{α_1} *if and only if* ω *is a volume form, that is,* $u(0) \neq 0$.

In [**He4**, ch. 5] a classifying space for Brieskorn lattices is studied. Its elements are subspaces $\mathcal{L}_0 \subset V^{>-1}$ with the following properties.

(α) The subspace $\mathcal{L}_0$ is a free $\mathbb{C}\{z\}$-module of rank μ.
(β) The subspace $\mathcal{L}_0$ is a free $\mathbb{C}\{\{\partial_z^{-1}\}\}$-module of rank μ.
(γ) The decreasing h_s-invariant filtration $F^{alg}_{n-\bullet}$ on H^∞ from (7.24) (cf. (10.88)) is in the classifying space D_{PMHS} for PMHSs from (10.74).
(δ) One has $K_f^{(-k)}(\mathcal{L}_0, \mathcal{L}_0) = 0$ for $-n \leq k \leq -1$.

Theorem 10.34 [**He4**] *The classifying space*

$$D_{BL} = \{\mathcal{L}_0 \subset V^{>-1} \mid \mathcal{L}_0 \text{ satisfies } (\alpha), (\beta), (\gamma), (\delta)\} \tag{10.94}$$

for Brieskorn lattices is a complex manifold and a locally trivial holomorphic bundle $pr_{BL} : D_{BL} \to D_{PMHS}$ *with fibres isomorphic to* $\mathbb{C}^{N_{BL}}$ *for some* $N_{BL} \in \mathbb{N}$. *The fibres have a natural* $\mathbb{C}^*$*-action with negative weights. The group* $G_{\mathbb{Z}} = \mathrm{Aut}(H^\infty_{\mathbb{Z}}, h, S)$ *acts properly discontinuously on* D_{PMHS} *and thus also on* D_{BL}.

More details on D_{BL} including a formula for N_{BL} can be found in [**He4**]. Essentially because of Theorem 10.32, there is a period map from the μ-constant stratum to the space D_{BL}. Such period maps are studied in [**He1**] [**He2**] [**He3**] and in section 12.2.

Chapter 11

Frobenius manifolds for hypersurface singularities

The construction of Frobenius manifolds for singularities is due to K. Saito and M. Saito, using results of Malgrange. The version presented in section 11.1 replaces the use of Malgrange's results by the solution of the Riemann–Hilbert–Birkhoff problem in section 7.4 and by the tools in section 8.2. All the other ingredients from the Gauß–Manin connection are provided in chapter 10.

Section 11.2 establishes series of functions which are close to Dubrovin's deformed flat coordinates. Some use of them is made in chapter 12. In view of some results of Dubrovin, Zhang, and Givental one can hope that much more can be found in these series of functions.

Sabbah generalized most of K. Saito and M. Saito's construction to the case of tame functions with isolated singularities on affine manifolds **[Sab3][Sab2] [Sab4]**. But the details are quite different, there one uses oscillating integrals, and the results are not as complete as in the local case. We discuss this at some length in section 11.4.

The case of tame functions is important for the following question within mirror symmetry: Are certain Frobenius manifolds from quantum cohomology isomorphic to certain Frobenius manifolds somehow coming from functions with isolated singularities? This is motivated by the results of Givental. A special case was looked at by Barannikov. In section 11.3 we make some remarks about this version of mirror symmetry.

11.1 Construction of Frobenius manifolds

Let $f : (\mathbb{C}^{n+1}, 0) \to (\mathbb{C}, 0)$ be a holomorphic function germ with an isolated singularity at 0, with Milnor number μ, and with $n \geq 1$. The base space of a semiuniversal unfolding is a germ $(M, 0) \cong (\mathbb{C}^{\mu}, 0)$. It can be equipped with the structure of a massive Frobenius manifold. The multiplication and the Euler field are unique, but the flat metric depends on some choice. Its existence is highly nontrivial and follows from the existence of a primitive form of K. Saito [**SK6**][**SK9**], which was proved in the general case by M. Saito [**SM3**], building on work of K. Saito and many other people. M. Saito used a result of Malgrange [**Mal3**][**Mal5**] on the extension of special bases in the microlocal Gauß–Manin system, whose proof involved the Fourier–Laplace transformation of this system. That made it difficult to use the construction of the metrics on $(M, 0)$ and to work with these Frobenius manifolds.

Below we give a simpler and more explicit version of the construction, without using Malgrange's result. We will also provide precise information as to which choices have to be made and what they yield. This makes the construction sufficiently transparent to be subsequently applied.

To describe the choices, we need the space H^∞ of multivalued global flat sections on the cohomology bundle of a Milnor fibration. It is defined as follows (cf. (7.2) and (7.3) and section 10.6). Choose $\varepsilon>0$ and $\delta>0$ as in section 10.1 such that $f: B_\varepsilon^{n+1} \cap f^{-1}(\Delta) \to \Delta$ for $\Delta := B_\delta^1 \subset \mathbb{C}$ and $B_\varepsilon^{n+1} \subset \mathbb{C}^{n+1}$ is a Milnor fibration.

The cohomology bundle $H^n|_{\Delta^*} := \bigcup_{z\in\Delta^*} H^n(f^{-1}(z), \mathbb{C})$ has rank μ and a flat structure. The universal covering $e : \mathbb{C} \to \mathbb{C}^*$, $\zeta \mapsto \mathrm{e}^{2\pi i\zeta}$ restricts to a universal covering $e : e^{-1}(\Delta^*) \to \Delta^*$. If $pr : e^*(H^n|_{\Delta^*}) \to H^n|_{\Delta^*}$ denotes for a moment the projection, then

$$H^\infty = \{pr \circ A \mid A \text{ is a global flat section of } e^*(H^n|_{\Delta^*})\} \tag{11.1}$$

is the μ-dimensional space of multivalued global flat sections of the cohomology bundle. It is independent of the choice of the Milnor fibration. It is equipped with a lattice $H^\infty_\mathbb{Z} \subset H^\infty$, a monodromy h with semisimple part h_s and unipotent part h_u, a polarizing form S (see section 10.6), and a polarized mixed Hodge structure with Steenbrink's Hodge filtration $F^\bullet$ (see Theorem 10.30).

The spectral numbers $\alpha_1, \ldots, \alpha_\mu \in (-1, n) \cap \mathbb{Q}$ come from the Brieskorn lattice H_0'', but they also encode the dimensions of the spaces $\mathrm{Gr}^p_F H^\infty_\lambda$ (see (10.50)), where $H^\infty_\lambda = \ker(h_s - \lambda)$. M. Saito's result (Theorem 10.33) that the smallest spectral number α_1 has multiplicity 1 implies

$$\dim F^{[n-\alpha_1]} H^\infty_{e^{-2\pi i\alpha_1}} = \dim \mathrm{Gr}_F^{[n-\alpha_1]} H^\infty_{e^{-2\pi i\alpha_1}} = 1. \tag{11.2}$$

The notion of an opposite filtration for $F^\bullet$ was defined in Definition 10.19. Monodromy invariant opposite filtrations $U_\bullet$ for $F^\bullet$ exist. They have to satisfy (10.61) with $m = n+1$ on H^∞_1 and with $m = n$ on $H^\infty_{\neq 1} = \bigoplus_{\lambda\neq 1}^{\infty} H^\infty_\lambda$. They form an algebraic manifold isomorphic to $\mathbb{C}^{N_{opp}}$ for some $N_{opp} \in \mathbb{N}$ (Lemma 10.21). Such a filtration satisfies

$$\dim \mathrm{Gr}^U_{[n-\alpha_1]} H^\infty_{e^{-2\pi i\alpha_1}} = 1. \tag{11.3}$$

Theorem 11.1 *Any choice $(U_\bullet, \gamma_1)$ of a monodromy invariant opposite filtration $U_\bullet$ for $F^\bullet$ on H^∞ and of a generator γ_1 of the 1-dimensional space $\mathrm{Gr}^U_{[n-\alpha_1]} H^\infty_{e^{-2\pi i\alpha_1}}$ induces the structure of a germ of a Frobenius manifold on $(M, 0)$. Multiplication $\circ$, unit field e, and Euler field E are unique. Different choices of $(U_\bullet, \gamma_1)$ give different metrics g. If $(U_\bullet, \gamma_1)$ gives a metric g then $(U_\bullet, c\gamma_1)$ for $c \in \mathbb{C}^*$ gives the metric cg.*

Let ∇^g be the Levi–Civita connection of g. The flat endomorphism $\nabla^g E$: $\mathcal{T}_{M,0} \to \mathcal{T}_{M,0}$, $X \mapsto \nabla^g_X E$, is semisimple with eigenvalues $d_i = 1 + \alpha_1 - \alpha_i$, $i = 1, \ldots, \mu$, and $D = 2 - (\alpha_\mu - \alpha_1) = 2 + 2\alpha_1 - (n-1)$.

This follows from the results of M. Saito **[SM3]** and K. Saito **[SK6][SK9]** on the Gauß–Manin connection and the primitive forms. In this section we give a simplified proof. It uses ingredients from all previous chapters. A rough idea of the proof was given in section 6.1.

Proof of Theorem 11.1: The Brieskorn lattice H_0'' (cf. (10.18), Theorem 10.5, (10.25), section 10.6) induces Steenbrink's Hodge filtration $F^\bullet = F^{alg}_{n-\bullet}$ on H^∞ (Theorem 10.30). Here $F^{alg}_{n-\bullet}$ is the increasing filtration which is associated to H_0'' by (7.27).

Let $(U_\bullet, \gamma_1)$ be as in Theorem 11.1. Define a monodromy invariant increasing filtration $\widetilde{U}_\bullet$ on $H_\mathbb{C}$ by $\widetilde{U}_p := U_{p+n}$ on H_1^∞ and $\widetilde{U}_p := U_{p+n+1}$ on $H^\infty_{\neq 1}$. Then $F^{alg}_\bullet$ and $\widetilde{U}_\bullet$ are opposite in H_1^∞, and $F^{alg}_\bullet$ and $\widetilde{U}_{\bullet+1}$ are opposite in $H^\infty_{\neq 1}$, in the sense of Definition 7.15.

The cohomology bundle $H^n|_{\Delta^*}$ extends uniquely to a flat bundle over $\mathbb{C}^*$. The Brieskorn lattice H_0'' gives an extension of its sheaf of sections over 0. The filtration $\widetilde{U}_\bullet$ induces an extension over ∞. By Theorem 7.17 this twofold extension $\overline{H''}$ is a free $\mathcal{O}_{\mathbb{P}^1}$-module of rank μ with a logarithmic pole at ∞. The residue endomorphism at ∞ has eigenvalues $-\alpha_1, \ldots, \alpha_\mu$ (Theorem 7.17) and is semisimple because of $N(U_p) \subset U_{p-1}$ and Theorem 7.10 ($N = \log h_u$ as usual).

Let $F : \mathcal{X} \to \Delta$ be a semiuniversal unfolding of f as in section 10.2 with discriminant $\check{\mathcal{D}} \subset \Delta \times M$ and cohomology bundle H^n over $\Delta \times M - \check{\mathcal{D}}$. The cohomology bundle H^n extends uniquely to a flat bundle over $\mathbb{C} \times M - \check{\mathcal{D}}$. The twofold extension $\overline{H''}$ embeds into a twofold extension $\overline{\mathcal{H}^{(0)}}$ of the sheaf of sections of H^n: The sheaf $\mathcal{H}^{(0)}$ from Lemma 10.2 is the unique locally free (Theorem 10.5) extension over $\check{\mathcal{D}}$ with a logarithmic pole along $\check{\mathcal{D}}_{reg}$ with (semisimple for $n > 1$) residue endomorphism with eigenvalues $(\frac{n-1}{2}, 0, \ldots, 0)$. By Theorem 8.7 $\widetilde{U}_\bullet$ or $(H_0'')_\infty$ induce a unique extension over $\{\infty\} \times M$ with a logarithmic pole along $\{\infty\} \times M$.

A priori $\overline{\mathcal{H}^{(0)}}$ is a locally free $\mathcal{O}_{\mathbb{P}^1 \times M}$-module of rank μ. But the restrictions of its sections to $\mathbb{P}^1 \times \{0\}$ give the free $\mathcal{O}_{\mathbb{P}^1}$-module $\overline{H''}$. By a classical theorem on families of vector bundles over $\mathbb{P}^1$ (e.g. [**Sab4**, I 5.b], [**Mal4**, §4]) then $\overline{\mathcal{H}^{(0)}}$ itself is a free $\mathcal{O}_{\mathbb{P}^1\times M}$-module if M is small enough. Choosing M arbitrarily small does no harm. So we will always suppose that it is small enough and often go from the germ $(M, 0)$ to such a representative M.

As in chapter 9, $pr : \mathbb{P}^1 \times M \to M$ denotes the projection. The sheaf $pr_*\overline{\mathcal{H}^{(0)}}$ of fibrewise global sections is a free $\mathcal{O}_M$-module of rank μ. The restriction to $\{\infty\} \times M$ yields a canonical isomorphism of $\mathcal{O}_M$-modules

$$\rho_1 : pr_*\overline{\mathcal{H}^{(0)}} \stackrel{\cong}{\longrightarrow} \left(\overline{\mathcal{H}^{(0)}} \Big/ \frac{1}{z}\overline{\mathcal{H}^{(0)}}\right)\Bigg|_{\{\infty\}\times M}. \tag{11.4}$$

The latter sheaf is equipped with the flat residual connection with respect to the coordinate $\frac{1}{z}$ (see section 8.2). The global sections whose restrictions to $\{\infty\} \times M$ are flat with respect to this residual connection form a μ-dimensional space. In order to study the features of these special sections we need some choices and notations.

The opposite filtration $U_\bullet$ induces a splitting of the Hodge filtration $F^\bullet$ on H^∞ with specific properties (cf. (10.59), (10.64), (10.65)). This can be shifted to the spaces C^α of elementary sections in $H^n|_{\Delta^*\times\{0\}}$ with the maps $\psi_\alpha : H^\infty_{e^{-2\pi i\alpha}} \to C^\alpha$ from (7.5). The subspaces

$$G^\alpha := \nabla_{\partial_z}^{[-\alpha]}\psi_{\alpha+[-\alpha]}\big(F^{[n-\alpha]} \cap U_{[n-\alpha]}H^\infty_{e^{-2\pi i\alpha}}\big) \subset C^\alpha \tag{11.5}$$

have dimensions $d(\alpha)$ = multiplicity of α as a spectral number. They satisfy (K_f is the pairing on $V^{>-1}$ from section 10.6):

$$\big(z\nabla_{\partial_z} - \alpha\big)G^\alpha = \frac{-N}{2\pi i}G^\alpha \subset \nabla_{\partial_z}G^{\alpha+1} \subset \bigoplus_{p>0} z^{-p}G^{\alpha+p}, \tag{11.6}$$

$$K_f(G^\alpha, G^\beta) = 0 \quad \text{for } \alpha+\beta \neq n-1, \tag{11.7}$$

$$K_f(G^\alpha, G^{n-1-\alpha}) = \mathbb{C}\cdot\partial_z^{-n-1}, \tag{11.8}$$

$$C^\alpha = \bigoplus_{p\in\mathbb{Z}} \nabla_{\partial_z}^p G^{\alpha+p} = \bigoplus_{p\in\mathbb{Z}} z^{-p}G^{\alpha+p} \quad \text{for } \alpha > -1, \tag{11.9}$$

$$\mathrm{Gr}_V^\alpha H_0'' = \bigoplus_{p\leq 0} \nabla_{\partial_z}^p G^{\alpha+p} = \bigoplus_{p\leq 0} z^{-p}G^{\alpha+p}. \tag{11.10}$$

In order to describe the situation along $\{\infty\} \times M$ we make the following observations:

(I) The space H^∞ is canonically isomorphic to the space of multivalued flat sections on the restriction to $(\mathbb{C} - \Delta) \times M$ of the (extended) cohomology bundle H^n.

(II) On $H^n|(\mathbb{C} - \Delta) \times M$ one has elementary sections with respect to the coordinate z. They are defined by the same formula, compare (7.5) and (8.13). An elementary section σ of order α satisfies

$$\left(z\nabla_{\partial_z} - \alpha\right)^{n+1}\sigma = 0 \quad \text{and} \quad \nabla_X \sigma = 0, \tag{11.11}$$

where $X \in \mathcal{T}_M$ is lifted to $\mathbb{P}^1 \times M$ in the canonical way with $X(z) = 0$.

(III) The elementary sections over $\Delta^* \times \{0\}$ and those in (II) glue to elementary sections over $\mathbb{C}^* \times \{0\} \cup (\mathbb{C} - \Delta) \times M$. From now on the spaces C^α denote spaces of such global elementary sections. But we will still take the liberty to embed the C^α into spaces of germs of sections at $(\Delta \times \{0\}, 0)$ or $(\mathbb{P}^1 \times M, (\infty, 0))$. One will usually see from the coefficients what is meant.

By the construction of the extension over $\{\infty\} \times M$, one has

$$\overline{\mathcal{H}^{(0)}} \mid (\mathbb{P}^1 - \Delta) \times M = \bigoplus_\alpha \mathcal{O}_{(\mathbb{P}^1 - \Delta)\times M} \cdot G^\alpha. \tag{11.12}$$

The restriction to $\{\infty\} \times M$ yields a canonical isomorphism of free $\mathcal{O}_M$-modules of rank μ

$$\rho_2 : \bigoplus_\alpha \mathcal{O}_M \cdot G^\alpha \xrightarrow{\cong} \left(\overline{\mathcal{H}^{(0)}} \Big/ \frac{1}{z}\overline{\mathcal{H}^{(0)}}\right)\Bigg|_{\{\infty\}\times M}. \tag{11.13}$$

By definition of the residual connection on the right module, the flat sections of this residual connection correspond to the sections in $\bigoplus_\alpha G^\alpha$ on the left.

Now we choose a basis $s_1, \ldots, s_\mu$ of $\bigoplus_\alpha G^\alpha$ with

$$s_i \in G^{\alpha_i}, \tag{11.14}$$

$$K_f(s_i, s_{\mu+1-j}) = \delta_{ij}\partial_z^{-n-1}, \tag{11.15}$$

$$\left[\psi^{-1}_{\alpha_1 + [-\alpha_1]}\left(\nabla_{\partial_z}^{-[-\alpha_1]} s_1\right)\right] = \gamma_1 \quad \text{in } \mathrm{Gr}^U_{[n-\alpha_1]} H^\infty_{e^{-2\pi i \alpha_1}}. \tag{11.16}$$

Note that $\dim G^{\alpha_1} = 1$ and that s_1 is uniquely determined by (11.16). The sections

$$v_i := \rho_1^{-1} \circ \rho_2(s_i) \in pr_* \overline{\mathcal{H}^{(0)}} \tag{11.17}$$

for $i = 1, \ldots, \mu$ form a basis of the space of global sections whose restrictions to $\{\infty\} \times M$ are flat with respect to the residual connection at $\{\infty\} \times M$. Their germs at $(\infty, 0)$ satisfy

$$v_i \in s_i + \widetilde{z} \cdot \overline{\mathcal{H}^{(0)}}_{(\infty,0)}, \tag{11.18}$$

here $\widetilde{z} := \frac{1}{z}$ with $\widetilde{z}\partial_{\widetilde{z}} = -z\partial_z$.

In the following, if $X \in \mathcal{T}_M$ we will also denote by X the canonical lift to $\mathbb{P}^1 \times M$ with $X(z) = 0$; this lift also satisfies $[X, \partial_z] = 0$.

The operator $\nabla_{\partial_z}^{-1}$ is welldefined for sections in C^α with $\alpha \neq -1$ and, by Theorem 10.7, for sections in $\mathcal{H}_0^{(0)}$ and $\pi_*\mathcal{H}^{(0)}$, where $\pi : \Delta \times M \to M$ is the projection. By the proof of Theorem 10.7, $\nabla_{\partial_z}^{-1}$ is also welldefined for fibrewise global sections of $\overline{\mathcal{H}^{(0)}}|_{\mathbb{C}\times M}$.

Lemma 11.2 *(a) For $X \in \mathcal{T}_M$*

$$\nabla_X \nabla_{\partial_z}^{-1} v_i \in \bigoplus_j \mathcal{O}_M \cdot v_j = pr_* \overline{\mathcal{H}^{(0)}}. \tag{11.19}$$

(b)

$$z \cdot v_i \in \bigoplus_j \mathcal{O}_M \cdot v_j + (\alpha_i + 1)\nabla_{\partial_z}^{-1} v_i. \tag{11.20}$$

Proof: (a) By Corollary 10.8, $\pi_*\mathcal{H}^{(0)}$ and the sheaf of fibrewise global sections of $\overline{\mathcal{H}^{(0)}}|_{\mathbb{C}\times M}$ are invariant under $\nabla_X \nabla_{\partial_z}^{-1}$.

For the germs at $(\infty, 0)$ one has to observe the following. The germ $\overline{\mathcal{H}^{(0)}}_{(\infty,0)}$ has a logarithmic pole along $\{\infty\} \times M$. It contains C^0, because its spectral numbers are $-\alpha_1, \ldots, -\alpha_\mu \in (-n, 1)$. If the monodromy has eigenvalue 1, i.e. $C^0 \neq 0$, then the germs in $\widetilde{z} \cdot \overline{\mathcal{H}^{(0)}}_{(\infty,0)}$ do not have unique preimages under ∇_{∂_z}, but any preimage is contained in $\overline{\mathcal{H}^{(0)}}_{(\infty,0)}$. Therefore the germs at $(\infty, 0)$ satisfy

$$\nabla_{\partial_z}^{-1} v_i \in \overline{\mathcal{H}^{(0)}}_{(\infty,0)} + \nabla_{\partial_z}^{-1} s_i, \tag{11.21}$$

$$\nabla_X \nabla_{\partial_z}^{-1} v_i \in \overline{\mathcal{H}^{(0)}}_{(\infty,0)}. \tag{11.22}$$

Thus $\nabla_X \nabla_{\partial_z}^{-1} v_i$ is a global section in $\overline{\mathcal{H}^{(0)}}$.

(b) One has to apply $z\nabla_{\partial_z} - (\alpha_i + 1)\,\mathrm{id}$ to (11.21) and observe (11.6). Then at $(\infty, 0)$

$$z \cdot v_i - (\alpha_i + 1)\nabla_{\partial_z}^{-1} v_i \in \overline{\mathcal{H}^{(0)}}_{(\infty,0)}. \tag{11.23}$$

This section is in any case in $\overline{\mathcal{H}^{(0)}}|_{\mathbb{C}\times M}$. Therefore it is a global section in $\overline{\mathcal{H}^{(0)}}$. □

In M. Saito's approach, the equations (11.19) and (11.20) follow directly from Malgrange's result. Formula (11.19) is used to prove Lemma 11.1 (b) (cf. [**SM3**, 4.3]). Again K_F is K. Saito's higher residue pairing (Theorem 10.11), and K_f is the pairing on $V^{>-1}$ defined in section 10.6.

Lemma 11.3 *(a) The restrictions $v_i^0 \in H_0''$ of the sections v_i satisfy*

$$v_i^0 \in s_i + \bigoplus_{\beta>\alpha_i} \bigoplus_{p>0} \nabla_{\partial_z}^p G^{\beta+p}, \tag{11.24}$$

$$K_f\big(v_i^0, v_{\mu+1-j}^0\big) = \delta_{ij}\partial_z^{-n-1}. \tag{11.25}$$

(b)

$$K_F(v_i, v_{\mu+1-j}) = \delta_{ij}\partial_z^{-n-1}. \tag{11.26}$$

Proof: (a) Formula (11.24) is obtained as (7.44) in the proof of Theorem 7.16: because of (11.9) and (11.10) H_0'' intersects the right hand side of (11.24) in a unique element. This element extends to a global section in $\overline{H''}$ and must coincide with the restriction of v_i.

The formulas (11.7), (11.8), and (11.24) show

$$K_f\big(v_i^0, v_{\mu+1-j}^0\big) \in \delta_{ij} \cdot \partial_z^{-n-1} + \sum_{k=1}^{n} \mathbb{C} \cdot \partial_z^{-k}. \tag{11.27}$$

On the other hand, $K_f(H_0'', H_0'') = \mathbb{C}\{\{\partial_z^{-1}\}\} \cdot \partial_z^{-n-1}$ (Theorem 10.7).

(b) One considers $K_F^{(-k)}(v_i, v_{\mu+1-j}) \in \mathcal{O}_{M,0}$ as a power series. The constant term is $K_f^{(-k)}(v_i^0, v_{\mu+1-j}^0) = \delta_{ij} \cdot \delta_{k0}$ because of Theorem 10.28 (i). The following calculation shows that the higher terms vanish. The third step uses (11.19) once. Here $X_1, \dots, X_m \in \mathcal{T}_{M,0}$ and $k \ge 0$.

$$\begin{aligned}
&\big(X_1 \dots X_m\, K_F^{(-k)}(v_i, v_j)\big)(0) \qquad\qquad (11.28)\\
&= K_F^{(-k)}\big(\nabla_{X_1} \dots \nabla_{X_m} v_i, v_j\big)(0) + \cdots\\
&= K_F^{(-k-m)}\big(\nabla_{X_1} \dots \nabla_{X_m} \nabla_{\partial_z}^{-m} v_i, v_j\big)(0) + \cdots\\
&\in K_F^{(-k-m)}\left(\bigoplus_{l=0}^{m-1} \nabla_{\partial_z}^{-l} \bigoplus_a \mathcal{O}_M \cdot v_a, v_j\right)(0) + \cdots\\
&= K_f^{(-k-m)}\left(\bigoplus_{l=0}^{m-1} \nabla_{\partial_z}^{-l} \bigoplus_a \mathbb{C} \cdot v_a^0, v_j^0\right) + \cdots = \{0\}.
\end{aligned}$$

□

Remark 11.4 In the case of a singularity with nondegenerate intersection form the description of K_F in Theorem 10.13 gives a better idea why (11.26) holds: $K_F^{(-k)}(v_i, v_j)$ is by (10.47) the residue along $\{\infty\} \times M$ of a certain meromorphic 1-form on $\mathbb{P}^1 \times M$. One can check that this residue is constant because of the properties of v_i and v_j along $\{\infty\} \times M$.

In the case of a degenerate intersection form one can embed the generalized Milnor fibration of the semiuniversal unfolding in a relative compactification and obtain an analogon of Theorem 10.13 (cf. Remarks 10.14 (b) and 10.29 (a)). I checked that one can make a refinement of the choice $(U_\bullet, \gamma_1)$ and that one can also explain (11.26) in this case with a residue along $\{\infty\} \times M$, but carrying out the details is quite intricate.

The section v_1 plays a special role. It turns out to be a primitive form in the sense of K. Saito. The germ $v_1^0 \in H_0''$ has the principal part in C^{α_1}. Then by Theorem 10.33 the germ $v_1 \in \mathcal{H}_0^{(0)}$ is represented by a volume form $u(x,t)$ $\mathrm{d}x_0 \ldots \mathrm{d}x_n$ with $u(0) \neq 0$. By Corollary 10.8 (iv) it induces an injective period map

$$v : \mathcal{T}_{M,0} \to \mathcal{H}_0^{(0)}, \quad X \mapsto -\nabla_X \nabla_{\partial_z}^{-1} v_1, \tag{11.29}$$

such that

$$\mathcal{H}_0^{(0)} = v(\mathcal{T}_{M,0}) \oplus \nabla_{\partial_z}^{-1} \mathcal{H}_0^{(0)}. \tag{11.30}$$

Because of (11.19), v is then an isomorphism

$$v : \mathcal{T}_{M,0} \to \bigoplus_{i=1}^{\mu} \mathcal{O}_{M,0} \cdot v_i. \tag{11.31}$$

Let

$$\delta_i := v^{-1}(v_i). \tag{11.32}$$

By the Kodaira–Spencer map $\mathbf{a} : \mathcal{T}_{M,0} \to \mathcal{O}_{C,0}$ from (10.3) we have the multiplication $\circ$ on $\mathcal{T}_{M,0}$ and the Euler field $E = \mathbf{a}^{-1}([F])$.

Lemma 11.5 concludes the proof of Theorem 11.1, except for the statement that different choices $(U_\bullet, \gamma_1)$ lead to different metrics g. This will be proved after Corollary 11.6.

Lemma 11.5 *(a) The δ_i satisfy $\delta_1 = e$ and $[\delta_i, \delta_j] = 0$ and*

$$-\nabla_{\delta_i} \nabla_{\partial_z}^{-1} v(\delta_j) = v(\delta_i \circ \delta_j). \tag{11.33}$$

(b) The composition of v and of the projection $r^{(0)} : \mathcal{H}_0^{(0)} \to \Omega_{F,0}$ from (10.38) coincides with the isomorphism

$$\mathcal{T}_{M,0} \to \Omega_{F,0} \quad X \mapsto \mathbf{a}(X) \cdot r^{(0)}(v_1). \tag{11.34}$$

The Grothendieck residue pairing J_F on Ω_F induces a flat metric g on M. The vector fields δ_i are flat with $g(\delta_i, \delta_{\mu+1-j}) = \delta_{ij}$. Let ∇^g denote the Levi–Civita connection of g.

(c) The metric g is multiplication invariant and satisfies together with the multiplication the potentiality condition.

(d) The Euler field E satisfies $[\delta_i, E] = (1 + \alpha_1 - \alpha_i)\delta_i$.

Proof: (a) Crucial for (a) and (b) is formula (10.39),

$$r^{(0)}\big(-\nabla_X \nabla_{\partial_z}^{-1}\omega\big) = \mathbf{a}(X) \cdot r^{(0)}(\omega) \tag{11.35}$$

for $X \in \mathcal{T}_{M,0}$, $\omega \in \mathcal{H}_0^{(0)}$. For $\omega = v_1$ and $X = \delta_1$ it shows $\mathbf{a}(\delta_1) = 1$, so $\delta_1 = e$. Together with $\mathbf{a}(\delta_i \circ \delta_j) = \mathbf{a}(\delta_i) \cdot \mathbf{a}(\delta_j)$ it shows that (11.33) holds modulo $\ker r^{(0)} = \nabla_{\partial_z}^{-1}\mathcal{H}_0^{(0)} = \mathcal{H}_0^{(-1)}$. Then (11.33) holds because of (11.19) and (11.30). The vector fields δ_i commute because of

$$\begin{aligned} -\nabla_{\partial_z}^{-1} v([\delta_i, \delta_j]) &= \nabla_{[\delta_i,\delta_j]}\nabla_{\partial_z}^{-2} v_1 \\ &= v(\delta_i \circ \delta_j) - v(\delta_j \circ \delta_i) = 0. \end{aligned} \tag{11.36}$$

(b) The first statement is (11.35) for $\omega = v_1$. The pairing $K_F^{(0)}$ induces J_F via the projection $r^{(0)}$ (Theorem 10.11). Then (11.26) shows $g(\delta_i, \delta_{\mu+1-j}) = \delta_{ij}$. The metric g is flat because the vector fields δ_i commute.

(c) The metric g is multiplication invariant because J_F satisfies the corresponding property (10.45). The manifold $(M, \circ, e)$ is an F-manifold and e is flat. In Theorem 2.15 it is proved that these three properties imply potentiality. But the potentiality in the form $\nabla^g_{\delta_i}(\delta_j \circ \delta_k) = \nabla^g_{\delta_j}(\delta_i \circ \delta_k)$ also follows from rewriting $0 = \nabla_{[\delta_i,\delta_j]}\nabla_{\partial_z}^{-2} v(\delta_k)$. Here one needs the $\mathcal{O}_{M,0}$-linear extension of (11.33),

$$-\nabla_X \nabla_{\partial_z}^{-1} v(Y) = v(X \circ Y) - \nabla_{\partial_z}^{-1} v\big(\nabla^g_X Y\big) \tag{11.37}$$

for $X, Y \in \mathcal{T}_{M,0}$.

(d) The vector fields $E - ze$ and $\partial_z + e$ and therefore also $E + z\partial_z$ are logarithmic along $\check{\mathcal{D}}$ by (10.9); in fact, $E + z\partial_z$ is also logarithmic along $\{\infty\} \times M$. The sheaf $\nabla_{\partial_z}^{-1}\mathcal{H}^{(0)} = \mathcal{H}^{(-1)}$ has a logarithmic pole along $\check{\mathcal{D}}$. Therefore

$$\big(z\nabla_{\partial_z} + \nabla_E\big)\nabla_{\partial_z}^{-1} v_i \in \nabla_{\partial_z}^{-1}\mathcal{H}^{(0)} = z \cdot v_i - v(E \circ \delta_i). \tag{11.38}$$

Then (11.20) shows

$$z \cdot v_i = v(E \circ \delta_i) + (\alpha_i + 1)\nabla_{\partial_z}^{-1} v_i. \tag{11.39}$$

We write $E = \sum_j \varepsilon_j \delta_j$ for some $\varepsilon_j \in \mathcal{O}_{M,0}$. Comparison of (11.39) and

$$\begin{aligned} z\nabla_{\partial_z} v_i &= -\nabla_{\delta_i} z \cdot v_1 \\ &= -\sum_j \delta_i(\varepsilon_j) v_j - \sum_j \varepsilon_j \cdot \nabla_{\delta_i} v_j + (\alpha_1 + 1) v_i \end{aligned} \tag{11.40}$$

shows $\delta_i(\varepsilon_j) = 0$ for $i \neq j$ and $\delta_i(\varepsilon_i) = 1 + \alpha_1 - \alpha_i$.

Therefore $\nabla^g_{\delta_i} E = [\delta_i, E] = (1 + \alpha_1 - \alpha_i)\delta_i$.

One also now obtains the following nice formula as an $\mathcal{O}_M$-linear extension of (11.39). For $X \in \mathcal{T}_M$

$$z \cdot v(X) = v(E \circ X) + \nabla_{\partial_z}^{-1} v\big((2 + \alpha_1)X - \nabla^g_X E\big). \qquad (11.41)$$

□

Corollary 11.6 *The period map v yields an isomorphism between the pairs $(\overline{\mathcal{H}^{(0)}}, \nabla)$ and $(pr^*\mathcal{T}_M, \check{\nabla}^{(-\frac{n}{2})})$ (the second one is a second structure connection of the Frobenius manifold $(M, \circ, e, E, g)$).*

Proof: One compares (9.10) and (9.11) with (11.37) and (9.13) with (11.41). One needs $\mathcal{V} - \frac{1}{2} - \frac{n}{2} = \nabla^g E - \frac{D}{2} - \frac{1}{2} - \frac{n}{2} = \nabla^g E - (2 + \alpha_1)$. □

Now we can prove the injectivity of the map {choices $(U_\bullet, \gamma_1)$} → {metrics g} in Theorem 11.1. If one fixes $U_\bullet$ then the construction of v_1 in (11.16) and (11.17) shows that the metric g varies linearly with the element γ_1.

Suppose that two choices of $(U_\bullet, \gamma_1)$ lead to the same metric. By Corollary 11.6, there exists an automorphism of the flat cohomology bundle which maps one extension $\overline{\mathcal{H}^{(0)}}$ of its sheaf of sections to the other. The automorphism commutes with the whole monodromy group.

The monodromy group is generated by the Picard–Lefschetz transformations of a distinguished basis of vanishing cycles. The automorphism commutes with them and must map each distinguished cycle to a multiple of itself. The Coxeter–Dynkin diagram is connected [**La**][**Ga**]. Therefore the multiples are all the same, and the automorphism is a multiple of the identity.

Thus the two extensions $\overline{\mathcal{H}^{(0)}}$ coincide and the two opposite filtrations coincide, because one can recover them from the extension at $\{\infty\} \times M$. Now the elements γ_1 must also coincide. This finishes the proof of Theorem 11.1.

Remarks 11.7 (a) For most of the singularities there also exist metrics g other than those constructed in Theorem 11.1 such that the F-manifold $(M, \circ, e, E)$ together with g is a Frobenius manifold. Even the eigenvalues of $\nabla^g E$ are different.

These other metrics arise in the same way as those in Theorem 11.1, but after changing a very fundamental choice: One replaces the natural order $<$ on the set $\{\alpha \mid e^{-2\pi i\alpha}$ is an eigenvalue of the monodromy $\} \cup \mathbb{Z}$ by another order $\prec$ which satisfies (7.9).

Then the spaces H^∞, C^α, $V^{>-\infty}$, and H_0'' are still the same, but the V-filtration, the spectral numbers of H_0'', and the corresponding filtrations $F_\bullet$ and $F_\bullet^{alg}$ from (7.24) and (7.27) usually change. Often opposite filtrations exist and the whole construction in this section can be carried out.

(b) Let us sketch this for an example of M. Saito [**SM3**, 4.4]. The semi-quasihomogeneous singularity $f = x_0^6 + x_1^6 + x_0^4 x_1^4$ has spectral numbers $\alpha_1 \le \ldots \le \alpha_{25}$ which are as an unordered tuple $(-1 + \frac{i+j+2}{6} \mid 0 \le i, j \le 4)$. One chooses $s_i \in C^{\alpha_i}$ as in (11.14) and (11.15) (this includes the choice of an opposite filtration). It then turns out that $v_i^0 = s_i$ for $i \ge 2$ and $v_1^0 = s_1 + a \cdot \nabla_{\partial_z} s_{25}$ for some $a \in \mathbb{C}^*$, with $s_1 \in C^{-2/3} = C^{\alpha_1}$, $\nabla_{\partial_z} s_{25} \in C^{-1/3} = C^{\alpha_{25}-1}$.

If one now chooses a new order $\prec$ as in (7.9) with $-\frac{1}{3} \prec -\frac{2}{3}$ then the spectral numbers for this order are (no longer with indices fitting to their order) $\widetilde{\alpha}_i = \alpha_i$ for $2 \le i \le 24$ and $\widetilde{\alpha}_1 = \alpha_{25} - 1 = -\frac{1}{3}, \widetilde{\alpha}_{25} = \alpha_1 + 1 = \frac{1}{3}$.

Now one can choose $\widetilde{s}_i = s_i$ for $2 \le i \le 24$ and $\widetilde{s}_1 = \nabla_{\partial_z} s_{25}, \widetilde{s}_{25} = -\nabla_{\partial_z}^{-1} s_1$. Then $\widetilde{v}_i^0 = v_i^0$ for $1 \le i \le 24$ and $\widetilde{v}_{25}^0 = \widetilde{s}_{25}$.

The whole construction in this section can be carried out. One obtains a Frobenius manifold with $[\widetilde{\delta}_i, E] = (1 + \widetilde{\alpha}_1 - \widetilde{\alpha}_i)\widetilde{\delta}_i$.

(c) The different choices of opposite filtrations and of orders $\prec$ seem to be related both to certain transformations of Frobenius manifolds of Dubrovin [**Du3**, Appendix B] [**Du4**, first part of Lecture 4] and to the Schlesinger transformations in [**JMU**][**JM**].

11.2 Deformed flat coordinates

In this section we want to establish series of functions $c_{ij}^{(p)} \in \mathcal{O}_{M,0}$ which are implicit in the construction of section 11.1 and which are very close to Dubrovin's *deformed flat coordinates*. These deformed flat coordinates play a major role in many of Dubrovin's papers on Frobenius manifolds. They lead to rich hidden structures [**Du3**, Lecture 6], [**DuZ1**], [**DuZ2**], [**Gi8**], whose meaning for singularities has still to be explored.

We will use our series of functions in chapter 12 in order to establish a canonical complex structure on the μ-constant stratum and study a period map.

Remark 11.8 The most detailed description of the deformed flat coordinates can be found in [**Du4**, Lecture 2]. We refrain from repeating that here and restrict ourselves to some comments.

The first structure connection $\hat{\nabla}^{(-\frac{1}{2})}$ (Definition 9.6) on the lift pr^*TM of the tangent bundle to $\mathbb{P}^1 \times M$ is flat and has a logarithmic pole along $\{0\} \times M$. The same holds for the induced connection on the dual bundle pr^*T^*M.

Dubrovin's deformed flat coordinates $\widetilde{t}_i(t,z)$ are multivalued functions on $\mathbb{C}^* \times M$ whose restrictions to slices $\{z\} \times M$ form locally flat coordinates with respect to $\hat{\nabla}^{(-\frac{1}{2})}$.

They also have the best possible behaviour with respect to ∂_z. To make that precise, let d_M denote the differential only with respect to the coordinates on M, not with respect to z. Then the differentials $\mathrm{d}_M\widetilde{t}_i$ are multivalued flat sections in $pr^*T^*M\,|\mathbb{C}^* \times M$ with respect to the induced connection.

The deformed flat coordinates $\widetilde{t}_i$ are usually written as a tuple

$$(\widetilde{t}_1(t,z),\ldots,\widetilde{t}_m(t,z)) = (\theta_1(t,z),\ldots,\theta_m(t,z))z^{\mu}z^{R} \tag{11.42}$$

[**Du4**, (2.83)], [**DuZ1**, (2.63)], [**DuZ2**, (3.13)]. The matrices μ and R are supposed to be chosen such that the following holds:

The functions $\theta_i(t,z)$ are holomorphic in $\mathbb{C}^* \times M$. Their differentials $\mathrm{d}_M\theta_i$ form a basis of elementary sections in $pr^*T^*M\,|\mathbb{C}^* \times M$. (Here $\pm\mu$ is the matrix for the (by assumption) semisimple residue endomorphism, the matrix R carries the Jordan block structure of the monodromy.)

The coefficients $\theta_{jp}(t)$ in the expansions $\theta_j(t,z) = \sum_{p\geq 0}\theta_{jp}(t)z^p$ are more or less equivalent to the coefficients $c_{1j}^{(p+1)}$ below. 'More or less', because our coefficients come from the second structure connection $\check{\nabla}^{(-\frac{n}{2})}$ which is close to, but in general not equal to the Fourier dual of $\hat{\nabla}^{(-\frac{1}{2})}$ (cf. Remark 9.9 (b)). The θ_{jp} are the real starting point for the profound structures in [**Du3**, Lecture 6], [**DuZ1**], [**DuZ2**], [**Gi8**].

We consider the same situation as in section 11.1 and use the notions established there. Most important are the global sections v_i and the elementary sections s_i. The germs in $\overline{\mathcal{H}^{(0)}}_{(\infty,0)}$ of the global sections v_i will be developed in series

$$v_i = \sum_{1\leq j\leq\mu}\sum_{p\geq 0} c_{ij}^{(p)}(-\nabla_{\partial_z})^p s_j \tag{11.43}$$

with coefficients $c_{ij}^{(p)} \in \mathcal{O}_{M,0}$. Because of (11.12) and (11.14) one has

$$\overline{\mathcal{H}^{(0)}}_{(\infty,0)} = \bigoplus_j \mathcal{O}_{M,0}\{\widetilde{z}\}\cdot s_j, \tag{11.44}$$

where $\widetilde{z} = \frac{1}{z}$. Therefore the germs v_i in $(\infty,0)$ can be written uniquely as sums of elementary sections $\widetilde{z}^p s_j$. But we want to write them with $(-\nabla_{\partial_z})^p s_j$.

For $\alpha \notin \mathbb{Z}_{<0}$ there is no problem. The sheaf $\overline{\mathcal{H}^{(0)}}$ has a logarithmic pole along $\{\infty\}\times M$, and $-\partial_z = \widetilde{z}(\widetilde{z}\partial_{\widetilde{z}})$. Therefore the sections $(-\nabla_{\partial_z})^p s_j$ with $\alpha_j - p = \alpha$ and $p \geq 0$ form a basis of the same subspace of C^α as the sections $\widetilde{z}^p s_j$. But for $\alpha \in \mathbb{Z}_{<0}$ there are two related problems:

(I) The sections $(-\nabla_{\partial_z})^p s_j$ with $\alpha_j - p = \alpha$ and $p \geq 0$ generate a strictly smaller subspace of C^α than the sections $\widetilde{z}^p s_j$. A priori it is not clear that the elementary part of v_i in C^α is in this smaller subspace. But it will turn out to be the case.

(II) $(-\nabla_{\partial_z})^p s_j = 0$ for certain j and p. The coefficients $c_{ij}^{(p)}$ in (11.43) for these j and p are not unique, but arbitrary.

In order to obtain unique coefficients $c_{ij}^{(p)}$ we have to introduce a dummy exponent $\xi \in \mathbb{N}$ and consider the analogon of (11.43) for $(-\nabla_{\partial_z})^{-\xi} v_i$ for all $\xi \in \mathbb{N}$. That shifts the orders of the elementary sections out of the bad domain $\mathbb{Z}_{<0}$. Each coefficient $c_{ij}^{(p)}$ will be uniquely defined by the expansion of $(-\nabla_{\partial_z})^{-\xi} v_i$ for sufficiently large ξ. The same trick will be used to establish the relations between the $c_{ij}^{(p)}$. It is clumsy, but worth the effort. With the first structure connections, one would not have this difficulty.

Of course, $(-\nabla_{\partial_z})^{-\xi} v_i$ is the section in $\overline{\mathcal{H}^{(0)}}|\mathbb{C} \times M$ which is welldefined by the proof of Theorem 10.7. The germ at $(\infty, 0)$ is a germ in $\widetilde{z}^{-\xi} \cdot \overline{\mathcal{H}^{(0)}}_{(\infty,0)}$.

Theorem 11.9 *(a) Let v_i and s_i be as in section 11.1. There exist unique coefficients $c_{ij}^{(p)} \in \mathcal{O}_{M,0}$ ($p \in \mathbb{Z}$, $i, j \in \{1, \dots, \mu\}$) such that the germs at $(\infty, 0)$ satisfy*

$$(-\nabla_{\partial_z})^{-\xi} v_i = \sum_{1 \leq j \leq \mu} \sum_{p \in \mathbb{Z}} c_{ij}^{(p)} (-\nabla_{\partial_z})^{p-\xi} s_j \tag{11.45}$$

for all $\xi \in \mathbb{N}$. The coefficients satisfy $c_{ij}^{(p)} = 0$ for $p < 0$ and $c_{ij}^{(0)} = \delta_{ij}$.

(b) Let $\delta_i = v^{-1}(v_i)$ be the flat vector fields from (11.32). Let $a_{ij}^k \in \mathcal{O}_{M,0}$ be the multiplication coefficients with $\delta_i \circ \delta_j = \sum_k a_{ij}^k \delta_k$. Then

$$\nabla_{\delta_i} v_j = (-\nabla_{\partial_z}) \sum_k a_{ij}^k v_k \tag{11.46}$$

and

$$\delta_i c_{jl}^{(p)} = \sum_k a_{ij}^k c_{kl}^{(p-1)}. \tag{11.47}$$

Especially $\delta_1 = e$, $a_{i1}^k = \delta_{ik}$, and

$$\delta_i c_{1l}^{(p)} = c_{il}^{(p-1)}, \tag{11.48}$$

$$\delta_i c_{jl}^{(1)} = a_{ij}^l, \tag{11.49}$$

$$\delta_i c_{1l}^{(1)} = \delta_{il}. \tag{11.50}$$

The last equation says that $\tau_i := c_{1i}^{(1)}$ are flat coordinates on M with $\frac{\partial}{\partial \tau_i} = \delta_i$.

(c) For any $i, j \in \{1, \dots, \mu\}$, $r \in \mathbb{Z}$,

$$\delta_{i\mu+1-j}\delta_{0r} = \sum_{p,q:\ p+q=r} \sum_k (-1)^p c_{ik}^{(p)} c_{j\mu+1-k}^{(q)}. \tag{11.51}$$

The right hand side has for $r \geq 1$ *the linear terms* $(-1)^r c_{i\mu+1-j}^{(r)} + c_{j\mu+1-i}^{(r)}$ *and for* $r \geq 2$ *additional quadratic terms. Especially*

$$c_{i\mu+1-j}^{(1)} = c_{j\mu+1-i}^{(1)}. \tag{11.52}$$

(d) Let $n_{ij} \in \mathbb{C}$ *be the coefficients with (cf.* (11.6)*)*

$$(z\nabla_{\partial_z} - \alpha_i)s_i = \frac{-N}{2\pi i} s_i = \sum_j n_{ij} \nabla_{\partial_z} s_j. \tag{11.53}$$

Then $n_{ij} = 0$ *for* $\alpha_j - 1 - \alpha_i \neq 0$ *and*

$$z \cdot v_i = (\alpha_i + 1)\nabla_{\partial_z}^{-1} v_i + \sum_j n_{ij} v_j - \sum_j c_{ij}^{(1)}(\alpha_j - 1 - \alpha_i)v_j, \tag{11.54}$$

and for $p \geq 2$

$$\begin{aligned} c_{ik}^{(p)}(\alpha_k - p - \alpha_i) = &\sum_j \left(c_{ij}^{(p-1)} n_{jk} - n_{ij} c_{jk}^{(p-1)}\right) \\ &+ \sum_j (\alpha_j - 1 - \alpha_i) c_{ij}^{(1)} c_{jk}^{(p-1)}. \end{aligned} \tag{11.55}$$

(e) The Euler field E satisfies

$$E \circ \delta_i = \sum_j n_{ij}\delta_j - \sum_j c_{ij}^{(1)}(\alpha_j - 1 - \alpha_i)\delta_j, \tag{11.56}$$

$$E = \sum_j n_{1j}\delta_j + \sum_j \tau_j(1 + \alpha_1 - \alpha_j)\delta_j, \tag{11.57}$$

$$E\, c_{ij}^{(p)} = (\alpha_i + p - \alpha_j)c_{ij}^{(p)} + \sum_k c_{ik}^{(p-1)} n_{kj}. \tag{11.58}$$

Proof: (a) Because of $v_i = \nabla_{\partial_z}^{\xi}(\nabla_{\partial_z}^{-\xi} v_i)$, the elementary part of v_i in C^α is contained in $\nabla_{\partial_z}^{\xi} C^{\alpha+\xi}$ for any $\xi \in \mathbb{N}$. The same holds for the elementary parts of $(-\nabla_{\partial_z})^{-\xi} v_i$. This solves the problem (I) before Theorem 11.9. For each ξ one has an expansion (11.45). The coefficients $c_{ij}^{(p)}$ are independent of ξ and uniquely determined by the expansions for sufficiently large ξ. Now $c_{ij}^{(p)} = 0$ for $p < 0$ is trivial and $c_{ij}^{(0)} = \delta_{ij}$ follows from (11.18).

(b) Formula (11.46) is (11.33) rewritten. One complements $(-\nabla_{\partial_z})^{-\xi}$ on both sides and inserts the expansions (11.45). One obtains (11.47) by comparison of the coefficients of $(-\nabla_{\partial_z})^{p-\xi} s_l$ for sufficiently large ξ. The rest is obvious.

(c) First we formulate an approach which does not work, but which helps to understand (11.51). One would like to ignore that certain $(-\nabla_{\partial_z})^p s_k$ vanish. One would like to extend the pairing K_f from section 10.6 $\mathcal{O}_{M,0}\{\{\partial_z^{-1}\}\}[\partial_z]$-sesquilinear and calculate $K_F(v_i, v_j)$ with this extension of K_f and with the expansions (11.45). One would obtain (11.51) because of (11.15) and (11.26).

The real proof is somewhat similar to the proof of (11.26). One has to consider the right hand side of (11.51) as a power series in $\mathcal{O}_{M,0}$ and discuss the coefficients separately.

Let $X_1, \dots, X_m \in \mathcal{T}_{M,0}$ (the empty set for $m = 0$), $r \geq 0$ and $\xi \geq r + m$. Then the number

$$\delta_{m0}\delta_{i\mu+1-j}\delta_{r0} = \big(X_1 \dots X_m\, K_F^{(r-4\xi)}\big(\big(-\nabla_{\partial_z}\big)^{-2\xi} v_i, \big(-\nabla_{\partial_z}\big)^{-2\xi} v_j\big)\big)\big)(0) \tag{11.59}$$

can be calculated as follows: One inserts the expansion (11.45), uses Theorem 10.11 (v), Theorem 10.28 (i), (11.15), and the properties of K_f. One obtains the value at 0 of the derivative by $X_1 \dots X_m$ of the right hand side of (11.51). Details are left to the reader.

(d) Formulas (11.54) and (11.55) will be proved simultaneously by the following calculation. The first step uses (11.45), (11.53) and $[z, (-\partial_z)^k] = k(-\partial_z)^{k-1}$. In the second step the expansion (11.45) is applied to the terms for $p = 0$, in the third step to a part of the terms for $p = 1$.

$$\begin{aligned}
&\big(-\nabla_{\partial_z}\big)^{-\xi} z \cdot v_i \\
&= \big(-\nabla_{\partial_z}\big)^{-\xi} \sum_{j,p} c_{ij}^{(p)} \big(-\nabla_{\partial_z}\big)^p \Big((\alpha_j + 1 - p)\nabla_{\partial_z}^{-1} s_j + \sum_k n_{jk} s_k\Big) \\
&= \big(-\nabla_{\partial_z}\big)^{-\xi} (\alpha_i + 1)\nabla_{\partial_z}^{-1} v_i + \big(-\nabla_{\partial_z}\big)^{-\xi} \sum_j n_{ij} v_j \\
&\quad + \big(-\nabla_{\partial_z}\big)^{-\xi} \sum_j \sum_{p\geq 1} c_{ij}^{(p)} \big(-\nabla_{\partial_z}\big)^p \Big((\alpha_j - p - \alpha_i)\nabla_{\partial_z}^{-1} s_j + \sum_k n_{jk} s_k\Big) \\
&\quad - \big(-\nabla_{\partial_z}\big)^{-\xi} \sum_{j,k} \sum_{p\geq 1} n_{ij} c_{jk}^{(p)} \big(-\nabla_{\partial_z}\big)^p s_k \\
&= \big(-\nabla_{\partial_z}\big)^{-\xi} (\alpha_i + 1)\nabla_{\partial_z}^{-1} v_i + \big(-\nabla_{\partial_z}\big)^{-\xi} \sum_j n_{ij} v_j \\
&\quad - \big(-\nabla_{\partial_z}\big)^{-\xi} \sum_j c_{ij}^{(1)} (\alpha_j - 1 - \alpha_i) v_j \\
&\quad + \big(-\nabla_{\partial_z}\big)^{-\xi} \sum_{j,k} \sum_{p\geq 1} \big(c_{ij}^{(p)} n_{jk} - n_{ij} c_{jk}^{(p)}\big)\big(-\nabla_{\partial_z}\big)^p s_k
\end{aligned}$$

$$- (-\nabla_{\partial_z})^{-\xi} \sum_k \sum_{p\geq 2} c_{ik}^{(p)} (\alpha_k - p - \alpha_i)(-\nabla_{\partial_z})^{p-1} s_k$$

$$+ (-\nabla_{\partial_z})^{-\xi} \sum_j c_{ij}^{(1)} (\alpha_j - 1 - \alpha_i) \sum_k \sum_{p\geq 1} c_{jk}^{(p)} (-\nabla_{\partial_z})^p s_k. \qquad (11.60)$$

The germ at $(\infty, 0)$ of $z \cdot v_i$ is contained in

$$z\overline{\mathcal{H}^{(0)}}_{(\infty,0)} = \overline{\mathcal{H}^{(0)}}_{(\infty,0)} \oplus \bigoplus_j \mathcal{O}_{M,0} \nabla_{\partial_z}^{-1} v_j. \qquad (11.61)$$

The restriction to $\mathbb{C}\times M$ of $z\cdot v_i$ is contained in $\overline{\mathcal{H}^{(0)}}|\mathbb{C}\times M$. Therefore $z\cdot v_i$ is contained in $\bigoplus_j \mathcal{O}_{M,0}\cdot v_j \oplus \bigoplus_j \mathcal{O}_{M,0}\cdot \nabla_{\partial_z}^{-1} v_j$. The difference between $(-\nabla_{\partial_z})^{-\xi} z\cdot v_i$ and the first two lines of (11.60) is an element of $(-\nabla_{\partial_z})^{-\xi}(\bigoplus_j \mathcal{O}_{M,0} \cdot v_j \oplus \bigoplus_j \mathcal{O}_{M,0} \cdot \nabla_{\partial_z}^{-1} v_j)$. If it were nonvanishing then the expansion into elementary parts would yield a nonvanishing part in $(-\nabla_{\partial_z})^{-\xi}(\bigoplus_j \mathcal{O}_{M,0} \cdot s_j \oplus \bigoplus_j \mathcal{O}_{M,0} \cdot \nabla_{\partial_z}^{-1} s_j)$. But the sum of the last three lines of (11.60) does not have such a part. Therefore the difference and this sum are both vanishing. This gives (11.54) and (11.55).

(e) Formulas (11.56) and (11.57) follow from (11.39) and (11.54). Again with (11.39) one obtains $\alpha_i v_i = \nabla_{E+z\partial_z} v_i$. One puts into both sides the expansion (11.45) and applies (11.53) on the right. Comparison of coefficients gives (11.58). □

Remarks 11.10 (a) There is a rich structure hidden in the equations in Theorem 11.9, as is clear from Dubrovin's and Zhang's work. The most important coefficients are the $c_{1l}^{(p)}$ in view of (11.48). Central for Dubrovin is equation (11.47), usually written as a second order equations for the $c_{1l}^{(p+1)}$. For the μ-constant stratum we will mainly regard the equations in (d). Equation (11.55) shows that the coefficients $c_{ij}^{(1)}$ and the $c_{ij}^{(p)}$ with $\alpha_j - p - \alpha_i = 0$ determine all coefficients $c_{ij}^{(p)}$.

(b) For fixed coefficients $c_{ij}^{(p)} \in \mathbb{C}$ with $c_{ij}^{(p)} = 0$ for $\alpha_j - p - \alpha_i \leq 0$ the equations (11.55) and (11.52) imply (11.51). This was proved in [**He4**, Proposition 5.5]. It was used to describe coordinates for the fibres of the classifying space D_{BL} for Brieskorn lattices as a bundle over a classifying space D_{PMHS} for polarized mixed Hodge structures.

But if only $c_{ij}^{(p)} = 0$ for $\alpha_j - p - \alpha_i < 0$ then the coefficients with $\alpha_j - p - \alpha_i = 0$ satisfy equations of the type (11.51) which do not follow from (11.55) and (11.52).

11.3 Remarks on mirror symmetry

Dubrovin's definition [**Du1**][**Du3**] of Frobenius manifolds formalized a part of the structures which the physicists Witten [**Wit**], Dijkgraaf, E. Verlinde, and H. Verlinde [**DVV**] had found studying (moduli spaces of) topological field theories.

Different ways in physics to establish such structures lead to phenomena which are now comprised in the famous notion *mirror symmetry*.

There are several very different versions of mirror symmetry. A vague paraphrase of one version is that one has two sides, the A-side and the B-side: on the A-side one looks at data related to the Kähler geometry of a manifold X, on the B-side one looks at data related to the complex geometry of a family of manifolds Y_t. These data should be isomorphic if the manifold on the A-side and the family of manifolds on the B-side are *mirror dual* to one another. See [**Vo**] for a detailed discussion and references.

The structure on the A-side of this version of mirror symmetry comes from genus 0 Gromov–Witten invariants and is called *quantum cohomology* . Within the frame of algebraic geometry it has been established by Kontsevich, Manin, and many others (see [**KM**], [**Man2**], and references therein). If $X \subset \mathbb{P}^N$ is a manifold with $H^{odd}(X, \mathbb{C}) = 0$, then its quantum cohomology can be encoded in a formal germ of a Frobenius manifold, where the formal germ of a manifold is $(H^*(X, \mathbb{C}), 0)$. The part of the structure which lives in some sense on $(H^2(X, \mathbb{C}), 0) \subset (H^*(X, \mathbb{C}), 0)$ is called *small quantum cohomology* (see [**Man2**] for any details).

The structure on the B-side of this version of mirror symmetry is related to period integrals, Picard–Fuchs equations, hypergeometric functions. If Y_t is a family mirror dual to X, then the parameter t should be in a space isomorphic to a neighbourhood of 0 in $H^2(X, \mathbb{C})$, and the data from Y_t should be isomorphic to the small quantum cohomology of X. The first example of this was proposed be Candelas, de la Ossa, Green, Parkes [**CDGP**]. It was proved together with many other cases by Givental [**Gi4**].

In view of the full quantum cohomology and the Frobenius manifold on the A-side, one may ask about an extension of the family Y_t and the data on the B-side, such that in that case one also obtains a Frobenius manifold, and such that it is isomorphic to the Frobenius manifold on the A-side. That would be a stronger version of mirror symmetry.

One approach to this is provided by the Barannikov-Kontsevich construction [**BaK**][**Ba1**][**Ba2**]. Using tools from formal deformation theory, there an extended moduli space $\mathcal{M}$ (a priori a formal germ, in good cases a manifold) is constructed such that a subspace $\mathcal{M}_{cs}$ (*cs* for complex structure) parameterizes a family Y_t of complex manifolds. Under certain assumptions the space $\mathcal{M}$ is equipped with the structure of a Frobenius manifold. Building on Givental's results, Barannikov proved isomorphy of this Frobenius manifold with that from quantum cohomology of X when X is a projective complete intersection Calabi–Yau manifold [**Ba1**]. A major and still unsolved problem is to find objects which are parameterized by points $t \in \mathcal{M} - \mathcal{M}_{cs}$.

Another approach to get Frobenius manifolds on the B-side is motivated by a proposal of Givental [**Gi3**] and Eguchi, Hori, Xiong [**EHX**]: if X on the A-side is not a Calabi–Yau manifold, but, for example, a Fano manifold, one should consider on the B-side as mirror dual a family (Y_t, f_t), where $f_t : Y_t \to \mathbb{C}$ is a function on Y_t. If X is Calabi–Yau then f_t should be constant, but in other cases it may be a function with isolated singularities on the (now not compact, but usually affine) manifold Y_t.

Then the data on the B-side for the family (Y_t, f_t) come from oscillating integrals. Givental proved in many cases that they are equivalent to the small quantum cohomology of the mirror dual manifold X [**Gi3**][**Gi4**][**Gi5**][**Gi6**]. In many cases the parameter space of the family is isomorphic to an open domain in $H^2(X, \mathbb{C})$, and at all isolated singularities of the single functions f_t, the family is a μ-constant deformation.

Now the extension of the family (Y_t, f_t) to a family with parameter space M with $\dim M = \dim H^*(X, \mathbb{C})$ is often not difficult: one has to consider also deformations of the functions f_t which are not μ-constant at the singular points. One arrives at a situation which generalizes the semiuniversal unfolding of a local singularity $f : (\mathbb{C}^{n+1}, 0) \to (\mathbb{C}, 0)$.

Sabbah saw that one can often establish the structure of a Frobenius manifold by a procedure which is similar to that in section 11.1. But one needs oscillating integrals instead of the Gauß–Manin connection. This is because the middle cohomology group of the smooth fibres $f_t^{-1}(z)$ usually has a dimension $\neq \dim M$. Also, the results on the existence of a Frobenius manifold are not as complete as in the case of a local singularity. This is discussed in [**Sab2**][**Sab4**]. The best positive results are due to him [**Sab3**][**NS**]. We give a sketch of our understanding of the situation and part of his work in the next section.

The family of functions mirror dual to $\mathbb{P}^n$ was proposed and studied in [**Gi4**] and [**EHX**]. The construction of its Frobenius manifold was carried out by Barannikov [**Ba3**]. It fits into the frame given by Sabbah.

11.4 Remarks on oscillating integrals

The version of mirror symmetry which was discussed in section 11.3 asks in its strongest form whether certain Frobenius manifolds from quantum cohomology of projective manifolds are isomorphic to Frobenius manifolds coming from suitable families of functions on suitable manifolds.

Semiuniversal unfoldings of local singularities are not suitable families of functions, simply because their spectral numbers never fit to those from quantum cohomology.

But one may hope to find suitable families of functions starting from functions which were studied by Sabbah [**Sab2**][**NS**]: *M-tame functions* on affine manifolds (definition at the end of this section). Sabbah indicated how one may proceed to construct Frobenius manifolds from them [**Sab3**][**Sab4**].

The construction is similar to that in section 11.1. But instead of the Gauß–Manin connection and a second structure connection one considers oscillating integrals and a first structure connection. The two most difficult steps are:

(a) to construct a meromorphic connection on a trivial vector bundle over $\mathbb{P}^1 \times M$ which looks like a first structure connection;

(b) to find a certain global section (a *primitive form*) which induces an isomorphism between the lifted tangent bundle π^*TM and this trivial bundle.

Step (a) involves (the spirit of) oscillating integrals and the choice of an opposite filtration. By some work of Sabbah, it can be carried out for a large class of functions, all M-tame functions on affine manifolds. But the existence of a primitive form in step (b) is for the moment only clear for a much smaller class of functions, for polynomials with nondegenerate Newton boundary.

In [**Ba3**], Barannikov treated (independently) the family of functions mirror dual to $\mathbb{P}^{n+1}$. Although the presentation is completely different, this fits into Sabbah's frame.

Not all steps in Sabbah's programme have been worked out in detail. And we will not do it here. We will restrict ourselves to formulate this programme in 5 steps and then comment upon some aspects of the steps.

Step 1: Construct a family of functions $f_t : X_t \to \Delta$, $t \in M \subset \mathbb{C}^\mu$, with the following properties.

(a) The set $\Delta = \{z \in \mathbb{C} \mid |z| < \eta\}$ is a disc, $X_t \subset \mathbb{C}^N$ is a Stein manifold of dimension $n+1$, $\mathcal{X} := \bigcup_t X_t \times \{t\} \subset \mathbb{C}^N \times M$ is a manifold, M is a (small) open subset of $\mathbb{C}^\mu$.

(b) The function $f_t : X_t \to \Delta$ has isolated singularities with $\mu = \sum_{x \in Sing(f_t)} \mu(f_t, x)$, and for any $\widetilde{t} \in M$ the restriction of the family of functions to the multigerm at $Sing(f_{\widetilde{t}}) \times \{\widetilde{t}\} \subset \mathcal{X}$ is a product of semiuniversal unfoldings of the singularities of $f_{\widetilde{t}}$.

(c) The map

$$\begin{aligned} \varphi : \mathcal{X} &\to \Delta \times M \\ (x, t) &\mapsto (f_t(x), t) \end{aligned} \tag{11.62}$$

is a C^∞-locally trivial fibration outside of a discriminant $\check{\mathcal{D}} \subset \Delta \times M$. For any $(z,t) \in \check{\mathcal{D}}$, the fibre $\varphi^{-1}(z,t) = f_t^{-1}(z) \times \{t\}$ is singular, and there exist arbitrarily small neighbourhoods $U_1 \subset \mathcal{X}$ of $Sing(\varphi^{-1}(z,t))$, $U_2 \subset \Delta$ of z, and $U_3 \subset M$ of t such that the restriction of φ to $\varphi^{-1}(U_2 \times U_3) - U_1$ is a C^∞-locally trivial fibration.

Step 2: Construct a flat bundle H_{Lef} of rank μ over $\mathbb{C}^* \times M$ (definition in (11.66)) with connection ∇^{Lef} such that the fibres of the dual bundle can be interpreted as spaces of Lefschetz thimbles. Let $\mathcal{H}_{Lef}$ be the sheaf of holomorphic sections of H_{Lef}. Define a certain pairing on the sheaf $\pi_*\mathcal{H}_{Lef}$ (here $\pi : \mathbb{C}^* \times M \to M$).

Step 3: Extend the sheaf $\mathcal{H}_{Lef}$ to a free $\mathcal{O}_{\mathbb{C}\times M}$-module $\mathcal{H}_{Lef}^{(0)}$ such that the pair $(\mathcal{H}_{Lef}^{(0)}, \nabla^{Lef})$ has a pole of Poincaré rank 1 along $\{0\} \times M$. This uses the Gauß–Manin connection of φ and a Fourier–Laplace transformation with parameters. A geometric interpretation imitates oscillating integrals on a (co)homological level.

Step 4: Try to extend the sheaf $\mathcal{H}_{Lef}^{(0)}$ to a free $\mathcal{O}_{\mathbb{P}^1\times M}$-module $\overline{\mathcal{H}_{Lef}^{(0)}}$ with a logarithmic pole along $\{\infty\} \times M$. This is a Birkhoff problem with parameters. Sabbah showed that this is solvable for similar reasons as in the case of a local singularity if one starts with an M-tame function on an affine manifold. He established a mixed Hodge structure on the space H_{Lef}^∞ of manyvalued flat global sections of H_{Lef} [**NS**]. An opposite filtration to its Hodge filtration induces a solution of the Birkhoff problem.

Step 5: Try to find a global section v_1 (a primitive form) in $\overline{\mathcal{H}_{Lef}^{(0)}}$ which is flat at $\{\infty\} \times M$ with respect to the residual connection there, which is an eigenvector of the residue endomorphism at $\{\infty\} \times M$, and which induces an isomorphism $\mathcal{T}_M \to \pi_*\overline{\mathcal{H}_{Lef}^{(0)}}$ by a period map. Then a part of the pairing in step 2 induces a flat metric in $\mathcal{T}_M$ such that M gets a Frobenius manifold. This isomorphism also identifies $(\mathcal{H}_{Lef}^{(0)}, \nabla^{Lef})$ and the first structure connection $(\hat{\nabla}^{(-\frac{n}{2}-1)}, pr^*\mathcal{T}_M)$ (with the coordinate $z = -\frac{1}{\hbar}$, where $\hbar$ is a coordinate on $\mathbb{C}^*$ used below).

Because of condition (b) in step 1, M will be a massive F-manifold with Euler field and smooth analytic spectrum.

The cohomology bundle of the fibration φ from (11.62) is

$$H^n := \bigcup_{(z,t)\in\Delta\times M-\check{\mathcal{D}}} H^n(f_t^{-1}(z), \mathbb{C}). \tag{11.63}$$

If the spaces X_t are not contractible then in general the fibres of H^n do not have dimension μ. In that case there is no chance of extending the Gauß–Manin connection to something isomorphic to a second structure connection. But by condition (c) in step 1, the spaces of Lefschetz thimbles have the right dimension μ. This is the reason why oscillating integrals will be used. To define Lefschetz thimbles, fix for a moment $\hbar \in \mathbb{C}^*$ and $t \in M$. For sufficiently small $\varepsilon > 0$ let $\Delta^{(\hbar)} := \{z \in \Delta \mid |z - \eta\frac{\hbar}{|\hbar|}| < \varepsilon\}$ be a neighbourhood in Δ of the boundary point $\eta\frac{\hbar}{|\hbar|}$. Then $X_t^{(\hbar)} := f_t^{-1}(\Delta^{(\hbar)})$ is homotopy equivalent to a smooth fibre of f_t.

A *Lefschetz thimble* in X_t with boundary in $X_t^{(\hbar)}$ is an $(n+1)$-cycle $\Gamma = \bigcup_{z\in\gamma([0,1])} \delta(z)$, where $\gamma : [0,1] \to \Delta$ is a path with $\gamma(0) \in f_t(Sing(f_t))$, $\gamma((0,1)) \cap f_t(Sing(f_t)) = \emptyset$, $\gamma(1) \in \Delta^{(\hbar)}$, and $\delta(z)$ for $z \in \gamma([0,1])$ is a continuous family of n-cycles in $f_t^{-1}(z)$ which vanish for $z \to \gamma(0)$.

Condition (c) in step 1 and standard arguments (e.g. [**AGV2**, ch. 2], [**Lo2**, (5.11)]) show that one obtains a space homotopy equivalent to X_t if one glues μ Lefschetz thimbles to $X_t^{(\hbar)}$. (First one chooses $t \in M$ such that f_t has μ A_1-singularities with different critical values and considers Lefschetz thimbles over μ nonintersecting paths from $f_t(Sing(f_t))$ to $\Delta^{(\hbar)}$.) Therefore

$$H_{n+1}\big(X_t, X_t^{(\hbar)}, \mathbb{Z}\big) \cong \mathbb{Z}^\mu, \tag{11.64}$$

and this space is generated by the relative homology classes of Lefschetz thimbles. There is an exact sequence

$$\begin{aligned} 0 \to H_{n+1}(X_t, \mathbb{Z}) &\to H_{n+1}\big(X_t, X_t^{(\hbar)}, \mathbb{Z}\big) \\ &\to H_n\big(X_t^{(\hbar)}, \mathbb{Z}\big) \to H_n(X_t, \mathbb{Z}) \to 0. \end{aligned} \tag{11.65}$$

The bundle H_{Lef} in step 2 is

$$H_{Lef} := \bigcup_{(\hbar,t)\in\mathbb{C}^*\times M} H_{n+1}\big(X_t, X_t^{(\hbar)}, \mathbb{C}\big)^*. \tag{11.66}$$

A pairing on $\pi_*\mathcal{H}_{Lef}$ can be defined using the intersection form for Lefschetz thimbles, which is a perfect pairing

$$< .,. >: H_{n+1}\big(X_t, X_t^{(\hbar)}, \mathbb{Z}\big) \times H_{n+1}\big(X_t, X_t^{(-\hbar)}, \mathbb{Z}\big) \to \mathbb{Z}. \tag{11.67}$$

It induces a pairing $< .,. >^*$ on the dual spaces. Then

$$\begin{aligned} K_{Lef} : \pi_*\mathcal{H}_{Lef} \times \pi_*\mathcal{H}_{Lef} &\to \pi*\mathcal{O}_{\mathbb{C}^*\times M} \\ (a,b) &\mapsto \big(\hbar \mapsto < a(\hbar), b(-\hbar) >^*\big) \end{aligned} \tag{11.68}$$

is $\mathcal{O}_M$-bilinear and sesquilinear in $\hbar$.

The definition of the extension $\mathcal{H}_{Lef}^{(0)}$ of $\mathcal{H}_{Lef}$ to $\mathbb{C} \times M$ in step 3 is not easy. One has to start with an extension $\mathcal{H}^{(0)}$ to $\Delta \times M$ of the sheaf of sections of the cohomology bundle H^n. This should be defined using relative $(n+1)$-forms as in chapter 10, but the definition in Lemma 10.2 probably gives the right object also in this more general situation. Then a Fourier–Laplace transformation with parameters (i.e. keeping t and ∂_t, changing $z \mapsto -\partial_\tau$, $\partial_z \mapsto \tau$, where $\tau := \frac{1}{\hbar}$) should give $\mathcal{H}^{(0)}_{Lef}$.

The best way to handle this has still to be fixed. A geometric interpretation looks roughly as follows. One extends the homology bundle and the cohomology bundle H^n of φ to flat bundles over $\mathbb{C} \times M - \check{\mathcal{D}}$. If $\Gamma = \bigcup_{z\in\gamma([0,1])} \delta(z)$ is a Lefschetz thimble as above, one extends the family $[\delta(z)]$ of homology classes of cycles to a continuous family in the extended homology bundle over $\gamma([0,\infty))$, where $\gamma([1,\infty))$ is the half line which starts at $\gamma(1)$ and goes in the direction $\frac{\hbar}{|\hbar|}$. If $\omega \in \pi_*\mathcal{H}^{(0)}$ now extends to a global section over $\mathbb{C} \times M$ with moderate growth along $\{\infty\} \times M$, one defines

$$[e^{-z/\hbar}\omega]([\Gamma]) := \int_{\gamma([0,\infty))} e^{-z/\hbar}\omega([\delta(z)])\mathrm{d}z. \tag{11.69}$$

This imitates an *oscillating integral*. One has to show that it gives a holomorphic section $[e^{-z/\hbar}\omega]$ in H_{Lef}. Then such sections generate $\mathcal{H}^{(0)}_{Lef}$ as an $\mathcal{O}_{\mathbb{C}\times M}$-module

As soon as one has proved formulas which establish the Fourier–Laplace correspondence between sections of $\mathcal{H}^{(0)}$ and $\mathcal{H}^{(0)}_{Lef}$ (i.e. $z \mapsto -\partial_\tau$, $\partial_z \mapsto \tau$, where $\tau = \frac{1}{\hbar}$), it is not hard to see that $\mathcal{H}^{(0)}_{Lef}$ is a free $\mathcal{O}_{\mathbb{C}\times M}$-module with a pole of Poincaré rank 1 along $\{0\} \times M$.

If the family $f_t : X_t \to \Delta$ is a semiuniversal unfolding of a local singularity then $(-1)^{\frac{n(n+1)}{2}} \frac{1}{(2\pi i)^{n+1}} K_{Lef}$ on $\pi_*\mathcal{H}^{(0)}_{Lef}$ is the Fourier dual of K. Saito's higher residue pairings K_F on $\pi_*\mathcal{H}^{(0)}$ from section 10.4. This follows from [**Ph5**, 2ème partie 4].

In order to find an extension to a free $\mathcal{O}_{\mathbb{P}^1\times M}$-module $\overline{\mathcal{H}^{(0)}_{Lef}}$ with a logarithmic pole along $\{\infty\} \times M$, one can restrict to a fixed parameter $\widetilde{t} \in M$. Let $\mathcal{H}^{(0)}_{Lef}|_{\widetilde{t}}$ be the free $\mathcal{O}_{\mathbb{C}}$-module of restrictions of sections in $\mathcal{H}^{(0)}_{Lef}$ to $H_{Lef}|_{\mathbb{C}^*\times\{\widetilde{t}\}}$. It has a pole of order ≤ 2 at 0. One wants to solve the Birkhoff problem for $\mathcal{H}^{(0)}_{Lef}|_{\widetilde{t}}$ in the spirit of sections 7.2 and 7.4, by defining a filtration $F^\bullet$ in H^∞_{Lef} from $\mathcal{H}^{(0)}_{Lef}|_{\widetilde{t}}$ and choosing an opposite filtration.

But the procedures in sections 7.2 and 7.4 do not apply, because $\mathcal{H}^{(0)}_{Lef}|_{\widetilde{t}}$ has a pole of order ≤ 2 at 0. Sabbah found a very nice variation of these procedures. One has to look at global sections in $\mathcal{H}^{(0)}_{Lef}|_{\widetilde{t}}$ with moderate growth at ∞ and has to define principal parts for them using the V-filtration at ∞. The principal parts induce a filtration $F^\bullet$ on H^∞_{Lef}. Any opposite filtration gives rise to a solution of the Birkhoff problem for $\mathcal{H}^{(0)}_{Lef}|_{\widetilde{t}}$. See [**Sab4**, III 2.b and IV 5.b] and [**Sab2**] for details.

A profound result of Sabbah is that $F^\bullet$ is a Hodge filtration of a mixed Hodge structure if $f_{\tilde{t}} : X_{\tilde{t}} \to \Delta$ is the restriction of an M-tame function $f_{\tilde{t}} : Y \to \mathbb{C}$ on an affine manifold $Y \supset X_{\tilde{t}}$ [**NS**][**Sab2**]. Then opposite filtrations exist and the Birkhoff problem is solvable.

Definition 11.11 ([**NZ**][**NS**]) The function $f_{\tilde{t}} : Y \to \mathbb{C}$ is *M-tame* if for some closed embedding $Y \subset \mathbb{C}^N$ of the affine manifold Y the following holds: for any $\eta > 0$ there exists an $R(\eta) > 0$ such that the spheres $S^{2N+1}_{R(\eta)} = \{x \in \mathbb{C}^N \mid |x| = R(\eta)\}$ are transversal to $f_{\tilde{t}}^{-1}(z)$ if $z \in \Delta$.

Then the restriction $f_{\tilde{t}} : B^{2N+1}_{R(\eta)} \cap f_{\tilde{t}}^{-1}(\Delta) \to \Delta$ is the analogon of a Milnor fibration of a local singularity, if Δ contains all critical values of $f_{\tilde{t}} : Y \to \mathbb{C}$. In [**Sab2**] a related more algebraic notion, *cohomologically tame*, is considered. But it is not clear whether it implies M-tame. Suppose now that $f_{\tilde{t}} : Y \to \mathbb{C}$ is M-tame.

Then one can realize the definition of $\mathcal{H}^{(0)}_{Lef}|_{\tilde{t}}$ and the formula (11.69) by proper oscillating integrals over Lefschetz thimbles in the sense of [**Ph4**][**Ph5**].

One can also realize the family $f_t : X_t \to \Delta$ with $t \in M$ by global functions $f_t : Y \to \mathbb{C}$. But in general these functions will not be tame in any sense and will have nonisolated singularities or too many isolated singularities. In order to obtain a family of functions as in step 2, in general one has to cut out bounded spaces $X_t \subset Y$, which contain the singularities of $f_{\tilde{t}}$, but exclude the bad phenomena coming from infinity for $t \neq \tilde{t}$.

In step 5, the existence of a primitive form is unfortunately at the moment only clear for a much smaller class of functions, for polynomials $f_{\tilde{t}} : \mathbb{C}^{n+1} \to \mathbb{C}$ with nondegenerate Newton boundary [**Sab3**][**Sab4**]. The problem is that the analogon of M. Saito's Theorem 10.33, that the smallest spectral number α_1 has multiplicity 1 and corresponds to volume forms, is not clear in the case of globalfunctions $f_{\tilde{t}} : Y \to \mathbb{C}$. But it is not difficult to establish this in concrete cases, and it may well hold for many classes of functions important for mirror symmetry.

Chapter 12

μ-constant stratum

Two applications of the results of chapter 11 are given here. In section 12.1 a canonical complex structure on the μ-constant stratum of a singularity is established. The most difficult part is to prove that it is independent of the choice $(U_\bullet, \gamma_1)$ in Theorem 11.1. In section 12.2 the period map from a μ-constant stratum to a classifying space for Brieskorn lattices from **[He1]**[**He2**][**He3**][**He4**] is taken up. It is shown that the map is an embedding. This strengthens a result of M. Saito.

12.1 Canonical complex structure

The μ-constant stratum of a singularity can be equipped with a canonical complex structure. This will follow from the construction of Frobenius manifolds in chapter 11 and from Varchenko's result that the spectral numbers are constant within a μ-constant family (Theorem 10.32, [**Va3**]). One can write down the complex structure quite explicitly for a choice of an opposite filtration. But the proof that it is independent of this choice is technical.

Let $f : (\mathbb{C}^{n+1}, 0) \to (\mathbb{C}, 0)$ be a holomorphic function germ with an isolated singularity at 0, with Milnor number μ, and with $n \geq 1$. We choose a representative $F : \mathcal{X} \to \Delta$ of a semiuniversal unfolding $F : (\mathbb{C}^{n+1} \times \mathbb{C}^\mu, 0) \to (\mathbb{C}, 0)$ as in section 10.1, that means, $\Delta = B^1_\delta \subset \mathbb{C}$, $M = B^\mu_\theta \subset \mathbb{C}^\mu$, $\mathcal{X} = F^{-1}(\Delta) \cap (B^{n+1}_\varepsilon \times M)$ for $\varepsilon, \delta, \theta > 0$ sufficiently small. The function $F_t : \mathcal{X} \cap (B^{n+1}_\varepsilon \times \{t\}) \to \Delta$ is the restrictionto the parameter $t = (t_1, \ldots, t_\mu) \in M$.

The μ-constant stratum $S_\mu \subset M$ is

$$S_\mu = \{t \in M \mid F_t \text{ has a singularity with Milnor number } \mu \text{ and critical value } 0\}. \tag{12.1}$$

It is a subset of the discriminant $\mathcal{D} \subset M$,

$$\mathcal{D} = \{t \in M \mid F_t \text{ has a singularity with critical value } 0\}. \tag{12.2}$$

Both can be described using the canonical structure of M as an F-manifold $(M, \circ, e, E)$ (see section 10.1 and section 5.1 for the definition of this structure). The discriminant is $\mathcal{D} = \det(\mathcal{U})^{-1}(0)$, where $\mathcal{U}$ is the multiplication with the Euler field.

Gabrielov [**Ga**], Lazzeri [**La**], and Lê [**Le**] proved that F_t has only one singularity if all its singularities are concentrated in one fibre. This is used in the following characterization of S_μ.

Theorem 12.1 *(a) The μ-constant stratum $S_\mu \subset M$ is an analytic subvariety. It is*

$$S_\mu = \mathcal{D} \cap \{t \in M \mid (T_t M, \circ, e) \text{ is irreducible}\} \tag{12.3}$$

$$= \{t \in M \mid \mathcal{U} : T_t M \to T_t M \text{ is nilpotent}\} \tag{12.4}$$

$$= \{t \in \mathcal{D} \mid \text{ the e-orbit through } t \text{ intersects} \mathcal{D} \text{ only in } t\}. \tag{12.5}$$

(b) The germ $((M, 0), \circ, e, E)$ of an F-manifold together with the germs $(\mathcal{D}, 0)$ and $(S_\mu, 0)$ is independent of the choice of F, it is unique up to canonical isomorphism.

Proof: (a) Equation (12.3) holds because $(T_t M, \circ, e)$ is isomorphic to the sum of the Jacobi algebras of the singularities of F_t. Both sets on the right are analytic (cf. Proposition 2.5 for the second). So S_μ is analytic. The endomorphism $\mathcal{U}$ corresponds to the multiplication by F_t on the Jacobi algebras. Equations (12.4) and (12.5) follow with the result of Gabrielov, Lazzeri and Lê.

(b) Theorem 5.4. □

Now we consider the construction in chapter 11. We will quite freely make use of the notations introduced there. The choice of an opposite filtration $U_\bullet$ and a vector γ_1 as in Theorem 11.1 and the choice of $s_1, \ldots, s_\mu$ as in (11.14)–(11.16) induces a flat metric g on M, a basis $\delta_1, \ldots, \delta_\mu$ of flat vector fields, and coefficients $c_{ij}^{(p)}$ (Theorem 11.9).

Theorem 12.2 *Let $(U_\bullet, \gamma_1)$, $s_1, \ldots, s_\mu$, $\delta_1, \ldots, \delta_\mu$, and $c_{ij}^{(p)}$ be as in chapter 11. Define coefficients $\varepsilon_{ij} \in \mathcal{O}_{M,0}$ by $E \circ \delta_i = \sum_j \varepsilon_{ij} \delta_j$.*

Then there is an equality of ideals

$$(\varepsilon_{ij} \mid \alpha_j - 1 - \alpha_i < 0) = \left(c_{ij}^{(1)} \mid \alpha_j - 1 - \alpha_i < 0\right) \tag{12.6}$$

$$= \left(c_{ij}^{(p)} \mid \alpha_j - p - \alpha_i < 0\right) \subset \mathcal{O}_{M,0}.$$

This ideal defines the μ-constant stratum $(S_\mu, 0) \subset (M, 0)$.

Proof: In (11.56) $n_{ij} = 0$ for $\alpha_j - 1 - \alpha_i \neq 0$. This shows $\varepsilon_{ij} = -(\alpha_j - 1 - \alpha_i)c_{ij}^{(1)}$ for $\alpha_j - 1 - \alpha_i \neq 0$ and the first equality. The second equality follows with (11.55). We have to show

$$S_\mu = \{t \in M \mid c_{ij}^{(1)}(t) = 0 \text{ for } \alpha_j - 1 - \alpha_i < 0\}. \tag{12.7}$$

The matrix $(\varepsilon_{ij}(t))$ for t as on the right hand side is nilpotent. Equation (12.4) gives the inclusion $\supset$. For the opposite inclusion $\subset$ we need Varchenko's result that the spectral numbers are constant within a μ-constant family (Theorem 10.32, [**Va3**]).

Fix a parameter $t \in S_\mu$. The restriction to $(\Delta \times \{t\}, (0, t))$ of the sections of $\mathcal{H}^{(0)}$ gives the Brieskorn lattice $H_0''(t)$ of the unique singularity of F_t (see Remark 12.3). Its spectral numbers are $\alpha_1, \ldots, \alpha_\mu$ by Theorem 10.32. Let $v_i^t \in H_0''(t)$ be the restriction of the global section v_i to this Brieskorn lattice. Consider the expansion (11.45) for v_i^t. It is finite, $c_{ij}^{(p)}(t) = 0$ for $\alpha_j - p < \alpha_1$, because of $H_0''(t) \subset V^{\alpha_1}$.

The inclusion $\subset$ in (12.7) is equivalent to the claim $v_i^t \in V^{\alpha_i}$. This holds obviously for v_1. If any v_i^t with $i \geq 2$ had a principal part of order $< \alpha_2$ then this would be independent of the principal part of v_1 because that contains s_1 and v_i^t with $i \geq 2$ does not contain s_1. The second spectral number would be smaller than α_2, a contradiction. Therefore $v_i^t \in V^{\alpha_2}$ for $i \geq 2$. The claim follows inductively. □

Remarks 12.3 The identification of the restrictions to $(\Delta \times \{t\}, (0, t))$ of the sections of $\mathcal{H}^{(0)}$ with the sections in the Brieskorn lattice $H_0''(t)$ of F_t for $t \in S_\mu$ is not as self-evident as it looks.

The restriction to $(\Delta \times \{t\}, (0, t))$ of the fibration $\varphi : \mathcal{X} \to \Delta \times M,\ (x, t) \mapsto (F(x, t), t)$ of the unfolding F is not necessarily a Milnor fibration for the unique singularity of F_t. It could happen that one has to choose $\varepsilon(t) \ll \varepsilon$ to make sure that all balls in $\mathbb{C}^{n+1}$ which have radius $\varepsilon' \leq \varepsilon(t)$ and are centred at the singular point of $F_t^{-1}(0)$ intersect $F_t^{-1}(0)$ transversally.

But according to [**LeR**, Lemma 2.2] the restriction $\varphi|(\Delta \times \{t\}, (0, t))$ is fibre-wise homotopy equivalent to a Milnor fibration for F_t. The cohomology bundles are canonically isomorphic. The restrictions to $(\Delta \times \{t\}, (0, t))$ of sections of $\mathcal{H}^{(0)}$ give the Brieskorn lattice $H_0''(t)$ of F_t.

Also, the space of multivalued flat global sections of the cohomology bundle of F_t is canonically isomorphic to H^∞. The spaces of elementary sections for this cohomology bundle can be identified with the spaces C^α of elementary sections near $\{\infty\} \times M$ for the extended cohomology bundle of φ. Compare the observations (I)–(III) before Lemma 11.2.

This allows the Brieskorn lattices $H_0''(t)$ for all $t \in S_\mu$ to be considered as sublattices of the same lattice $V^{>-1}$. In section 12.2 we will consider the period map $S_\mu \to D_{BL}$ to a classifying space for such lattices.

Theorem 12.4 *The ideal in* (12.6) *is independent of the choices of* $s_1, \dots, s_\mu$ *and* $(U_\bullet, \gamma_1)$. *It defines a canonical complex structure on the* μ*-constant stratum* $(S_\mu, 0)$.

Proof: 1st step: We fix an opposite filtration $U_\bullet$. That does not induce a unique metric, but a unique flat structure on M. Then $\operatorname{ad} E$ acts semisimple on the space of flat vector fields with eigenvalues $-(1+\alpha_1-\alpha_i)$. Let $\widetilde{\delta}_1, \dots, \widetilde{\delta}_\mu$ be any basis of flat vector fields with $[\widetilde{\delta}_i, E] = (1+\alpha_1-\alpha_i)\widetilde{\delta}_i$. Define $\widetilde{\varepsilon}_{ij} \in \mathcal{O}_{M,0}$ by $E \circ \widetilde{\delta}_i = \sum_j \widetilde{\varepsilon}_{ij} \cdot \widetilde{\delta}_j$. Then the ideal $(\widetilde{\varepsilon}_{ij} \mid \alpha_j - 1 - \alpha_i < 0)$ is the same for all such choices of $\widetilde{\delta}_1, \dots, \widetilde{\delta}_\mu$, because a base change mixes only $\widetilde{\varepsilon}_{ij}$ with the same orders α_i and α_j. Therefore the ideal in (12.6) depends at most on $U_\bullet$, not on $\gamma_1, s_1, \dots, s_\mu$.

2nd step: Let $U_\bullet$ and $\widetilde{U}_\bullet$ be two opposite filtrations. They induce subspaces G^α and $\widetilde{G}^\alpha$ of C^α as in (11.5). They satisfy

$$G^\alpha \subset \bigoplus_{p\le 0} \nabla_{\partial_z}^p \widetilde{G}^{\alpha+p}$$

and vice versa. Choose sections $\widetilde{s}_1, \dots, \widetilde{s}_\mu$ as in (11.14) and (11.15) for $\widetilde{U}_\bullet$. There exist unique sections $s_i \in G^{\alpha_i}$ of the form

$$s_i = \sum_{j,p:\ p\le 0, \alpha_j - p = \alpha_i} d_{ij}^{(p)} (-\nabla_{\partial_z})^p \widetilde{s}_j \tag{12.8}$$

with $d_{ij}^{(0)} = \delta_{ij}$. They satisfy (11.15). Let v_i and $\widetilde{v}_i$ be the corresponding global sections for the two extensions of $\mathcal{H}^{(0)}$ by $U_\bullet$ and $\widetilde{U}_\bullet$. Over $\mathbb{C} \times M$ they are bases of the same sheaf, along $\{\infty\} \times M$ they generate different extensions with logarithmic pole along $\{\infty\} \times M$, but both with spectral numbers $-\alpha_1, \dots, -\alpha_\mu \in (-n, 1)$. Therefore the coefficients in

$$\widetilde{v}_i = \bigoplus_j \beta_{ij}(z,t) \cdot v_j \tag{12.9}$$

satisfy

$$\beta_{ij}(z,t) \in \bigoplus_{m=0}^{n} \mathcal{O}_M \cdot z^m. \tag{12.10}$$

Because of (11.54) one can rewrite this with $\nabla_{\partial_z}^{-1}$,

$$\widetilde{v}_i = \sum_j \sum_{m=-n}^{0} b_{ij}^{(m)} \cdot \left(-\nabla_{\partial_z}\right)^m v_j \tag{12.11}$$

with unique $b_{ij}^{(m)} \in \mathcal{O}_M$.

Now one can calculate the coefficients $\widetilde{c}_{ij}^{(p)}$ in terms of the $c_{ij}^{(p)}$. One uses (11.45) for $\widetilde{v}_i$ and for v_i, (12.11) and (12.8). One obtains

$$\widetilde{c}_{il}^{(r)} = \sum_{j,k,m,p,q:\ p+q+m=r} b_{ij}^{(m)} c_{jk}^{(p)} d_{kl}^{(q)}. \tag{12.12}$$

Here note that $b_{ij}^{(m)} = 0$ for $m < -n$ and for $m > 0$; $d_{kl}^{(q)} = 0$ for $q > 0$ and for $\alpha_l - q - \alpha_k \neq 0$; $d_{kl}^{(0)} = \delta_{kl}$; $c_{jk}^{(p)} = 0$ for $p < 0$; $c_{jk}^{(0)} = \delta_{jk}$.

We want to show

$$\widetilde{c}_{il}^{(r)} \in \left(c_{jk}^{(p)} \mid \alpha_k - p - \alpha_j < 0\right) \quad \text{for } \alpha_l - r - \alpha_i < 0.$$

Because of (12.12) it is sufficient to show the following claim:

$$b_{il}^{(r)} \in \left(c_{jk}^{(p)} \mid \alpha_k - p - \alpha_j < 0\right) \quad \text{for } \alpha_l - r - \alpha_i < 0. \tag{12.13}$$

Claim (12.13) will be proved inductively over the size of $\alpha_l - r - \alpha_i$. Fix $\beta < 0$ and suppose that (12.13) is true for $\alpha_l - r - \alpha_i < \beta$. Fix a triple (i, l, r) with $\alpha_l - r - \alpha_i = \beta$ and $r \leq 0$. Then $\widetilde{c}_{il}^{(r)} = 0$. Equation (12.12) gives a relation for the monomials on the right hand side. Modulo the ideal $(c_{jk}^{(p)} \mid \alpha_k - p - \alpha_j < 0)$ the only monomials left are those with

$$p > 0, \quad \alpha_k - p - \alpha_j = 0, \quad \alpha_j - m - \alpha_i = \beta \tag{12.14}$$

and those with

$$p = 0, \quad \alpha_j - m - \alpha_i = \beta. \tag{12.15}$$

Because of (11.24) $c_{jk}^{(p)}(0) = 0$ for $p > 0,\ \alpha_k - p - \alpha_j = 0$.

So, (12.12) modulo the ideal is a linear relation between $b_{ij}^{(m)}$ with $\alpha_j - m - \alpha_i = \beta$; the only eventually invertible coefficients are 1 for $b_{il}^{(r)}$ and $d_{jl}^{(r-m)}$ for $b_{ij}^{(m)}$ with $m > r$.

Putting together the relations (12.12) for all triples (i, l, r) as above, the coefficient matrix is triangular and unipotent for $t=0$. Therefore the claim holds for (i, l, r) as above. This finishes the induction and the whole proof.

□

The Kodaira–Spencer map yields a canonical isomorphism (cf. (10.1))

$$\mathbf{a}_0 : T_0M \to \mathcal{O}_{\mathbb{C}^{n+1},0}/J_f, \quad \frac{\partial}{\partial t_i} \mapsto \left[\frac{\partial F}{\partial t_i}\bigg|_{t=0} \right] \tag{12.16}$$

between the tangent space at 0 and the Jacobi algebra $\mathcal{O}/J_f$. The space

$$\Omega_f = \Omega^{n+1}_{\mathbb{C}^{n+1},0}/\mathrm{d}f \wedge \Omega^{n}_{\mathbb{C}^{n+1},0} = H_0''/\nabla_{\partial_z}^{-1} H_0'' \tag{12.17}$$

(cf. (10.87)) is a free module of rank 1 over the Jacobi algebra. Also, the V-filtration on H_0'' induces the V-filtration with $V^\alpha\Omega_f := pr(V^\alpha \cap H_0'')$ on Ω_f.

Varchenko and Chmutov [**ChV**] considered the tangent cone to the μ-constant stratum. It is estimated by the space

$$V^1(\mathcal{O}/J_f) := \left\{ g \in \mathcal{O}/J_f \mid g \cdot V^\alpha\Omega_f \subset V^{\alpha+1}\Omega_f \right\}. \tag{12.18}$$

In fact, they took not only the V-filtration into account, but also some weight filtration. Some remarks on that will come in section 12.2. Here we will see that the space $V^1(\mathcal{O}/J_f)$ fits very well to the canonical complex structure on $(S_\mu, 0)$.

Lemma 12.5 *The space $V^1(\mathcal{O}/J_f) \subset \mathcal{O}/J_f$ is canonically isomorphic to the Zariski tangent space in T_0M of the μ-constant stratum $(S_\mu, 0)$ with the canonical complex structure from* (12.6).

Proof: Choose $(U_\bullet, \gamma_1)$ and $v_1, \ldots, v_\mu$ as in section 11.1. One arrives at the following identity in Ω_f for $X \in \mathcal{T}_{M,0}$ for example through (11.35), (11.33), (11.49)

$$\mathbf{a}_0(X|_{t=0}) \cdot [v_j^0] = \sum_k X(c_{jk}^{(1)})\big|_{t=0} \cdot [v_k^0]. \tag{12.19}$$

Because of $V^\alpha\Omega_f = \bigoplus_{\alpha_i \geq \alpha} \mathbb{C}[v_i^0]$, the space $V^1(\mathcal{O}/J_f)$ is isomorphic (via $\mathbf{a}_0$) to

$$\left\{ X|_{t=0} \in T_0M \mid X \in \mathcal{T}_{M,0},\ X\left(c_{ij}^{(1)}\right)\big|_{t=0} = 0 \text{ for } \alpha_j - 1 - \alpha_i < 0 \right\}. \tag{12.20}$$

This is the Zariski tangent space of $(S_\mu, 0)$ with its canonical complex structure. □

Remarks 12.6 (a) Choose $(U_\bullet, \gamma_1)$ as in section 11.1. The functions $\tau_i := c_{1i}^{(1)}$ are flat coordinates on M with $\delta_i = \frac{\partial}{\partial \tau_i}$ (Theorem 11.9 (b)). The μ-constant stratum is contained in the flat subspace of M

$$\{t \in M \mid \tau_j(t) = 0 \text{ for } \alpha_j - 1 - \alpha_1 < 0\}. \tag{12.21}$$

In general this is only a coarse estimate. But for a quasihomogeneous singularity f this flat subspace is the μ-constant stratum. In that case $E|_{t=0} = 0$, and $E = \sum_j (1+\alpha_1-\alpha_j)\tau_j\delta_j$ by (11.57). All coefficients $c_{ij}^{(p)}$ are weighted homogeneous with weights $\alpha_i + p - \alpha_j$ because of (11.58). Therefore then

$$\left(c_{ij}^{(1)} \mid \alpha_j - 1 - \alpha_i < 0\right) = (\tau_j \mid \alpha_j - 1 - \alpha_1 < 0) \subset \mathcal{O}_{M,0}.$$

It coincides also with the ideal $(t_j \mid \alpha_j - 1 - \alpha_1 < 0)$. Varchenko [**Va4**] was the first person to see that this last ideal gives the μ-constant stratum.

(b) In the curve case ($n = 1$) the μ-constant stratum is smooth ([**Wah**], see also [**Matt**]). Unfortunately, it is not clear whether the canonical complex structure is then the reduced structure. Results on the Zariski tangent space in [**ChV**] indicate that this should be the case for (generic?) curve singularities with nondegenerate Newton boundary.

(c) For $n \geq 2$ the μ-constant stratum can have singularities [**Lu**][**Stv**]. The construction in [**Lu**] is elegant, but it makes use of a part of the Zariski conjecture, that the multiplicity should be constant along the μ-constant stratum. At the time of [**Lu**] a proof of it was announced which later turned out to be wrong. So [**Lu**] and an also elegant generalization in [**VaS**] concern the sets $S_\mu \cap \{\text{multiplicity constant}\}$.

In [**Stv**] it is proved for one example of [**Lu**] that this (singular) subset is at least a component of S_μ, if it is not the whole space S_μ. Results in [**VaS**] indicate that this subset (and probably S_μ itself) can have arbitrary singularities (but the Milnor numbers in examples in [**VaS**] become astronomically high).

(d) One may also expect that the canonical complex structure of S_μ can be nonreduced. But examples are not known. Computing this canonical structure for higher Milnor numbers is difficult.

12.2 Period map and infinitesimal Torelli

Let f be a singularity as in section 12.1 with semiuniversal unfolding F, base space M, and μ-constant stratum $S_\mu \subset M$. We argued in Remark 12.3 that the Brieskorn lattices $H_0''(t)$ for all $t \in S_\mu$ can be considered as sublattices of the same lattice $V^{>-1}$. Also the Hodge filtrations $F^\bullet(t)$ can be considered as filtrations on the same space H^∞.

By Varchenko's result (Theorem 10.32, [**Va3**]) that the spectral pairs are constant within the μ-constant stratum, these Hodge filtrations are contained in the same classifying space D_{PMHS} for polarized mixed Hodge structures.

By definition, the classifying space D_{BL} from (10.94) for Brieskorn lattices is a fibre bundle over D_{PMHS}, with projection $pr_{PMHS} : D_{BL} \to D_{PMHS}$. Therefore we obtain two period maps

$$\Psi : S_\mu \to D_{BL}, \quad t \mapsto H_0''(t), \tag{12.22}$$

$$\Psi_{PMHS} = pr_{PMHS} \circ \Psi : S_\mu \to D_{PMHS}, \quad t \mapsto F^\bullet(t). \tag{12.23}$$

There is also a global period map. In section 13.3 we will discuss the space

$$\mathcal{M}_\mu := \{\text{singularities in one } \mu\text{-homotopy class}\}/\text{right equivalence} \tag{12.24}$$

and prove that it has a natural structure of a complex space. The group $G_{\mathbb{Z}} = \mathrm{Aut}(H_{\mathbb{Z}}^\infty, h, S)$ acts on $V^{>-1}$, respecting the pairing K_f and the actions of z and ∂_z^{-1}. It acts on D_{PMHS} and D_{BL} properly discontinuously (Theorem 10.34).

The quotient $D_{BL}/G_{\mathbb{Z}}$ is the moduli space for Brieskorn lattices, or more precisely, for tuples $(H_{\mathbb{Z}}^\infty, h, S, V^{>-1}, H_0'')$ up to isomorphism. One obtains a global period map

$$\mathcal{M}_\mu \to D_{BL}/G_{\mathbb{Z}}. \tag{12.25}$$

In [**He2**] the following global Torelli type conjecture was formulated.

Conjecture 12.7 *The period map $\mathcal{M}_\mu \to D_{BL}/G_{\mathbb{Z}}$ is injective for any μ-homotopy class of singularities.*

It was proved in [**He1**][**He2**] for all unimodal and most of the bimodal singularities. The only exceptions were some subseries of the 8 bimodal series where the period map could not be determined precisely enough to see whether it is injective. In [**He3**] it was proved for the semiquasihomogeneous singularities with weights $(\frac{1}{3}, \frac{1}{3}, \frac{1}{3}, \frac{1}{3})$ (with $\dim S_\mu = 5$) and for all semiquasihomogeneous singularities with weights $(\frac{1}{a_0}, \ldots, \frac{1}{a_n})$ such that $\gcd(a_i, a_j) = 1$ for all $i \neq j$ (with $\dim S_\mu$ arbitrarily high). At to the present time no counterexamples are known.

In section 13.3 we will see that the space $\mathcal{M}_\mu$ is locally the quotient of a μ-constant stratum by a finite group. The period map $\Psi : S_\mu \to D_{BL}$ is a local lift of the global period map. M. Saito did not have the classifying space D_{BL}, but nevertheless he considered essentially the period map Ψ, as a map into a manifold which is a classifying space for subspaces of $V^{>-1}/V^{n-1}$ with the correct spectral numbers.

In [**SM3**, 2.10] he proved that it is injective if S_μ is smooth. In [**SM4**, Theorem 3.3] he used this to show that for any S_μ it is finite-to-one if one chooses S_μ sufficiently small. He did not use the construction in section 11.1. With that and with the precise knowledge of coordinates on D_{BL} we obtain the following infinitesimal Torelli type result.

Theorem 12.8 *The period map $\Psi : S_\mu \to D_{BL}$ is an embedding of S_μ with its reduced complex structure into D_{BL} if S_μ is chosen small enough.*

Proof: Choose an opposite filtration $U_\bullet$ for the singularity f and elementary sections $s_i \in C^{\alpha_i}$ as in section 11.1. We claim that any lattice $K \in D_{BL}$ in a neighbourhood of the Brieskorn lattice $H_0'' \in D_{BL}$ has a unique element of the type

$$v_i^K = s_i + \sum_{j,p:\ p\geq 1, \alpha_j - p \geq \alpha_i} \gamma_{ij}^{(p)}(K)(-\nabla_{\partial_z})^p s_j \tag{12.26}$$

for $i = 1, \ldots, \mu$. One constructs it in two steps, first the principal part, then the rest. The filtration $U_\bullet$ is also opposite to the Hodge filtration $pr_{PMHS}(K) \in D_{PMHS}$. So one obtains spaces G_K^α as in (11.5). They satisfy the analoga of (11.6)–(11.10) and also

$$G_K^\alpha \subset \bigoplus_{p\geq 0} \nabla_{\partial_z}^p G_{H_0''}^{\alpha+p} \tag{12.27}$$

and vice versa. There is a unique elementary section $s_i^K \in G_K^{\alpha_i}$ with $s_i^K - s_i \in \bigoplus_{p>0} \nabla_{\partial_z}^p G_{H_0''}^{\alpha_i+p}$. It is

$$s_i^K = s_i + \sum_{j,p:\ p\geq 1, \alpha_j - p = \alpha_i} \gamma_{ij}^{(p)}(K)(-\nabla_{\partial_z})^p s_j. \tag{12.28}$$

The element v_i^K is built from the sections s_j^K by the same procedure as in (11.24) and the proof of Theorem 7.16. Uniqueness is clear from the construction.

The sections v_i^K generate K as a $\mathbb{C}\{\{\partial_z^{-1}\}\}$- module and as a $\mathbb{C}\{z\}$-module. The lattice K and the coefficients $\gamma_{ij}^{(p)}(K)$ determine one another uniquely.

The period map $\Psi : S_\mu \to D_{BL}$ sends $t \in S_\mu$ to the lattice $K \in D_{BL}$ with $\gamma_{ij}^{(p)}(K) = c_{ij}^{(p)}(t)$. Now the injectivity of the period map follows simply from the facts, that the coefficients $c_{1j}^{(1)}$ for $j = 1, \ldots, \mu$ are flat coordinates on M (Theorem 11.9 (b)) and that $S_\mu \subset \{t \mid c_{1j}^{(1)} = 0 \text{ for } \alpha_j - 1 - \alpha_1 < 0\}$ (Theorem 12.2).

To see that the period map is an embedding is much more difficult. We will show the existence of an index set $I \subset \{(i, j, p) \mid \alpha_j - p - \alpha_i \geq 0,\ p > 0\}$ and of a submanifold $\widetilde{M} \subset M$ with $S_\mu \subset \widetilde{M}$ such that the coefficients $\{c_{1j}^{(1)} \mid (1, j, 1) \in I\}$ serve as coordinates on $\widetilde{M}$ and that the coefficients $\{\gamma_{ij}^{(p)} \mid (i, j, p) \in I\}$ serve locally as coordinates on D_{BL}. That is obviously sufficient.

The existence of an index set I such that $\{\gamma_{ij}^{(p)} \mid (i, j, p) \in I\}$ serve locally as coordinates on D_{BL} follows from the construction in [**He4**]. It decomposes into the index sets $I_1 = \{(i, j, p) \in I \mid \alpha_j - p - \alpha_i = 0\}$ and $I_2 = I - I_1$. The set I_2 corresponds to coordinates for the fibres of $pr_{PMHS} : D_{BL} \to D_{PMHS}$. It can be chosen explicitly as

$$I_2 = \{(i, j, 1) \mid i \leq \mu + 1 - j,\ \alpha_j - 1 - \alpha_i > 0\}, \tag{12.29}$$

compare [**He4**, Theorem 5.6] and (11.51), (11.52), (11.55), and Remark 11.10 (b). It especially contains all triples $(1, j, 1)$ with $\alpha_j - 1 - \alpha_i > 0$.

The set I_1 corresponds to coordinates for the base D_{PMHS}. Unfortunately, an explicit choice is difficult and not provided in [**He4**]. We now have to construct it.

Let us first forget about the s_i and $\gamma_{ij}^{(p)}$ and find a basis a_i^0 of H^∞ and coefficients $\beta_{ij}^{(p)}$ with $\alpha_j - p - \alpha_i = 0$, which are better suited.

Because of the proof of Lemma 10.21 there exist subspaces $\widetilde{I}^{p,q}$ and $\widetilde{I}_0^{p,q}$ with all the properties in Lemma 10.18 except (10.51)+(10.52) and which give the opposite filtration $U_\bullet$ by the analogon of (10.66). One can choose bases of all the primitive subspaces $(\widetilde{I}_0^{p,q})_\lambda$ such that they together with their images under powers of N form a basis $a_1^0, \ldots, a_\mu^0$ with the following properties: set $a_{\mu+1}^0 := 0$; there exists a map $\nu : \{1, \ldots, \mu\} \to \{1, \ldots, \mu, \mu + 1\}$ and an involution $\kappa : \{1, \ldots, \mu\} \to \{1, \ldots, \mu\}$ with

$$N\, a_i^0 = a_{\nu(i)}^0, \tag{12.30}$$

$$S\big(a_i^0, a_j^0\big) = \pm\delta_{i\kappa(j)}, \tag{12.31}$$

$$a_i^0 \in F_0^{[n-\alpha_i]} \cap U_{[n-\alpha_i]} H_{e^{-2\pi i\alpha_i}}^\infty, \tag{12.32}$$

here $F_0^\bullet$ denotes the Hodge filtration for f. The basis $a_1^0, \ldots, a_\mu^0$ is a basis of Jordan blocks which are dual or selfdual with respect to the polarizing form S and which fit to the splittings and strict morphisms in Lemma 10.18. Note that N is an infinitesimal isometry of S.

For any Hodge filtration $F^\bullet \in D_{PMHS}$ sufficiently close to $F_0^\bullet$, there exists a unique basis $a_1, \ldots, a_\mu$ of H^∞ with

$$a_i \in F^{[n-\alpha_i]} \cap U_{[n-\alpha_i]} H_{e^{-2\pi i\alpha_i}}^\infty, \tag{12.33}$$

$$a_i - a_i^0 \in U_{[n-\alpha_i]-1}. \tag{12.34}$$

It has the form

$$a_i = a_i^0 + \sum_{j,p:\ p>0,\alpha_j-p-\alpha_i=0} \beta_{ij}^{(p)} a_j^0. \tag{12.35}$$

One checks as in the proof of Lemma 10.21 that the $a_1, \dots, a_\mu$ satisfy the same relations (12.30) and (12.31) with respect to S and N as the $a_1^0, \dots, a_\mu^0$. This gives relations between the coefficients $\beta_{ij}^{(p)}$. The relations from N say:

(α) each coefficient $\beta_{ij}^{(p)}$ with $a_i^0 \in ImN$ is zero or equal to a coefficient $\beta_{kl}^{(p)}$ with $a_k^0 \notin ImN$,

(β) such a coefficient $\beta_{kl}^{(p)}$ with $a_k^0 \notin ImN$ is zero if $a_k^0 \in \ker N^m$ and $a_l^0 \notin \ker N^m$ for some m.

One eliminates with (α) all coefficients $\beta_{ij}^{(p)}$ with $a_i \in ImN$. With respect to S, observe that N is an infinitesimal isometry. One checks easily that the relations from S satisfy:

(γ) All the nonvanishing different relations have one or two coefficients as linear terms and otherwise quadratic terms; in one relation the upper indices or sums of two upper indices are all equal; any coefficient turns up as linear term in at most one relation.

In Definition 10.16 of a polarized mixed Hodge structure, the conditions (i) and (iv) (β) are open conditions. The others are satisfied for a filtration which is defined by a basis $a_1, \dots, a_\mu$ as in (12.35) whose coefficients $\beta_{ij}^{(p)}$ satisfy the relations from the above N and S. In view of (α)–(γ), certain of the coefficients $\beta_{ij}^{(p)}$ form local coordinates for D_{PMHS}. We are especially interested in the behaviour of the coefficients $\beta_{1j}^{(1)}$:

(δ) certain of them might be zero because of (β),

(ε) some others might be zero or linearly related only to one another because of (γ),

(ζ) all others can be chosen as coordinates.

Now we return to the coefficients $\gamma_{ij}^{(p)}$. The relations in (δ) and (ε) induce linear relations between the $\gamma_{1j}^{(1)}$ for $\alpha_j - 1 - \alpha_1 = 0$. The index set I_1 can be chosen to contain the indices of all $\gamma_{1j}^{(1)}$ with $\alpha_j - 1 - \alpha_1 = 0$ except those which can be eliminated by these relations.

The corresponding linear relations between the $c_{1j}^{(1)}$ with $\alpha_j - 1 - \alpha_1 = 0$ define the submanifold $\widetilde{M} \subset \{t \in M \mid c_{1j}^{(1)} = 0 \text{ for } \alpha_j - 1 - \alpha_1 < 0\}$. Obviously $S_\mu \subset \widetilde{M}$. □

Remarks 12.9 (a) It is not clear whether the period map $\Psi : S_\mu \to D_{BL}$ is an embedding of S_μ with its canonical complex structure into D_{BL}. The reason is that it is not clear whether S_μ with its canonical complex structure is embedded in the flat submanifold $\widetilde{M}$ which was constructed in the proof. One does not know whether the corresponding relations between the $c_{1j}^{(1)}$ with $\alpha_j - 1 - \alpha_1 = 0$ from (δ) and (ε) in the proof are contained in the ideal in Theorem 12.2.

(b) It might be that such extra relations do not turn up or that they are always contained in the ideal in Theorem 12.2. It might also be possible to define in a natural way an a priori bigger ideal which takes such relations into account. But one would need more elaborate choices and independence of choices would be a much bigger problem.

An infinitesimal refinement of $V^1(\mathcal{O}/J_f)$ which takes into account relations from the Jordan blocks and the weight filtration (i.e. type (β) and (δ) in the proof) has been discussed in [**ChV**].

(c) The sections v_i^K in (12.26) are very similar to the sections v_i in chapter 11. One has to replace M by D_{BL}. The sections v_i^K glue to generators of a free $\mathcal{O}_{\mathbb{P}^1 \times D_{BL}}$-module over $\mathbb{P}^1 \times D_{BL}$, which is the sheaf of sections of a flat vector bundle over $\mathbb{C}^* \times D_{BL}$. The restrictions to $(\Delta \times \{K\}, (0, K))$ generate K. Along $\{\infty\} \times D_{BL}$ the sheaf has a logarithmic pole, and the restrictions to $\{\infty\} \times D_{BL}$ of the sections v_i^K are flat with respect to the residual connection along $\{\infty\} \times D_{BL}$.

Chapter 13

Moduli spaces for singularities

The choice $(U_\bullet, \gamma_1)$ in Theorem 11.1 induces isomorphisms between spaces which are not canonically isomorphic. Section 13.1 fixes this and the compatibilities of these isomorphisms with respect to coordinate changes and μ-constant deformations.

This is used in section 13.2 for a discussion of symmetries of singularities and in section 13.3 for the (up to now) best application of the construction of Frobenius manifolds in singularity theory: the existence of global moduli spaces for singularities. This really comes from an interplay between polarized mixed Hodge structures and Frobenius manifolds, with the construction in section 11.1 as the link.

13.1 Compatibilities

A choice $(U_\bullet, \gamma_1)$ as in Theorem 11.1 for a singularity f does not only induce the metric of a Frobenius manifold. It also induces isomorphisms between several μ-dimensional spaces which are not canonically isomorphic (Lemma 13.3). These spaces and isomorphisms behave naturally with respect to μ-constant families (Lemma 13.4) and with respect to isomorphisms of singularities (Lemma 13.6). A part of this is elementary, a part is a direct consequence of the construction in section 11.1. All of it fits to the expectations. But to use it in sections 13.2 and 13.3 we need precise notations and statements. First, in Theorem 13.1 and Theorem 13.2 some results from part 1 are summarized which will be needed in sections 13.2 and 13.3.

Theorem 13.1 *Let $((M,t),\circ,e)$ be the germ of a massive F-manifold with multiplication $\circ$ and unit field e. The group* $\mathrm{Aut}((M,t),\circ,e)$ *of automorphisms of the germ of the F-manifold is finite. If E is an Euler field of the F-manifold (i.e.* $\mathrm{Lie}_E(\circ) = 1\cdot\circ$*) such that the endomorphism $T_tM \to T_tM$, $X \mapsto E\circ X$, is nilpotent, then* $\mathrm{Aut}((M,t),\circ,e) = \mathrm{Aut}((M,t),\circ,e,E)$.

Proof: The finiteness is Theorem 4.14. An Euler field is mapped to an Euler field by an automorphism of an F-manifold. If the germ (M, t) has an Euler field, then there is a unique Euler field whose action on $T_t M$ is nilpotent. This follows from Theorem 3.3. Any automorphism of the germ maps this Euler field to itself. □

Theorem 13.2 *Let $f(x_0, \ldots, x_n)$ and $\widetilde{f}(x_0, \ldots, x_n)$ be two isolated hypersurface singularities with semiuniversal unfoldings F and $\widetilde{F}$ and germs of base spaces $(M, 0)$ and $(\widetilde{M}, 0)$ as in section 10.1. The germs $(M, 0)$ and $(\widetilde{M}, 0)$ are germs of massive F-manifolds.*

(a) A germ of a biholomorphic map $\varphi : (\mathbb{C}^{n+1}, 0) \to (\mathbb{C}^{n+1}, 0)$ (i.e. a coordinate change) with $f = \widetilde{f} \circ \varphi$ can be extended to an isomorphism of the unfoldings, that is, a pair of isomorphisms (Φ, φ_M) such that the diagram commutes

$$\begin{array}{ccc} (\mathbb{C}^{n+1} \times M, 0) & \stackrel{\Phi}{\longrightarrow} & (\mathbb{C}^{n+1} \times \widetilde{M}, 0) \\ \downarrow pr_M & & \downarrow pr_{\widetilde{M}} \\ (M, 0) & \stackrel{\varphi_M}{\longrightarrow} & (\widetilde{M}, 0), \end{array} \tag{13.1}$$

and

$$\Phi|_{(\mathbb{C}^{n+1} \times \{0\}, 0)} = \varphi, \tag{13.2}$$

$$F = \widetilde{F} \circ \Phi \tag{13.3}$$

hold. The map φ_M is an isomorphism of germs of F-manifolds. The map Φ is not unique, but φ_M is uniquely determined by φ.

(b) For any isomorphism $\varphi_M : (M, 0) \to (\widetilde{M}, 0)$ of the germs of F-manifolds there exists a Φ such that the pair (Φ, φ_M) is an isomorphism of the unfoldings, that is, it satisfies (13.1) and (13.3) (Φ is not at all unique).

Proof: (a) The existence of an isomorphism (Φ, φ_M) of the unfoldings which extends φ is classic, it is essentially part of the definition of a semiuniversal unfolding (e.g. [**Was**], [**AGV1**, ch. 8]). The base map φ_M is an isomorphism of germs of F-manifolds, because the Kodaira–Spencer maps, which define the multiplications, are compatible with the pair (Φ, φ_M) ((5.10) and Remark 5.2 (v)). It is unique because the tangent map at 0 is unique and the germs of F-manifolds are so rigid by the above Theorem 13.1 (Theorem 5.4).

(b) Theorem 5.6 (b)). □

Now we turn to the construction in section 11.1. Let $f(x_0, \dots, x_n)$ with $n \geq 1$ be an isolated hypersurface singularity as in chapters 10 and 11 and $F : (\mathbb{C}^{n+1} \times M, 0) \to (\mathbb{C}, 0)$ a semiuniversal unfolding, H^n the cohomology bundle over $\Delta \times M - \check{\mathcal{D}}$ (for a representative of F), $\mathcal{H}^{(0)}$ the extension from Lemma 10.2 to $\Delta \times M$ of its sheaf of holomorphic sections, H_0'' the Brieskorn lattice, H^∞ the space of global flat multivalued sections on $H^n|(\Delta \times \{0\}, 0)$, and $F^\bullet$ Steenbrink's Hodge filtration on it.

Consider the four μ-dimensional spaces T_0M, $\mathcal{O}/J_f$, Ω_f, and H^∞. Here $\mathcal{O}/J_f$ is the Jacobi algebra and (cf. (10.87))

$$\Omega_f = \Omega^{n+1}_{\mathbb{C}^{n+1},0} / \mathrm{d}f \wedge \Omega^n_{\mathbb{C}^{n+1},0} = H_0'' / \nabla_{\partial_z}^{-1} H_0''.$$

The reduced Kodaira–Spencer map $\mathbf{a}_0 : T_0M \to \mathcal{O}/J_f$ (cf. (10.1)) is a canonical isomorphism. The space Ω_f is a free $\mathcal{O}/J_f$-module of rank 1; generators are the classes in Ω_f of volume forms. One has to choose such a class of a volume form to obtain an isomorphism $\mathcal{O}/J_f \cong \Omega_f$. There are no obvious natural isomorphisms between H^∞ and the other three spaces.

Lemma 13.3 *A choice $(U_\bullet, \gamma_1)$ as in Theorem 11.1 induces the isomorphisms $\psi_1[U_\bullet]$, $\psi_3[U_\bullet]$, $\psi_4[U_\bullet, \gamma_1]$ in the following diagram*

$$T_0M \xrightarrow{\mathbf{a}_0} \mathcal{O}/J_f \xrightarrow{\psi_4} \Omega_f \xleftarrow{\psi_3} \bigoplus_p \mathrm{Gr}_p^U H^\infty \xleftarrow{\psi_1} H^\infty. \tag{13.4}$$

The map ψ_1 comes from the splitting $H^\infty = \bigoplus_p F^p \cap U_p$, the map ψ_4 is the $\mathcal{O}/J_f$-module isomorphism with $\psi_4(1) = [v_1^0]$ (cf. (11.16) and (11.17)), the map ψ_3 is the composition of the projection $H_0'' \to \Omega_f$ with an embedding

$$\psi_2[U_\bullet] : \bigoplus_p \mathrm{Gr}_p^U H^\infty \to H_0'', \tag{13.5}$$

which will be explained in the proof. Denote by

$$\psi_5[U_\bullet, \gamma_1] : \bigoplus_p \mathrm{Gr}_p^U H^\infty \to T_0M \tag{13.6}$$

the induced isomorphism. The maps ψ_1, ψ_2, and ψ_3 depend only on $U_\bullet$, the maps ψ_4 and ψ_5 depend also linearly on γ_1, that means, for $c \in \mathbb{C}^*$ one has

$$\psi_5[U_\bullet, c \cdot \gamma_1] = c \cdot \psi_5[U_\bullet, \gamma_1]. \tag{13.7}$$

Proof: For the definition of $\psi_2[U_\bullet]$ one has to be well aware of the construction in section 11.1 up to (11.17). The map ψ_2 is composed of the isomorphisms $\psi_1^{-1} : \mathrm{Gr}_p^U H^\infty \to F^p \cap U_p H^\infty$, (11.5), (11.13), (11.4), and the restriction of global sections in $\overline{\mathcal{H}^{(0)}}$ to sections in the Brieskorn lattice H_0''. Properties of ψ_2 will be discussed in the proof of Lemma 13.4

The dependence on γ_1 of the isomorphisms follows from $\psi_3(\gamma_1) = [v_1^0]$. □

Lemma 13.4 *Fix a choice $(U_\bullet, \gamma_1)$ as in Theorem 11.1. Consider two points t_1 and t_2 in the μ-constant stratum $S_\mu \subset M$. The spaces H^∞ of the corresponding singularities are canonically isomorphic and will be identified (cf. Remark 12.3).*

By Theorem 11.1, the opposite filtration $U_\bullet$ (without γ_1) induces a flat structure on M. This yields an isomorphism

$$\sigma[U_\bullet, t_1, t_2] : T_{t_1}M \to T_{t_2}M. \tag{13.8}$$

One has for the singularities for t_1 and t_2 isomorphisms as in (13.6). *Denote them by $\psi_5[(U_\bullet, \gamma_1), t_1]$ and $\psi_5[(U_\bullet, \gamma_1), t_2]$. They are compatible with* (13.8)*:*

$$\sigma[U_\bullet, t_1, t_2] \circ \psi_5[(U_\bullet, \gamma_1), t_1] = \psi_5[(U_\bullet, \gamma_1), t_2]. \tag{13.9}$$

Proof: One has to review the definition of ψ_5. The isomorphism ψ_1^{-1} and the spaces G^α of elementary sections in (11.5) depend on the Hodge filtration, which may vary for the singularities in S_μ. But the composition of ψ_1^{-1}, (11.5), and (11.13) is independent of the Hodge filtration. With (11.4) one arrives at the same isomorphism

$$\bigoplus_p \mathrm{Gr}_p^U H^\infty \to \bigoplus_i \mathbb{C} \cdot v_i \tag{13.10}$$

for all singularities in S_μ. One now has to see that for any $t \in S_\mu$ the isomorphism

$$\bigoplus \mathbb{C} \cdot v_i \to T_t M, \quad v_i \mapsto \delta_i | T_t M \tag{13.11}$$

coincides with the composition of the restriction of v_i to the Brieskorn lattice $H_0''(t)$ and the isomorphisms ψ_4^{-1} and $\mathbf{a}_t^{-1}$ for the singularity corresponding to $t \in S_\mu$. This follows with Lemma 11.5 (b). □

Remarks 13.5 (a) The isomorphism ψ_1 depends on the Hodge filtration of a singularity. Therefore, in general, one does not have a compatibility of the isomorphisms $\psi_5 \circ \psi_1[t_i] : H^\infty \to T_{t_i}M, i = 1, 2$, for two parameters $t_1, t_2 \in S_\mu$ with (13.8).

(b) In the diagram (13.4) of isomorphisms the semiuniversal unfolding is needed only for the Kodaira–Spencer isomorphism $\mathbf{a}_0$. In the definition of ψ_2 one can replace (11.13) and (11.4) by the corresponding restrictions to the slice $\mathbb{P}^1 \times \{0\} \subset \mathbb{P}^1 \times M$, or one can define the section v_i^0 directly as the unique section in H_0'' with principal part s_i and higher elementary parts as in formula (11.24).

Lemma 13.6 *As well as f consider a second isolated hypersurface singularity $\widetilde{f} = \widetilde{f}(x_0, \dots, x_n)$ which is right equivalent to f. Choose a semiuniversal unfolding $\widetilde{F}$ with base $(\widetilde{M}, 0)$ for $\widetilde{f}$ and write all the other associated objects for $\widetilde{f}$ with a tilde. Fix a coordinate change $\varphi : (\mathbb{C}^{n+1}, 0) \to (\mathbb{C}^{n+1}, 0)$ with $f = \widetilde{f} \circ \varphi$.*

(a) It induces isomorphisms between the corresponding objects for f and $\widetilde{f}$, which respect all the canonical additional structures,

$$\begin{aligned}
\varphi_M &: ((M,0), \circ, e, E) \to ((\widetilde{M},0), \circ, \widetilde{e}, \widetilde{E}),\\
\varphi_{Jac} &: (\mathcal{O}/J_f,\ mult., [f]) \to (\mathcal{O}/J_{\widetilde{f}},\ mult., [\widetilde{f}]),\\
\varphi_{Gro} &: (\Omega_f, J_f) \to (\Omega_{\widetilde{f}}, J_{\widetilde{f}}),\\
\varphi_{Bri} &: \left(H_0'', z, \partial_z^{-1}, K_f\right) \to \left(\widetilde{H_0''}, z, \partial_z^{-1}, K_{\widetilde{f}}\right),\\
\varphi_{coh} &: \left(H_{\mathbb{Z}}^{\infty}, h, S, F^\bullet\right) \to \left(\widetilde{H_{\mathbb{Z}}^{\infty}}, \widetilde{h}, \widetilde{S}, \widetilde{F}^\bullet\right),
\end{aligned}$$

(cf. section 10.6 for the residue pairing J_f, the polarizing form S and the form K_f).

(b) Choose $(U_\bullet, \gamma_1)$ as in Theorem 11.1 and set $\widetilde{U}_\bullet := \varphi_{coh}(U_\bullet)$ in H^∞, $\widetilde{\gamma}_1 := (Gr^U \varphi_{coh})(\gamma_1)$. The diagram (13.4) for $(f, (U_\bullet, \gamma_1))$ and the diagram (13.4) for $(\widetilde{f}, (\widetilde{U}_\bullet, \widetilde{\gamma}_1))$ are compatible with the above isomorphisms and with the induced isomorphisms.

The same holds for the embeddings ψ_2 and $\widetilde{\psi}_2$ from (13.5). The map φ_M is an isomorphism of Frobenius manifolds with respect to the metrics on M and $\widetilde{M}$ induced by $(U_\bullet, \gamma_1)$ and $(\widetilde{U}_\bullet, \widetilde{\gamma}_1)$.

Proof: This holds, because all the considered objects and structures and also the construction in section 11.1 are essentially coordinate independent. Theorem 13.2 provides the uniqueness of φ_M. For the other isomorphisms one does not even need to consider the semiuniversal unfoldings because of Remark 13.5 (b). A nontrivial point is to only formulate one or several natural definitions for the isomorphisms $\varphi_M, \dots, \varphi_{coh}$ and to check that all are compatible. See Remark 13.7. □

Remark 13.7 For example, φ_{Bri} can be defined by pulling back differential forms with $(\varphi^{-1})^*$. It can also be recovered from φ_{coh}: By sections 7.1 and 10.6, the tuple (H^∞, h, S) determines the structure $(V^{>-1}, z, \partial_z^{-1}, K_f)$. The isomorphism φ_{coh} induces an isomorphism from $(V^{>-1}, z, \partial_z^{-1}, K_f)$ to the corresponding structure for $\widetilde{f}$. It maps H_0'' to $\widetilde{H_0''}$.

Another example: $\mathcal{O}/J_f$ is naturally embedded in $\text{End}_{self-adjoint}(\Omega_f, J_f)$ by the action on Ω_f. Knowing this embedding for f and $\widetilde{f}$, one can recover φ_{Jac} from φ_{Gro}.

13.2 Symmetries of singularities

The following is motivated by the results of Slodowy [**Sl**] and Wall [**Wal1**] and an extension of them. Remarks on the relations will be made below.

Let $f : (\mathbb{C}^{n+1}, 0) \to (\mathbb{C}, 0)$ be an isolated hypersurface singularity with $n \geq 1$, $\mathcal{R} := \{\varphi : (\mathbb{C}^{n+1}, 0) \to (\mathbb{C}^{n+1}, 0) \text{ biholomorphic}\}$ the group of all coordinate changes, and $\mathcal{R}^f = \{\varphi \in \mathcal{R} \mid f \circ \varphi = f\}$ the group of *symmetries* of f.

The group of symmetries acts on all the objects associated to the singularity. The interrelations presented in section 13.1 result from heavy machinery and profound facts, but now they make it easy to analyze the actions of the symmetry group.

The group $\mathcal{R}^f$ is ∞-dimensional, but the group of k-jets $j_k\mathcal{R}^f$ is an algebraic group for any $k \geq 1$. Let

$$R_f := j_1\mathcal{R}^f/(j_1\mathcal{R}^f)_0 \tag{13.12}$$

be the finite group of components of $j_1\mathcal{R}^f$. The following is classical.

Lemma 13.8 *The kernel* $\ker(j_k\mathcal{R}^f \to j_1\mathcal{R}^f)$ *is unipotent. The groups* $j_k\mathcal{R}^f$ *and* $j_1\mathcal{R}^f$ *have the same number of components.*

Proof: The group $j_k\mathcal{R}^f$ acts faithfully on $\mathcal{O}_{\mathbb{C}^{n+1},0}/\mathbf{m}^{k+1}$, where $\mathbf{m}$ is the maximal ideal. An element of the kernel acts trivially on $\mathbf{m}/\mathbf{m}^2$ and thus on $\mathbf{m}^l/\mathbf{m}^{l+1}$ for any $l \leq k$. It is unipotent. The kernel is unipotent and thus connected. □

We use the following abbreviations for the groups of automorphisms of different objects associated to the singularity.

$$\text{Aut}_M = \text{Aut}((M, 0), \circ, e, E), \tag{13.13}$$

$$\text{Aut}_{Jac} = \text{Aut}(\mathcal{O}/J_f, \text{mult.}, [f]) \cong \text{Aut}(T_0M, \circ, E|_0), \tag{13.14}$$

$$G_{\mathbb{Z}} = \text{Aut}\big(H_{\mathbb{Z}}^{\infty}, h, S\big), \tag{13.15}$$

$$\text{Stab}_{G_{\mathbb{Z}}}(F^\bullet) = \text{Aut}\big(H_{\mathbb{Z}}^{\infty}, h, S, F^\bullet\big), \tag{13.16}$$

$$\text{Stab}_{G_{\mathbb{Z}}}\big(H_0''\big) = \text{Aut}\big(H_{\mathbb{Z}}^{\infty}, h, S, V^{>-1}, H_0''\big). \tag{13.17}$$

Here $(M, 0)$ is the base of a semiuniversal unfolding, a germ of an F-manifold with Euler field, $\mathcal{O}/J_f$ is the Jacobi algebra, $H^\infty_{\mathbb{Z}}$ the lattice in the space H^∞ of global flat multivalued sections in the cohomology bundle over a punctured disc, h its monodromy, S the polarizing form from section 10.6, H_0'' the Brieskorn lattice, and $F^\bullet$ Steenbrink's Hodge filtration on H^∞.

One can see that $G_{\mathbb{Z}}$ is isomorphic to the automorphism group of the Milnor lattice with Seifert form [**He4**]. The group $G_{\mathbb{Z}}$ acts on $(V^{>-1}, z, \partial_z^{-1}, K_f)$ (compare sections 7.1 and 10.6). It acts properly discontinuously on the classifying spaces D_{PMHS} and D_{BL} (Theorems 10.22, 10.34).

Therefore the groups $\mathrm{Stab}_{G_{\mathbb{Z}}}(H_0'') \subset \mathrm{Stab}_{G_{\mathbb{Z}}}(F^\bullet)$ are finite. The group Aut_M is also finite (Theorem 13.1). Because of Theorem 13.2 and Lemma 13.6 there are canonical homomorphisms from the group $\mathcal{R}^f$ of symmetries of f to the groups in (13.13)–(13.17). Denote

$$\det_0 : \mathcal{R}^f \to \mathbb{C}^*, \quad \varphi \mapsto \det\left(\frac{\partial \varphi_i}{\partial x_j}\right)(0). \tag{13.18}$$

Because of the splitting lemma one can transform any singularity by a coordinate change to the form in Theorem 13.9.

Theorem 13.9 *Suppose that* $f = g(x_0, \dots, x_m) + x_{m+1}^2 + \dots + x_n^2$ *with* $g \in \mathbf{m}^3$ *for some* $m \le n$.

(a) The canonical homomorphisms from $\mathcal{R}^f$ *to the groups in* (13.13)–(13.17) *factor through* R_f.

(b) If $n = m$ *then the map* $R_f \to \mathrm{Aut}_M$ *is an isomorphism. If* $n > m$ *then* $R_f \cong R_g \times \mathbb{Z}_2$, *the map* $R_f \to \mathrm{Aut}_M$ *is two-to-one, and the kernel is generated by* $(x_0, \dots, x_{n-1}, x_n) \mapsto (x_0, \dots, x_{n-1}, -x_n)$. *In both cases* $(n \ge m)$, *the map* $R_f \to \mathrm{Aut}_{Jac}$ *is induced by the embedding of* Aut_M *in* $\mathrm{Aut}(T_0M, \circ, E|_0) \cong \mathrm{Aut}_{Jac}$.

(c) If $n = m$ *then* $R_f \cong j_1\mathcal{R}^f$ *and any maximal reductive subgroup of* $j_k\mathcal{R}^f$ *is isomorphic to it. If* $n > m$ *then any maximal reductive subgroup of* $j_k\mathcal{R}^f$ *is isomorphic to* $R_g \times O(n-m, \mathbb{C})$.

(d) The map $\det_0$ *factors through* R_f. *The action of* R_f *on* H^∞ *is obtained from the action on* T_0M *by twist with* $\det_0^{-1}$. *More precisely, there is an isomorphism* $\psi_6 : H^\infty \to T_0M$ *such that for all* $\varphi \in \mathcal{R}^f$

$$\varphi_{coh} = \psi_6^{-1} \circ \det_0^{-1}(\varphi) \cdot d\varphi_M|_0 \circ \psi_6. \tag{13.19}$$

Here $\varphi_{coh} \in G_{\mathbb{Z}}$ *and* $\varphi_M \in \mathrm{Aut}_M$ *are the induced isomorphisms.*

(e) The map $R_f \to \mathrm{Stab}_{G_{\mathbb{Z}}}(H_0'')$ *is injective, so* R_f *acts faithfully on the Milnor lattice.*

Proof: (a) The group Aut_M is embedded in $\mathrm{Aut}(T_0M, \circ, E|_0) \cong \mathrm{Aut}_{Jac}$, because it is finite. The actions of $\mathcal{R}^f$ on T_0M and on $\mathcal{O}/J_f$ are isomorphic via the Kodaira–Spencer isomorphism $\mathbf{a}_0$. The action on $\mathcal{O}/J_f$ depends only on sufficiently large k-jets and depends continuously on them, the action on T_0M is finite. Therefore both factor through R_f.

The same argument works for H^∞ and Ω_f with a special choice of $U_\bullet$: The opposite filtration $U_\bullet^{(0)}$ from Deligne's $I^{p,q}$ on H^∞ (cf. (10.66))

$$U_p^{(0)} = \bigoplus_{i,q:\, i\le p} I^{p,q} \tag{13.20}$$

is invariant under φ_{coh} for any $\varphi \in \mathcal{R}^f$, because the $I^{p,q}$ are invariant by their definition (cf. (10.50)). It induces an isomorphism $\psi_3 \circ \psi_1 : H^\infty \to \Omega_f$ by Lemma 13.3. The actions of $\mathcal{R}^f$ on Ω_f and on H^∞ are compatible with it by Lemma 13.6. But $\mathrm{Stab}_{G_\mathbb{Z}}(H_0'')$ is finite and the action on Ω_f depends continuously on sufficiently high k-jets. So both factor through R_f.

(b) and (c) First, suppose $n = m$. Then $f \in \mathbf{m}^3$ and there is a natural surjective map $\mathbf{m}/J_f \to \mathbf{m}/\mathbf{m}^2$. The group action of $j_1\mathcal{R}^f$ on $\mathbf{m}/\mathbf{m}^2$ is faithful. It factors through R_f by (a). Thus $j_1\mathcal{R}^f \cong R_f$ and R_f acts faithfully on $\mathcal{O}/J_f$. Therefore $R_f \to \mathrm{Aut}_M$ is injective. It is surjective because of Theorem 13.2 (b). Because the kernel $\ker(j_k\mathcal{R}^f \to j_1\mathcal{R}^f)$ is unipotent (Lemma 13.8), any reductive subgroup of $j_k\mathcal{R}^f$ is mapped injectively to R_f.

Now, suppose $n > m$. A symmetry $\varphi \in \mathcal{R}^f$ acts on $\mathcal{O}_{\mathbb{C}^{n+1},0}$ by $h \mapsto h \circ \varphi^{-1}$. It leaves the Jacobian ideal invariant. The ideal $\mathbf{m}^2 + J_f = \mathbf{m}^2 + (x_{m+1}, \dots, x_n)$ is invariant under the action of $\mathcal{R}^f$. There is a natural homomorphism

$$j_1\mathcal{R}^f \to \mathrm{Aut}\left(\frac{\mathbf{m}}{\mathbf{m}^2 + (x_{m+1}, \dots, x_n)}\right) \times \mathrm{Aut}\left(\frac{\mathbf{m}^2 + (x_{m+1}, \dots, x_n)}{\mathbf{m}^2}\right). \tag{13.21}$$

The kernel is unipotent, the image is isomorphic to $R_g \times O(n-m, \mathbb{C})$, because $j_2 f = x_{m+1}^2 + \cdots + x_n^2$. Therefore $R_f \cong R_g \times \mathbb{Z}_2$. The other statements are also now clear.

(d) Choose any generator γ_1 of the 1-dimensional space $\mathrm{Gr}^{U^{(0)}}_{[n-\alpha_1]} H^\infty_{e^{-2\pi i\alpha_1}}$. By Lemma 13.3 $(U_\bullet^{(0)}, \gamma_1)$ induces a class of a volume form $[v_1^0] \in \Omega_f$ and an isomorphism $\psi_4 : \mathcal{O}/J_f \to \Omega_f$, $[1] \mapsto [v_1^0]$, of $\mathcal{O}/J_f$-modules. Also $\psi_3(\gamma_1) = [v_1^0]$.

The actions on $\mathrm{Gr}^{U^{(0)}} H^\infty$ and Ω_f of a symmetry $\varphi \in \mathcal{R}^f$ are compatible with ψ_3 and have γ_1 and $[v_1^0]$ as eigenvectors. The eigenvalue is $\det_0^{-1}(\varphi)$, because $[v_1^0]$ is represented by a volume form. By Lemma 13.6 the actions φ_{Jac} and φ_{Ω_f} are compatible with ψ_4 with the twist $\det_0^{-1}(\varphi)$,

$$\det{}_0^{-1}(\varphi) \cdot \psi_4 \circ \varphi_{Jac} = \varphi_{\Omega_f} \circ \psi_4. \tag{13.22}$$

Consider (13.4) and define $\psi_6 := \mathbf{a}_0^{-1} \circ \psi_4^{-1} \circ \psi_3 \circ \psi_1$. The compatibilities of the actions of φ with $\mathbf{a}_0$ and $\psi_3 \circ \psi_1$ show that the twist extends to (13.19). The map $\det_0$ factors automatically through R_f.

(e) The kernel $\ker(R_f \to G_{\mathbb{Z}})$ is contained in the kernel of $\det_0$ because of the action on $\mathbb{C} \cdot \gamma_1$ in (d). Then by (13.19) it is the intersection of the kernels of $\det_0$ and $R_f \to \mathrm{Aut}_M$. But this is trivial. □

Remarks 13.10 (a) Slodowy [**Sl**] considered a real singularity with a given compact group of symmetries and showed that one can construct a semiuniversal unfolding on which the group acts. He also showed that such a group is finite if $f \in \mathbf{m}^3$. Compact corresponds to reductive in the complex case. So his results are close to Theorem 13.9 (b) and (c). But he did not have the uniqueness of the action on the basis $(M, 0)$, which follows from the rigidity of massive F-manifolds (Theorem 13.1), and he did not consider R_f and Aut_M.

(b) Part (d) in Theorem 13.9 is due to Wall [**Wal1**], also the way in which this is proved. But he did not have the up to a scalar canonical isomorphism $\psi_6 : H^\infty \to T_0M$, which comes from the construction in section 11.1 and from Deligne's $I^{p,q}$.

(c) In [**Sl**] and [**Wal1**] one starts with a given (compact or) finite subgroup of $\mathcal{R}^f$. The group R_f and its liftability to a subgroup of $\mathcal{R}^f$ are not considered. But this liftability can be shown easily, following arguments in [**Wal2**] (see also [**Mue1**]):

It is known that the group R_f can be lifted to a subgroup $\widetilde{R}_f$ in the algebraic group $j_k\mathcal{R}^f$ for any k. Let $k \geq \mu+1$. Within the k-jets one carries out the usual averaging procedure to find coordinates on which the group $\widetilde{R}_f$ acts linearly and obtains a k-jet $j_k\varphi$ of a coordinate change such that $j_k\varphi^{-1} \circ \widetilde{R}_f \circ j_k\varphi$ acts linearly on $\mathcal{O}/\mathbf{m}^{k+1}$ and respects $j_k f \circ j_k\varphi$. The last group lifts to a group of linear coordinate changes and respects $j_k f \circ j_k\varphi$, considered as a polynomial. This polynomial is right equivalent to f, because f is $\mu + 1$-determined.

In the case of a quasihomogeneous singularity f one can calculate the group R_f using the following characterization.

Theorem 13.11 *Let $f \in \mathbb{C}[x_0, \dots, x_n]$ be a quasihomogeneous isolated singularity with weights $w_0, \dots, w_n \in (0, \frac{1}{2}] \cap \mathbb{Q}$ and degree 1. Suppose that $w_0 \leq \dots \leq w_{n-1} < \frac{1}{2}$ (then $f \in \mathbf{m}^3$ if and only if $w_n < \frac{1}{2}$). Let G_w be the algebraic group of quasihomogeneous coordinate changes, that means, those which respect $\mathbb{C}[x_0, \dots, x_n]$ and the grading by the weights $w_0, \dots, w_n$ on it. Then*

$$R_f \cong \mathrm{Stab}_{G_w}(f). \tag{13.23}$$

Proof: The group $\text{Stab}_{G_w}(f)$ is finite by [**GrHP**, Proposition 2.7] (the proof is similar to that in [**Sl**, (4.6)]). In [**GrHP**, Theorem 2.1] it is proved that any symmetry $\varphi \in \mathcal{R}^f$ has weighted degree ≥ 0, that means, the i-th component $\varphi_i \in \mathcal{O}$ does not contain monomials of weighted degree $< w_i$. The degree 0 part of any symmetry is an element of $\text{Stab}_{G_w}(f)$.

One can rewrite Lemma 13.8 for weighted jets and for $\text{Stab}_{G_w}(f)$ instead of $j_1\mathcal{R}^f$. But the groups of jets and of weighted jets of symmetries are contained in one another for suitable high degrees. They have the same number of components. This shows (13.23). □

Conjecture 13.12 *The map $R_f \to \text{Stab}_{G_\mathbb{Z}}(H_0'')$ is an isomorphism for any isolated hypersurface singularity $f(x_0, \ldots, x_n)$ with $n \geq 1$ and multiplicity 2.*

Theorem 13.13 *Suppose that the map $R_f \to \text{Stab}_{G_\mathbb{Z}}(H_0'')$ is an isomorphism for one singularity $f(x_0, \ldots, x_n)$ with $n \geq 1$ and multiplicity 2.*

Then the corresponding map is an isomorphism for any singularity in the (sufficiently small) μ-constant stratum S_μ of f.

Proof: We may choose a (sufficiently small) representative F of the semiuniversal unfolding, a representative in $\mathcal{R}^f$ for each element of R_f, and extensions of these representatives to automorphisms of the unfolding F in the sense of Theorem 13.2 (a).

Consider the period map $S_\mu \to D_{BL}$, $t \mapsto H_0''(t)$, from section 12.2. If S_μ is sufficiently small, then $\text{Stab}_{G_\mathbb{Z}}(H_0''(t)) \subset \text{Stab}_{G_\mathbb{Z}}(H_0''(0))$ for any $t \in S_\mu$ because $G_\mathbb{Z}$ acts properly discontinuously on D_{BL} (Theorem 10.34).

By assumption, an element $\varphi_{coh} \in \text{Stab}_{G_\mathbb{Z}}(H_0''(t)) \subset \text{Stab}_{G_\mathbb{Z}}(H_0''(0)) \cong R_f$ is induced by an element of $\mathcal{R}^f$. It induces an automorphism φ_M of M. By the compatibility of the morphisms in Lemma 13.6, $\varphi_{coh}(H_0''(t)) = H_0''(\varphi_M(t))$.

But the period map is injective (Theorem 12.8). Therefore $\varphi_M(t) = t$. The automorphism of the unfolding restricts to a symmetry of the singularity F_t. This symmetry induces φ_{coh}. Therefore $R_{F_t} \to \text{Stab}_{G_\mathbb{Z}}(H_0''(t))$ is an isomorphism. □

Remarks 13.14 (a) Conjecture 13.12 complements the global Torelli type conjecture 12.7. Together with the infinitesimal Torelli type theorem 12.8 and Theorem 13.15 below it would say that locally the period map $\mathcal{M}_\mu \to D_{BL}/G_\mathbb{Z}$ is an embedding. The only obstructions to Conjecture 12.7 would be possible intersections of the images of disjoint pieces of $\mathcal{M}_\mu$.

(b) The conjecture is true for the weighted homogeneous singularities with weights $(\frac{1}{a_0}, \ldots, \frac{1}{a_n})$ where $gcd(a_i, a_j) = 1$ for $i \neq j$.

For these singularities all eigenspaces of the monodromy are 1-dimensional and the orders of their eigenvalues can be put into certain chains with one biggest order. This is used in [**He3**, Proposition 6.3] to show that $\mathrm{Stab}_{G_\mathbb{Z}}(H_0'') = G_\mathbb{Z} = \{\pm h^k \mid k \in \mathbb{Z}\}$. This is isomorphic to $\mathrm{Stab}_{G_w}(f)$ if $a_n = 2$.

(c) The same reasoning as in (b) applies to the A_k, D_{2k+1}, E_k, and to 22 of the 28 quasihomogeneous unimodal and bimodal exceptional singularities, but not to D_{2k}, Z_{12}, Q_{12}, U_{12}, Z_{18}, Q_{16}, U_{16}.

Starting with a Coxeter–Dynkin diagram, I checked Conjecture 13.12 for Q_{12}, which has some 2-dimensional eigenspaces and $\mathrm{Stab}_{G_\mathbb{Z}}(H_0'') = G_\mathbb{Z} \cong \{\pm h^k \mid k \in \mathbb{Z}\} \times \mathbb{Z}_2$ ($\cong R_f$ for $n \geq 3$, $\cong \mathrm{Stab}_{G_w}(f)$ for $n = 3$).

13.3 Global moduli spaces for singularities

In this section we present our best application of the construction of Frobenius manifolds for hypersurface singularities: Theorem 13.15 gives a global moduli space for the singularities in one μ-homotopy class.

The three main ingredients in the proof are the F-manifold structure on a base M of a semiuniversal unfolding of a singularity, the construction of flat structures on M by choice of opposite filtrations, and the period map $S_\mu \to D_{PMHS}$ from a μ-constant stratum to the classifying space for polarized mixed Hodge structures. So it grows from an interplay between Frobenius manifolds and polarized mixed Hodge structures with the Gauß–Manin connection as the intermediary.

The construction of moduli spaces for singularities starts with Mather's theory of jets of singularities ([**Math1**][**Math2**], cf. [**BrL**]). The k-jet of a function $f \in \mathcal{O} = \mathcal{O}_{\mathbb{C}^{n+1},0}$ is the class $j_k f \in \mathcal{O}/\mathbf{m}^{k+1}$; here $\mathbf{m} \subset \mathcal{O}$ is the maximal ideal.

The action of the group $\mathcal{R} = \{\varphi : (\mathbb{C}^{n+1}, 0) \to (\mathbb{C}^{n+1}, 0)$ biholomorphic $\}$ of coordinate changes on $\mathbf{m}^2 \subset \mathcal{O}$ pulls down to an action of the algebraic group $j_k\mathcal{R}$ of k-jets of coordinate changes on $\mathbf{m}^2/\mathbf{m}^{k+1}$.

An isolated hypersurface singularity $f \in \mathbf{m}^2$ with Milnor number μ is $\mu+1$-determined, that means, any function g with $j_{\mu+1}g = j_{\mu+1}f$ is right equivalent to f. Therefore for $k \geq \mu+1$, the set of all $\mathcal{R}$-orbits in $\mathbf{m}^2$ of singularities with Milnor number μ is in one-to-one correspondence with the set of $j_k\mathcal{R}$-orbits of their k-jets in $\mathbf{m}^2/\mathbf{m}^{k+1}$.

Fix μ and $k \geq \mu+1$. The codimension of the orbit $j_k\mathcal{R}\cdot j_k f$ in $\mathbf{m}^2/\mathbf{m}^{k+1}$ for a singularity f with $\mu(f) = \mu$ is $\mu - 1$. The union of all orbits with codimension $\geq \mu - 1$ is an algebraic subvariety in $\mathbf{m}^2/\mathbf{m}^{k+1}$. The set $\{j_k f \mid \mu(f) = \mu\}$ is Zariski open in it and thus a quasiaffine variety. It decomposes into algebraically irreducible components and into (possibly bigger) topological components.

Each topological component corresponds to a *μ-homotopy class* of singularities. The singularities f and g in $\mathbf{m}^2$ with $\mu(f) = \mu(g) = \mu$ are *μ-homotopic*, $f \sim_\mu g$, if and only if there is a family $f_t \in \mathbf{m}^2$, $t \in [0, 1]$, with $\mu(f_t) = \mu$, $f_0 = f$, $f_1 = g$, such that the coefficients of the power series f_t depend continuously (or, equivalently, C^∞ or even piecewise real analytic) on t. A singularity f with $\mu(f) = \mu$ is μ-homotopic to its k-jet by the proof of the finite determinacy, and k-jets in one topological component of the set above $\{j_k f \mid \mu(f) = \mu\}$ are obviously μ-homotopic. Finally, the group $j_k\mathcal{R}$ is connected and acts on each topological component.

For a singularity $f \in \mathbf{m}^2$ with $\mu(f) = \mu$, denote by $C(k, f)$ the topological component of $\{j_k f \mid \mu(f) = \mu\}$ which contains f. We summarize: the map

$$\{g \in \mathbf{m}^2 \mid g \sim_\mu f\}/\mathcal{R} \to C(k, f)/j_k\mathcal{R} \tag{13.24}$$

is bijective, the algebraic group $j_k\mathcal{R}$ acts on the quasiaffine variety $C(k, f) \subset \mathbf{m}^2/\mathbf{m}^{k+1}$.

Theorem 13.15 *Fix μ, $k \geq \mu + 1$ and a singularity $f \in \mathbf{m}^2 \subset \mathcal{O}_{\mathbb{C}^{n+1},0}$ with $\mu(f) = \mu$.*

(a) The quotient $\mathcal{M}_\mu$ of the map $\pi : C(k, f) \to C(k, f)/j_k\mathcal{R} =: \mathcal{M}_\mu$ is an analytic geometric quotient. That means, the quotient topology is Hausdorff and the sheaf

$$(\pi_* \mathcal{O}_{C(k,f)})^{j_k\mathcal{R}} =: \mathcal{O}_{\mathcal{M}_\mu} \tag{13.25}$$

induces a reduced complex structure on $\mathcal{M}_\mu$.

(b) The germ at $[j_k f]$ is

$$(\mathcal{M}_\mu, [j_k f]) \cong (S_\mu, 0)/\mathrm{Aut}((M, 0), \circ, e, E). \tag{13.26}$$

Here $((M, 0), \circ, e, E)$ is the base space of a semiuniversal unfolding of f with its structure as a germ of an F-manifold (with Euler field), $(S_\mu, 0) \subset (M, 0)$ is the μ-constant stratum.

(c) The canonical complex structure on the μ-constant strata from Theorem 12.4 induces a (possibly nonreduced) canonical complex structure on $\mathcal{M}_\mu$.

The proof will be given after Remark 13.19. Results of Gabrielov [**Ga**] and Teissier [**Te1**] will allow semiuniversal unfoldings to be considered instead of spaces of k-jets. Then Theorem 13.17 shows that the quotient topology is Hausdorff. Theorem 13.18 gives the reduced complex structure and (13.26).

Remark 13.16 (a) All orbits in $C(k, f)$ have the same dimension and are closed. But this by no means implies that the quotient is Hausdorff. For example, the quotient of $\mathbb{C}^2 - \{0\}$ with the action of $\mathbb{C}^* \hookrightarrow GL(2, \mathbb{C})$, $x \mapsto \begin{pmatrix} x & 0 \\ 0 & x^{-1} \end{pmatrix}$, on it is not Hausdorff, the punctured coordinate planes cannot be separated.

(b) One may ask whether $\mathcal{M}_\mu$ is even an algebraic geometric quotient. But it is not at all clear how one could approach this question. In general, the group $j_k\mathcal{R}$ is not reductive and the variety $C(k, f)$ not affine.

(c) I expect that $\mathcal{M}_\mu$ with the canonical complex structure in Theorem 13.15 (c) is a coarse moduli space for an appropriate notion of μ-constant deformations over arbitrary bases. But this has still to be worked out and checked.

Theorem 13.17 *Let f and $\widetilde{f} \in \mathcal{O}_{\mathbb{C}^{n+1},0}$ be two isolated hypersurface singularities with (representatives of) base spaces M and $\widetilde{M}$ of semiuniversal unfoldings and with μ-constant strata $S_\mu \subset M$ and $\widetilde{S}_\mu \subset \widetilde{M}$. Suppose that there are two sequences $(t_i)_{i\in\mathbb{N}}$ and $(\widetilde{t}_i)_{i\in\mathbb{N}}$ with $t_i \in S_\mu$, $\widetilde{t}_i \in \widetilde{S}_\mu$, $t_i \to 0$ and $\widetilde{t}_i \to 0$ for $i \to \infty$, such that for each $i \in \mathbb{N}$ the singularities which correspond to t_i and $\widetilde{t}_i$ are right equivalent.*

Then f and $\widetilde{f}$ are right equivalent.

Proof: We may suppose $n \geq 1$, because for $n = 0$ there are only the A_μ-singularities. Let F and $\widetilde{F}$ be (representatives of) semiuniversal unfoldings of f and $\widetilde{f}$ over M and $\widetilde{M}$. Denote by $F_{t_i} = F|(\mathbb{C}^{n+1} \times \{t_i\}, (x^{(i)}, t_i))$ the singularity which corresponds to t_i; here $x^{(i)}$ is its singular point and $F_{t_i}((x^{(i)}, t_i)) = 0$ by definition of S_μ. Define $\widetilde{F}_{\widetilde{t}_i}$ analogously and choose a coordinate change $\varphi_i : (\mathbb{C}^{n+1}, x^{(i)}) \to (\mathbb{C}^{n+1}, \widetilde{x}^{(i)})$ with $F_{t_i} = \widetilde{F}_{\widetilde{t}_i} \circ \varphi_i$.

There is no possibility of controlling directly the sequence $(\varphi_i)_{i\in\mathbb{N}}$ of coordinate changes and finding a limit coordinate change for f and $\widetilde{f}$.

But we have induced sequences of isomorphisms on several related objects which can be controlled when they are seen together and which give the desired information.

First, the germ $F : (\mathbb{C}^{n+1} \times M, (x^{(i)}, t_i)) \to (\mathbb{C}^{n+1}, 0)$ is a semiuniversal unfolding for F_{t_i}, the same holds for $\widetilde{F}$ and $\widetilde{F}_{\widetilde{t}_i}$. By Theorem 13.2 (a) φ_i induces an isomorphism of germs of F-manifolds

$$\varphi_{i,M} : (M, t_i) \to (\widetilde{M}, \widetilde{t}_i). \tag{13.27}$$

We will show that a subsequence tends to a limit isomorphism

$$\varphi_{\infty,M} : (M, 0) \to (\widetilde{M}, 0) \tag{13.28}$$

of germs of F-manifolds. With Theorem 13.2 (b) or with Scherk's result ([**Sche2**], cf. Remark 5.5 (iv)) one concludes that f and $\widetilde{f}$ are right equivalent.

In order to control the sequence $(\varphi_{i,M})_{i\in\mathbb{N}}$, we need the flat structures and their construction on M and $\widetilde{M}$. Let H^∞ denote as usual the space of global flat multivalued sections in the cohomology bundle over a punctured disc for f. It is canonically isomorphic to the corresponding spaces for the singularities in the μ-constant stratum S_μ and will be identified with them. It is equipped with the lattice $H^\infty_{\mathbb{Z}}$, the monodromy h, the polarizing form S (cf. section 10.6) and for each $t \in S_\mu$ with Steenbrink's Hodge filtration $F^\bullet(t)$ of the corresponding singularity. The period map (cf. section 12.2)

$$\Psi_{PMHS} : S_\mu \to D_{PMHS}, \quad t \mapsto F^\bullet(t) \tag{13.29}$$

to a classifying space for polarized mixed Hodge structures on H^∞ is holomorphic.

Denote the corresponding objects for $\widetilde{f}$ by $\widetilde{\mathbb{H}^\infty}$, $\widetilde{\mathbb{H}^\infty_{\mathbb{Z}}}$, $\widetilde{h}$, $\widetilde{S}$, and $\widetilde{F}^\bullet(\widetilde{t})$ for $\widetilde{t} \in \widetilde{S}_\mu$. Choose any isomorphism

$$\chi : (\widetilde{H^\infty_{\mathbb{Z}}}, \widetilde{h}, \widetilde{S}) \to (H^\infty_{\mathbb{Z}}, h, S). \tag{13.30}$$

One obtains a holomorphic period map

$$\widetilde{\Psi}_{PMHS} : \widetilde{S}_\mu \to D_{PMHS}, \quad \widetilde{t} \mapsto \chi(\widetilde{F}^\bullet(\widetilde{t})). \tag{13.31}$$

The coordinate change φ_i induces an isomorphism $\varphi_{i,coh} : H^\infty \to \widetilde{H^\infty}$ which maps $F^\bullet(t_i)$ to $\widetilde{F}^\bullet(\widetilde{t}_i)$. Therefore $\chi \circ \varphi_{i,coh} \in G_{\mathbb{Z}} = \mathrm{Aut}(\mathbb{H}^\infty_{\mathbb{Z}}, h, S)$ acts on D_{PMHS} with

$$\chi \circ \varphi_{i,coh}(\Psi_{PMHS}(t_i)) = \widetilde{\Psi}_{PMHS}(\widetilde{t}_i). \tag{13.32}$$

Now first note that $\Psi_{PMHS}(t_i) \to \Psi_{PMHS}(0)$ and $\widetilde{\Psi}_{PMHS}(\widetilde{t}_i) \to \widetilde{\Psi}_{PMHS}(0)$ for $i \to \infty$, and second that the group $G_{\mathbb{Z}}$ acts properly discontinuously on D_{PMHS} (Theorem 10.22). Therefore there exists an infinite subset $I \subset \mathbb{N}$ and an isomorphism $\varphi_{coh} : H^\infty \to \widetilde{H^\infty}$ with

$$\varphi_{i,coh} = \varphi_{coh} \quad \text{for } i \in I \tag{13.33}$$

and

$$\varphi_{coh}(F^\bullet(0)) = \widetilde{F}^\bullet(0). \tag{13.34}$$

So f and $\widetilde{f}$ have isomorphic polarized mixed Hodge structures. (One can apply the same arguments to D_{BL} instead of D_{PMHS}. If the global Torelli type conjecture 12.7 were proved one could stop here.)

Choose a monodromy invariant filtration $U_\bullet$ on H^∞ which is opposite to $F^\bullet(0)$ and a generator γ_1 of the 1-dimensional space $\mathrm{Gr}^U_{[n-\alpha_1]} H^\infty_{e^{-2\pi i \alpha_1}}$. By Theorem 11.1 the pair $(U_\bullet, \gamma_1)$ induces a flat metric g on M.

The filtration $U_\bullet$ is opposite to $F^\bullet(t)$ for all $t \in S_\mu$ (sufficiently close to 0). The construction in section 11.1 gives the same metric g for all germs (M, t) with $t \in S_\mu$.

The image $(\widetilde{U}_\bullet, \widetilde{\gamma}_1)$ under φ_{coh} satisfies the same properties with respect to $\widetilde{F}^\bullet(\widetilde{t})$ for $\widetilde{t} \in \widetilde{S}_\mu$ and induces a flat metric $\widetilde{g}$ on $\widetilde{M}$.

By Lemma 13.6 (b), the maps $\varphi_{i,M}$, $i \in I$, are isomorphisms of germs of Frobenius manifolds

$$\varphi_{i,M} : ((M, t_i), \circ, e, E, g) \to ((\widetilde{M}, \widetilde{t}_i), \circ, \widetilde{e}, \widetilde{E}, \widetilde{g}). \tag{13.35}$$

Even more, by combining Lemma 13.4 and Lemma 13.6 (b) one sees that they differ at most by translations (with respect to the flat structures).

Thus for sufficiently large $i \in I$, the map $\varphi_{i,M}$ extends to a neighbourhood of $0 \in M$, and $\varphi_{i,M}(0) \to 0$ for $i \to \infty$.

If one had $\varphi_{i,M}(0) \neq 0$ for an arbitrarily large $i \in I$ then the set $\{\varphi_{i,M}(0) \mid i \in I\}$ would cluster around $0 \in \widetilde{M}$. The germs $((\widetilde{M}, \varphi_{i,M}(0)), \circ, \widetilde{e}, \widetilde{E})$ of F-manifolds with Euler field would all be isomorphic. This is not possible by Corollary 4.16.

(Also, the singularities for the parameters $\varphi_{i,M}(0) \in \widetilde{S}_\mu$ would all be right equivalent. With the proof below of Theorem 13.15 and with a closer look at the action of the algebraic group $j_k\mathcal{R}$ on the algebraic variety $C(k, f)$ one finds that this is also not possible.)

Therefore $\varphi_{i,M}(0) = 0$ for large $i \in I$. This is the limit isomorphism from (13.28). One finishes the proof with Theorem 13.2 (b) or Scherk's result. □

Theorem 13.18 *Let $f : (\mathbb{C}^{n+1}, 0) \to (\mathbb{C}, 0)$ be an isolated hypersurface singularity. There exists a representative M of the base space $(M, 0)$ of a semiuniversal unfolding of f with the following properties.*

The finite group $\mathrm{Aut}_M := \mathrm{Aut}((M, 0), \circ, e, E)$ of automorphisms of the germ of an F-manifold acts on M. If t and $\widetilde{t}$ are in the μ-constant stratum $S_\mu \subset M$ and $\varphi : ((M, t), \circ, e, E) \to ((M, \widetilde{t}), \circ, e, E)$ is an isomorphism of germs of F-manifolds, then $\varphi \in \mathrm{Aut}_M$.

Proof: We may suppose $n \geq 1$. Assume that such a representative does not exist. Then one can choose (for some representative M) two sequences $(t_i)_{i\in\mathbb{N}}$ and $(\widetilde{t}_i)_{i\in\mathbb{N}}$ with $t_i, \widetilde{t}_i \in S_\mu$, $t_i \to 0$ and $\widetilde{t}_i \to 0$ for $i \to \infty$, and a sequence of isomorphisms

$$\varphi_i : ((M, t_i), \circ, e, E) \to ((M, \widetilde{t}_i), \circ, e, E) \tag{13.36}$$

with $\varphi_i \notin \mathrm{Aut}_M$ for any i.

By Theorem 13.2 (b), for each i the singularities which correspond to t_i and $\widetilde{t}_i$ are right equivalent and there exists a coordinate change between them which induces φ_i. But then the proof of Theorem 13.17 provides an infinite subset $I \subset \mathbb{N}$ such that all φ_i for $i \in I$ are equal and contained in Aut_M. A contradiction. □

Remark 13.19 This result does not extend to $t, \widetilde{t} \in M - S_\mu$. For example, in the case of an A_μ-singularity, the *Lyashko–Looijenga map* (section 3.5 and (5.49)) $M \to \mathbb{C}^\mu$ is finite of degree $(\mu+1)^{\mu-1}$. This implies that a generic germ of a semisimple F-manifold with Euler field turns up at $(\mu+1)^{\mu-1}$ different points in M. But the group Aut_M only has order $\mu+1$. So most of the isomorphisms of the germs of F-manifolds do not extend to $0 \in M$.

Proof of Theorem 13.15: The $j_k\mathcal{R}$-orbit of $j_k f$ in $\mathbf{m}^2/\mathbf{m}^{k+1}$ has codimension $\mu - 1$. A *transversal disc* for f is an unfolding $\check{F} : (\mathbb{C}^{n+1} \times \check{M}, 0) \to (\mathbb{C}, 0)$ of f with the following properties:

$\check{M} \subset \mathbb{C}^{\mu-1}$ is a neighbourhood of 0;
for any parameter $a = (a_1, \ldots, a_{\mu-1}) \in \check{M}$ the function germ $\check{F}_a$ is

$$\check{F}_a = \check{F}|(\mathbb{C}^{n+1} \times \{a\}, (0, a)) \in \mathbf{m}^2; \tag{13.37}$$

of course $\check{F}_0 = f$;
the map $T_k : \check{M} \to \mathbf{m}^2/\mathbf{m}^{k+1}, a \mapsto j_k\check{F}_a$ is an embedding and the image $T_k(\check{M})$ intersects $j_k\mathcal{R} \cdot j_k f$ transversally in $j_k f$.

If we choose a smooth germ $(R, id) \subset (j_k\mathcal{R}, id)$ which is transversal to the stabilizer of $j_k f$ then the natural map of germs

$$(R, id) \times (T_k(\check{M}) \cap C(k, f), j_k f) \to (C(k, f), j_k f) \tag{13.38}$$

is an isomorphism. We need the germ $(T_k(\check{M}) \cap C(k, f), j_k f)$ to be isomorphic to the μ-constant stratum $(S_\mu, 0) \subset (M, 0)$ of f. It will follow from results of Gabrielov and Teissier.

Gabrielov [**Ga**] constructed explicitly a semiuniversal unfolding $F : (\mathbb{C}^{n+1} \times M, 0) \to (\mathbb{C}, 0)$ of f, a transversal disc $\check{F}$ for f, and an embedding

$$(\xi, \tau) = (\xi_0, \ldots, \xi_n, \tau_1, \ldots, \tau_\mu) : (\check{M}, 0) \to (\mathbb{C}^{n+1} \times M, 0) \tag{13.39}$$

with

$$\check{F}_a(x) = F_{\tau(a)}(x + \xi(a)). \tag{13.40}$$

The critical space $(C, 0) \subset (\mathbb{C}^{n+1} \times M, 0)$ (cf. section 10.1) and the intersection $C \cap F^{-1}(0)$ are smooth, $\mathcal{D} = pr_M(C \cap F^{-1}(0)) \subset M$ is the discriminant. Gabrielov's construction yields an isomorphism

$$(\xi, \tau) : (\check{M}, 0) \to (C \cap F^{-1}(0), 0) \tag{13.41}$$

which maps $T_k^{-1}(T_k(\check{M}) \cap C(k, f), j_k f)$ to the preimage in $(C \cap F^{-1}(0), 0)$ of the μ-constant stratum $(S_\mu, 0)$.

The projection to the μ-constant stratum is clearly a homeomorphism. Teissier [**Te1**, §6] showed that it is an isomorphism. He used the fact that the map $C \cap F^{-1}(0) \to \mathcal{D}$ is isomorphic to the *development* of the discriminant (cf. for example section 3.5) to construct a section $(M, 0) \to (\mathbb{C}^{n+1} \times M, 0)$ which contains the preimage of $(S_\mu, 0)$ in $(C \cap F^{-1}(0), 0)$.

Therefore the germs $(T_k(\check{M}) \cap C(k, f), j_k f)$ and $(S_\mu, 0)$ are isomorphic.

Now Theorem 13.17 shows that the quotient topology of $\mathcal{M}_\mu = C(k, f)/j_k\mathcal{R}$ is Hausdorff. This is the first of two conditions in a criterion of Holmann [**Hol**, Satz 17] for the existence of an analytic geometric quotient. The second condition is the existence of holomorphic functions in a neighbourhood of $j_k f$ in $C(k, f)$, which are constant on the $j_k\mathcal{R}$-orbits and which separate points in different orbits.

By (13.38) and the above isomorphism of germs it is sufficient to show the existence of such functions on S_μ. But in that case it follows from Theorem 13.18 and from the classical result of Cartan [**Ca**] that a quotient by a finite group is a reduced complex space. This shows parts (a) and (b) in Theorem 13.15.

The canonical complex structure on S_μ from Theorem 12.4 is invariant under Aut_M, for example because these automorphisms respect the Frobenius manifold structure from $U_\bullet^{(0)}$ and the canonical complex structure is completely determined by this Frobenius manifold structure (1st step in the proof of Theorem 12.4). One can apply locally [**Kaup**, 49 A.16] and glue the complex structures. This gives part (c). □

Remark 13.20 (a) Consider the local period map $\Psi : S_\mu \to D_{BL}$ for a singularity f. If $\varphi \in \mathcal{R}^f$ is a symmetry then the action of φ_M on $S_\mu \subset M$ and the action of φ_{coh} on $\Psi(S_\mu) \subset D_{BL}$ are compatible. Now Theorem 13.15 shows that the global period map $\mathcal{M}_\mu \to D_{BL}/G_{\mathbb{Z}}$ is a map between reduced complex varieties.

(b) A part of the splitting lemma says that two singularities $f(x_0, \dots, x_n)$ and $g(x_0, \dots, x_n)$ are right equivalent if and only if $f + x_{n+1}^2$ and $g + x_{n+1}^2$ are right equivalent (e.g. Remark 5.7 (i)). Therefore the global moduli space $\mathcal{M}_\mu(f)$ embeds into $\mathcal{M}_\mu(f + x_{n+1}^2)$.

It would be very desirable to know the answer to the following very weak form of the Zariski conjecture. Do the singularities in a μ-homotopy class have either all multiplicity 2 or all multiplicity ≥ 3?

If the answer is yes, $\mathcal{M}_\mu(f)$ and $\mathcal{M}_\mu(f + x_{n+1}^2)$ are isomorphic, if no, the second one would be larger in some cases.

(c) The answer to the question is yes for semiquasihomogeneous singularities and for curve singularities.

An interesting related result has been shown by Navarro-Aznar [**Nav**]: suppose $f(x_0, \ldots x_n)$ and $g(x_0, \ldots, x_n)$ are singularities with the same topological type and suppose that the rank of the Hessian of f, $\mathrm{rank}(\frac{\partial^2 f}{\partial x_i \partial x_j})(0)$, is *odd* (this implies that its multiplicity is 2). Then also the multiplicity of g is two.

This implies that two surface singularities with the same topological type both have multiplicity two or both have multiplicity ≥ 3.

(But precisely in the surface case it is not yet clear whether μ-homotopy implies the same topological type, cf. [**LeR**]).

(d) If the answer is yes for some μ-homotopy class then the groups R_f for singularities f in this μ-homotopy class satisfy a semicontinuity property. Then by Theorem 13.18 and Theorem 13.9 the groups R_{F_t} for the singularities in the μ-constant stratum of a fixed singularity f are (isomorphic to) subgroups of R_f.

For semiquasihomogeneous singularities the semicontinuity property of R_f has been proved previously in [**Mue2**].

Chapter 14

Variance of the spectral numbers

Section 14.3 gives a surprising statement on the spectral numbers of quasi-homogeneous singularities. It comes from properties of the G-function of a semisimple or massive Frobenius manifold. General remarks, the definition of Dubrovin, Zhang, and Givental and some properties of it are presented in section 14.2. The socle field of a Frobenius manifold is related to the simpler part of this G-function. It is more or less known, but not treated systematically in the literature. This is provided in section 14.1.

14.1 Socle field

A Frobenius manifold has another distinguished vector field besides the unit field and the Euler field. It will be discussed in this section. We call it the *socle field*. It is used implicitly in [**Du4**, Theorem 1.1] and [**Gi7**].

Let $(M, \circ, e, g)$ be a manifold with a multiplication $\circ$ on the tangent bundle, with a unit field, and with a multiplication invariant metric g. We do not need flatness and potentiality and an Euler field in the moment.

Each tangent space T_tM is a Frobenius algebra and splits uniquely into a direct sum of Gorenstein rings (cf. section 2.1)

$$T_tM = \bigoplus_{k=1}^{l(t)} (T_tM)_k \tag{14.1}$$

with maximal ideals $\mathbf{m}_{t,k} \subset (T_tM)_k$ and units e_k such that $e = \sum e_k$,

$$(T_tM)_j \circ (T_tM)_k = \{0\} \quad \text{for } j \neq k, \tag{14.2}$$

and thus

$$g((T_tM)_j, (T_tM)_k) = \{0\} \quad \text{for } j \neq k. \tag{14.3}$$

The socle $\mathrm{Ann}_{(T_tM)_k}(\mathbf{m}_{t,k})$ is 1-dimensional and has a unique generator $H_{t,k}$ which is normalized such that

$$g(e_k, H_{t,k}) = \dim(T_tM)_k. \tag{14.4}$$

The following lemma shows that the vectors $\sum_k H_{t,k}$ glue to a holomorphic vector field, the *socle field* of $(M, \circ, e, g)$.

Lemma 14.1 *For any dual bases $X_1, \ldots, X_m$ and $\widetilde{X}_1, \ldots, \widetilde{X}_m$ of T_tM, that means, $g(X_i, \widetilde{X}_j) = \delta_{ij}$, one has*

$$\sum_{k=1}^{l(t)} H_{t,k} = \sum_{i=1}^{m} X_i \circ \widetilde{X}_i. \tag{14.5}$$

Proof: One easily sees that the sum $\sum X_i \circ \widetilde{X}_i$ is independent of the choice of the basis $X_1, \ldots, X_m$. One can suppose that $l(t) = 1$ and that $X_1, \ldots, X_m$ are chosen such that they yield a splitting of the filtration $T_tM \supset \mathbf{m}_{t,1} \supset \mathbf{m}_{t,1}^2 \supset \ldots$. Then $g(e, X_i \circ \widetilde{X}_i) = 1$ and

$$g(\mathbf{m}_{t,1}, X_i \circ \widetilde{X}_i) = g(X_i \circ \mathbf{m}_{t,1}, \widetilde{X}_i) = 0.$$

Thus $X_i \circ \widetilde{X}_i = \frac{1}{m} H_{t,1}$. □

It will be useful to fix the multiplication and vary the metric.

Lemma 14.2 *Let $(M, \circ, e, g)$ be a manifold with multiplication $\circ$ on the tangent bundle, unit field e and multiplication invariant metric g. For each multiplication invariant metric $\widetilde{g}$ there exists a unique vector field Z such that the multiplication with it is invertible everywhere and for all vector fields X, Y*

$$\widetilde{g}(X, Y) = g(Z \circ X, Y). \tag{14.6}$$

The socle fields H and $\widetilde{H}$ of g and $\widetilde{g}$ satisfy

$$H = Z \circ \widetilde{H}. \tag{14.7}$$

Proof: The situation for one Frobenius algebra is described in Lemma 2.2. It yields (14.6) immediately. Formula (14.7) follows from the comparison of (14.4) and (14.6). □

Denote by

$$H_{op} : \mathcal{T}_M \to \mathcal{T}_M, \quad X \mapsto H \circ X \tag{14.8}$$

the multiplication with the socle field H of $(M, \circ, e, g)$ as above. The socle field is especially interesting if the multiplication is generically semisimple, that means, generically $l(t) = m$. Then the caustic $\mathcal{K} = \{t \in M \mid l(t) < m\}$ is the set where the multiplication is not semisimple. It is the hypersurface

$$\mathcal{K} = \det(H_{op})^{-1}(0). \tag{14.9}$$

In an open subset of $M - \mathcal{K}$ with basis $e_1, \dots, e_m$ of idempotent vector fields the socle field is

$$H = \sum_{i=1}^{m} \frac{1}{g(e_i, e_i)} e_i. \tag{14.10}$$

It determines the metric g everywhere because (14.10) determines the metric at semisimple points.

If $(M, \circ, e)$ is an F-manifold then each germ $((M, t), \circ, e)$ decomposes uniquely into a product of irreducible germs of F-manifolds (Theorem 2.11). This extends the infinitesimal decomposition (14.1).

At semisimple points one has the product A_1^m if m 1-dimensional germs of F-manifolds. The 2-dimensional irreducible germs are classified in Theorem 4.7, those with generically semisimple multiplication form a series $I_2(n)$, $n \geq 3$, with $I_2(3) = A_2$.

Theorem 14.3 *Let $(M, \circ, e, g)$ be a massive F-manifold with multiplication invariant metric g. Suppose that at generic points of the caustic the germ of the F-manifold is of the type $I_2(n)A_1^{m-2}$.*

Then the function $\det(H_{op})$ *vanishes with multiplicity $n-2$ along the caustic.*

Proof: The manifold $M = \mathbb{C}^m$ with coordinate fields $\delta_i = \frac{\partial}{\partial t_i}$ and multiplication defined by

$$\delta_1 \circ \delta_2 = \delta_2, \tag{14.11}$$

$$\delta_2 \circ \delta_2 = t_2^{n-2} \delta_1, \tag{14.12}$$

$$\delta_i \circ \delta_j = \delta_{ij} \delta_i \quad \text{if } (i, j) \notin \{(1, 2), (2, 1), (2, 2)\} \tag{14.13}$$

is an F-manifold with a global decomposition $\mathbb{C}^2 \times \mathbb{C} \times \cdots \times \mathbb{C}$ of the type $I_2(n)A_1^{m-2}$. The unit fields for the components are $\delta_1, \delta_3, \dots, \delta_m$, the global unit field is $e = \delta_1 + \delta_3 + \cdots + \delta_m$, the caustic is $\mathcal{K} = \{t \mid t_2 = 0\}$. The idempotent vector fields in a simply connected subset of $M - \mathcal{K}$ are

$$e_{1/2} = \frac{1}{2}\delta_1 \pm \frac{1}{2} t_2^{-\frac{n-2}{2}} \delta_2, \tag{14.14}$$

$$e_i = \delta_i \quad \text{for } i \geq 3, \tag{14.15}$$

canonical coordinates there are

$$u_{1/2} = t_1 \pm \frac{2}{n} t_2^{\frac{n}{2}}, \tag{14.16}$$

$$u_i = t_i \quad \text{for } i \geq 3. \tag{14.17}$$

A multiplication invariant metric g is uniquely determined by the 1-form $\varepsilon = g(e, .)$. Because of (14.7) it is sufficient to prove the claim for one metric. We choose the metric with 1-form

$$\varepsilon(\delta_i) = 1 - \delta_{i1}. \tag{14.18}$$

The bases $\delta_1, \delta_2, \delta_3, \ldots, \delta_m$ and $\delta_2, \delta_1, \delta_3, \ldots, \delta_m$ are dual with respect to this metric. Its socle field is by Lemma 14.1

$$H = 2\delta_2 + \delta_3 + \cdots + \delta_m \tag{14.19}$$

and satisfies $\det(H_{op}) = -4t_2^{n-2}$.

If an F-manifold M has at generic points of the caustic germs of the type $I_2(n)A_1^{m-2}$ then the set of nongeneric points is empty or has codimension ≥ 2 in M. □

14.2 G-function of a massive Frobenius manifold

Associated to any simply connected semisimple Frobenius manifold is a fascinating and quite mysterious function. Dubrovin and Zhang **[DuZ1]**[**DuZ2**] called it the G-function and proved the most detailed results for it. But Givental [**Gi7**] studied it slightly earlier, and it originates in much older work. It takes the form

$$G(t) = \log \tau_I - \tfrac{1}{24} \log J \tag{14.20}$$

and is determined only up to addition of a constant. First we explain the simpler part, $\log J$. Let $(M, \circ, e, E, g)$ be a semisimple Frobenius manifold with canonical coordinates $u_1, \ldots, u_m$ and flat coordinates $\widetilde{t}_1, \ldots, \widetilde{t}_m$. Then

$$J = \det\left(\frac{\partial \widetilde{t}_i}{\partial u_j}\right) \cdot \text{constant} \tag{14.21}$$

is the base change matrix between flat and idempotent vector fields.

One can rewrite it with the socle field. Denote $\eta_i := g(e_i, e_i)$ and consider the basis $v_1, \ldots, v_m$ of vector fields with

$$v_i = \frac{1}{\sqrt{\eta_i}} e_i \tag{14.22}$$

(for some choice of the square roots). The matrix $\det(g(v_i, v_j)) = 1$ is constant as is the corresponding matrix for the flat vector fields. Therefore

$$\text{constant} \cdot J = \prod_{i=1}^{m} \sqrt{\eta_i} = \det(H_{op})^{-\frac{1}{2}}. \tag{14.23}$$

Here $H = \sum v_i \circ v_i$ is the socle field.

One of the origins of the first part $\log \tau_I$ is the geometry of isomonodromic deformations. The second structure connections and the first structure connections of the semisimple Frobenius manifold are isomonodromic deformations over $\mathbb{P}^1 \times M$ of restrictions to a slice $\mathbb{P}^1 \times \{t\}$. The function τ_I is their τ*-function* in the sense of [**JMMS**][**JMU**][**JM**][**Mal4**]. See [**Sab4**] for other general references on this.

The situation for Frobenius manifolds is discussed and put into a Hamiltonian framework in [**Du3**, Lecture 3], [**Man2**, II§2], and in [**Hi**]. The coefficients H_i of the 1-form $d \log \tau_I = \sum H_i du_i$ are certain Hamiltonians and motivate the definition of this 1-form. Hitchin [**Hi**] compares the realizations of this for the first and the second structure connections.

Another origin of the whole G-function comes from quantum cohomology. Getzler [**Ge**] studied the relations between cycles in the moduli space $\overline{\mathcal{M}}_{1,4}$ and derived from it recursion relations for genus one Gromov–Witten invariants of projective manifolds and differential equations for the genus one Gromov–Witten potential.

Dubrovin and Zhang [**DuZ1**, chapter 6] investigated these differential equations for any semisimple Frobenius manifold and found that they have always one unique solution (up to addition of a constant), the G-function, a function which was proposed in [**Gi7**].

They also proved the major part of the conjectures in [**Gi7**] concerning $G(t)$. Finally, they found that the potential of the Frobenius manifold (for genus zero) and the G-function (for genus one) are the basements of full free energies in genus zero and one and give rise to Virasoro constraints [**DuZ2**]. Givental found formulas for the full free energies at higher genus [**Gi8**] and proved Virasoro constraints for them [**Gi9**]. Exploiting this for singularities will be a big task for the future.

For our application in section 13.3 we need only the definition of $\log \tau_I$ and the behaviour of $G(t)$ with respect to the Euler field and the caustic in a massive Frobenius manifold. We have to summarize some known formulas related to the canonical coordinates of a semisimple Frobenius manifold ([**Du3**], [**Man2**], also [**Gi7**]).

The 1-form $\varepsilon = g(e, .)$ is closed and can be written as $\varepsilon = \mathrm{d}\eta$. One defines

$$\eta_i := e_i\eta = g(e_i, e) = g(e_i, e_i), \tag{14.24}$$

$$\eta_{ij} := e_i e_j \eta = e_i\eta_j = e_j\eta_i, \tag{14.25}$$

$$\gamma_{ij} := \frac{1}{2}\frac{\eta_{ij}}{\sqrt{\eta_i}\sqrt{\eta_j}}, \tag{14.26}$$

$$V_{ij} := -(u_i - u_j)\gamma_{ij}, \tag{14.27}$$

$$\mathrm{d}\log\tau_I := \frac{1}{2}\sum_{i\neq j}\frac{V_{ij}^2}{u_i - u_j}\mathrm{d}u_i = \frac{1}{2}\sum_{i\neq j}(u_i - u_j)\gamma_{ij}^2\mathrm{d}u_i \tag{14.28}$$

$$= \frac{1}{8}\sum_{i\neq j}(u_i - u_j)\frac{\eta_{ij}^2}{\eta_i\eta_j}\mathrm{d}u_i. \tag{14.29}$$

Theorem 14.4 *Let $(M, \circ, e, E, g)$ be a semisimple Frobenius manifold with global canonical coordinates $u_1, \ldots, u_m$.*

(a) The rotation coefficients γ_{ij} (for $i \neq j$) satisfy the Darboux–Egoroff equations

$$e_k\gamma_{ij} = \gamma_{ik}\gamma_{kj} \quad \textit{for } k \neq i \neq j \neq k, \tag{14.30}$$

$$e\gamma_{ij} = 0 \quad \textit{for } i \neq j. \tag{14.31}$$

(b) The connection matrix of the flat connection for the basis $v_1, \ldots, v_m$ from (14.22) *is the matrix $(\gamma_{ij}d(u_i - u_j))$. The Darboux–Egoroff equations are equivalent to the flatness condition*

$$\mathrm{d}(\gamma_{ij}\mathrm{d}(u_i - u_j)) + (\gamma_{ij}\mathrm{d}(u_i - u_j)) \wedge (\gamma_{ij}d(u_i - u_j)) = 0. \tag{14.32}$$

(c) The 1-form $\mathrm{d}\log\tau_I$ *is closed and comes from a function* $\log\tau_I$.

(d) The Euler field E satisfies $E(\eta_i) = (D - 2)\eta_i$ and $E(\gamma_{ij}) = -\gamma_{ij}$.

(e) If the canonical coordinates are chosen such that $E = \sum u_i e_i$ then the matrix $-(V_{ij})$ is the matrix of the endomorphism $\mathcal{V} = \nabla E - \frac{D}{2}\mathrm{id}$ on $\mathcal{T}_M$ with respect to the basis $v_1, \ldots, v_m$.

Proof: (a) and (b) See [**Du3**, pp. 200–201] or [**Man2**, I§3].

(c) This can be checked easily with the Darboux–Egoroff equations.

(d) It follows from $\mathrm{Lie}_E(g) = D \cdot g$ and from $[e_i, E] = e_i$.

(e) This is implicit in [**Du3**, pp. 200–201]. One can check it with (a) and (b) and (d). □

The endomorphism $\mathcal{V}$ is skewsymmetric with respect to g and flat with eigenvalues $d_i - \frac{D}{2}$; the numbers d_i can be ordered such that $d_1 = 1$, $d_i + d_{m+1-i} = D$ (cf. Remark 9.2 (e)).

Corollary 14.5 [**DuZ1**, Theorem 3] *Suppose that $E = \sum u_i e_i$. Then*

$$E \log \tau_I = -\frac{1}{4} \sum_{i=1}^{m} \left(d_i - \frac{D}{2} \right)^2, \tag{14.33}$$

$$E\ G(t) = -\frac{1}{4} \sum_{i=1}^{m} \left(d_i - \frac{D}{2} \right)^2 + \frac{m(2-D)}{48} =: \gamma. \tag{14.34}$$

Proof:

$$E \log \tau_I = \frac{1}{2} \sum_{i \neq j} \frac{u_i V_{ij}^2}{u_i - u_j} = \frac{1}{2} \sum_{i<j} V_{ij}^2 \tag{14.35}$$

$$= -\frac{1}{4} \sum_{ij} V_{ij} V_{ji} = -\frac{1}{4} \text{trace}(V^2)$$

$$= -\frac{1}{4} \sum_{i=1}^{m} \left(d_i - \frac{D}{2} \right)^2.$$

Formula (14.23) shows $E(J) = m\frac{D-2}{2} J$. Now (14.34) follows from the definition of the G-function. □

If M is a massive Frobenius manifold with caustic $\mathcal{K}$, one may ask which kind of poles the 1-form $d \log \tau_I$ has along $\mathcal{K}$ and when the G-function extends over $\mathcal{K}$.

In [**DuZ1**, chapter 6] the G-function is calculated for the 2-dimensional Frobenius manifolds $I_2(n)$, $n \geq 3$, on $M = \mathbb{C}^2$ with coordinates (t_1, t_2) and $e = \frac{\partial}{\partial t_1}$. It turns out to be

$$G(t) = -\frac{1}{24} \frac{(2-n)(3-n)}{n} \log t_2. \tag{14.36}$$

Especially, for the case $I_2(3) = A_2$ the G-function is $G(t) = 0$. This was also checked in [**Gi7**]. Givental concluded that in the case of singularities the G-function of the base space of a semiuniversal unfolding with some Frobenius manifold structure extends holomorphically over the caustic. This is a good guess, but it does not follow from the case A_2, because a Frobenius manifold structure on a germ of an F-manifold of type $A_2 A_1^{m-2}$ for $m \geq 3$ is never the product of the Frobenius manifolds A_2 and A_1^{m-2} (the numbers $d_1, \ldots, d_m$ would not be symmetric). However, it is true, as the following result shows.

Theorem 14.6 *Let $(M, \circ, e, E, g)$ be a simply connected massive Frobenius manifold. Suppose that at generic points of the caustic $\mathcal{K}$ the germ of the underlying F-manifold is of type $I_2(n) A_1^{m-2}$ for one fixed number $n \geq 3$.*

(a) The form $\mathrm{d}\log\tau_I$ *has a logarithmic pole along* $\mathcal{K}$ *with residue* $-\frac{(n-2)^2}{16n}$ *along* $\mathcal{K}_{reg}$.

(b) The G-function extends holomorphically over $\mathcal{K}$ *if and only if* $n = 3$.

Proof: Theorem 14.3 and (14.23) say that the form $-\frac{1}{24}\mathrm{d}\log J$ has a logarithmic pole along $\mathcal{K}$ with residue $\frac{n-2}{48}$ along $\mathcal{K}_{reg}$. This equals $\frac{(n-2)^2}{16n}$ if and only if $n = 3$. So (b) follows from (a).

It is sufficient to show (a) for the F-manifold $M = \mathbb{C}^m$ in the proof of Theorem 14.3, equipped with some metric which makes a Frobenius manifold out of it (we do not need an Euler field here). Unfortunately we do not have an identity for $\mathrm{d}\log\tau_I$ as (14.7) for the socle field which would allow only a most convenient metric to be considered.

We use (14.11)–(14.17) and (14.24)–(14.29) and consider a neighbourhood of $0 \in \mathbb{C}^m = M$. Denote for $j \geq 3$

$$T_{1j} := (u_1 - u_j)\frac{\eta_{j1}^2}{\eta_j\eta_1} + (u_2 - u_j)\frac{\eta_{j2}^2}{\eta_j\eta_2}, \tag{14.37}$$

$$T_{2j} := (u_1 - u_j)\frac{\eta_{j1}^2}{\eta_j\eta_1} - (u_2 - u_j)\frac{\eta_{j2}^2}{\eta_j\eta_2},$$

$$T_{12} := (u_1 - u_2)\frac{\eta_{12}^2}{\eta_1\eta_2}\mathrm{d}(u_1 - u_2).$$

With $\eta_j(0) \neq 0$ for $j \geq 3$, (14.29) and (14.16) one calculates

$$\begin{aligned} 8\mathrm{d}\log\tau_I &= \text{holomorphic 1-form} + T_{12} \\ &\quad + \sum_{j\geq 3} T_{1j}\mathrm{d}t_1 + \sum_{j\geq 3} T_{2j}t_2^{\frac{n-2}{2}}\,\mathrm{d}t_2 - \sum_{j\geq 3} T_{1j}\mathrm{d}u_j. \end{aligned} \tag{14.38}$$

From (14.14) one obtains

$$\eta_{1/2} = \frac{1}{2}\delta_1(\eta) \pm \frac{1}{2}\delta_2(\eta)t_2^{-\frac{n-2}{2}}, \tag{14.39}$$

$$\eta_1 \cdot \eta_2 = \frac{1}{4}t_2^{-n+2}\big(-\delta_2(\eta)^2 + t_2^{n-2}\delta_1(\eta)^2\big), \tag{14.40}$$

$$\eta_{12} = \frac{1}{4}\delta_1\delta_1(\eta) + \frac{1}{4}\frac{n-2}{2}t_2^{-n+1}\delta_2(\eta) - \frac{1}{4}t_2^{-n+2}\delta_2\delta_2(\eta). \tag{14.41}$$

The vector $\delta_2|_0$ is a generator of the socle of the subalgebra in T_0M which corresponds to $I_2(n)$. Therefore $\delta_2(\eta)(0) \neq 0$. It is not hard to see with (14.39) and (14.16) that the terms T_{1j} and $T_{2j}t_2^{\frac{n-2}{2}}$ for $j \geq 3$ are holomorphic at 0. The term T_{12} is

$$\begin{aligned} T_{12} &= \frac{4}{n} \cdot t_2^{\frac{n}{2}} \cdot \frac{\eta_{12}^2}{\eta_1 \eta_2} \cdot \mathrm{d}\left(\frac{4}{n} \cdot t_2^{\frac{n}{2}}\right) \\ &= \frac{8}{n} \cdot t_2^{n-1} \cdot \frac{\eta_{12}^2}{\eta_1 \eta_2} \cdot \mathrm{d}t_2 \\ &= -\frac{(n-2)^2}{2n} \cdot \frac{\mathrm{d}t_2}{t_2} + \text{holomorphic 1-form.} \end{aligned} \tag{14.42}$$

This proves part (a). □

Remarks 14.7 (a) The base spaces of semiuniversal unfoldings meet the case $n = 3$ in Theorem 14.6. Closely related are base spaces of certain unfoldings of tame functions. The germs of F-manifolds are isomorphic to products of the germs of F-manifolds from hypersurface singularities. But Sabbah [**Sab3**] [**Sab2**][**Sab4**] equipped them with a metric such that the Frobenius manifold structure is in general not a product (cf. section 11.4).

(b) It might be interesting to look for massive Frobenius manifolds which meet the case $n = 3$ in Theorem 14.6, but where the underlying F-manifolds are not locally products of those from hypersurface singularities. In view of Theorem 5.6 the analytic spectrum of such F-manifolds would have singularities, but only in codimension ≥ 2, as the analytic spectrum of A_2 is smooth.

The analytic spectrum is Cohen–Macaulay and even Gorenstein and a Lagrange variety. P. Seidel (Ecole Polytechnique) showed me a normal and Cohen–Macaulay Lagrange surface. But it seems to be unclear whether there exist normal and Gorenstein Lagrange varieties which are not smooth.

14.3 Variance of the spectrum

By Theorem 14.6 the germ $(M, 0)$ of a Frobenius manifold as in Theorem 11.1 for an isolated hypersurface singularity f has a holomorphic G-function $G(t)$, unique up to addition of a constant. By Corollary 14.5 and Theorem 11.1 this function satisfies

$$EG(t) = -\frac{1}{4}\sum_{i=1}^{\mu}\left(\alpha_i - \frac{n-1}{2}\right)^2 + \frac{\mu(\alpha_\mu - \alpha_1)}{48} =: \gamma. \tag{14.43}$$

So it has a very peculiar strength: it gives a hold at the squares of the spectral numbers $\alpha_1, \ldots, \alpha_\mu$ of the singularity. Because of the symmetry $\alpha_i + \alpha_{\mu+1-i} = n-1$, the spectral numbers are scattered around their *expectation value* $\frac{n-1}{2}$. One may ask about their *variance* $\frac{1}{\mu}\sum_{i=1}^{\mu}(\alpha_i - \frac{n-1}{2})^2$.

Conjecture 14.8 *The variance of the spectral numbers of an isolated hypersurface singularity is*

$$\frac{1}{\mu}\sum_{i=1}^{\mu}\left(\alpha_i - \frac{n-1}{2}\right)^2 \le \frac{\alpha_\mu - \alpha_1}{12}, \tag{14.44}$$

or, equivalently,

$$\gamma \ge 0. \tag{14.45}$$

Theorem 14.9 *In the case of a quasihomogeneous singularity* f

$$\frac{1}{\mu}\sum_{i=1}^{\mu}\left(\alpha_i - \frac{n-1}{2}\right)^2 = \frac{\alpha_\mu - \alpha_1}{12}, \tag{14.46}$$

and

$$\gamma = 0. \tag{14.47}$$

Proof: One has the isomorphism $(\mathcal{O}/J_f, mult., [f]) \cong (T_0M, \circ, E|_0)$. Here $f \in J_f$ and $E|_0 = 0$ and therefore $E\ G(t) = 0$. □

Remarks 14.10 (a) When I presented Theorem 14.9 at the summer school on singularity theory in Cambridge in August 2000, I asked for an elementary proof of it. This was found by A. Dimca. It uses the characteristic function

$$\chi_f := \frac{1}{\mu}\sum_{i=1}^{\mu} T^{\alpha_i+1} \tag{14.48}$$

of the spectral numbers. The variance is

$$\frac{\mathrm{d}}{\mathrm{d}t}\left(T \cdot \frac{\mathrm{d}}{\mathrm{d}t}(\chi_f \cdot T^{-\frac{n+1}{2}})\right)|_{T=1}. \tag{14.49}$$

In the case of a quasihomogeneous singularity with weights $w_0, \ldots, w_n \in (0, \frac{1}{2}]$ and degree 1 the characteristic function is

$$\chi_f = \frac{1}{\mu}\prod_{i=0}^{n}\frac{T - T^{w_i}}{T^{w_i} - 1}, \tag{14.50}$$

as is well known. Using this product formula A. Dimca [**Di**] showed that in the case of a quasihomogeneous singularity the variance is $\sum_{i=0}^{n}\frac{1-2w_i}{12} = \frac{\alpha_\mu - \alpha_1}{12}$. He also made a conjecture dual to Conjecture 14.8 for the case of tame polynomials: there the inverse inequality to (14.44) should hold. The conjectures intersect in the case of quasihomogeneous singularities and give there the equality (14.46).

(b) M. Saito in September 2000 proved Conjecture 14.8 in the case of irreducible plane curve singularities [**SM5**].

(c) T. Brélivet in May 2001 proved it in the case of plane curve singularities with nondegenerate Newton polyeder [**Bre**].

(d) The only unimodal or bimodal families of not semiquasihomogeneous singularities are the cusp singularities T_{pqr} and the 8 bimodal series. The spectral numbers are given in [**AGV2**]. One finds

$$\gamma(T_{pqr}) = \frac{1}{24}\left(1 - \frac{1}{p} - \frac{1}{q} - \frac{1}{r}\right) \geq 0 \tag{14.51}$$

with equality only for the simple elliptic singularities. In the case of the 8 bimodal families one obtains

$$\gamma = \frac{p}{48 \cdot \kappa} \cdot \left(1 - \frac{1}{p + \kappa}\right) \geq 0 \tag{14.52}$$

with $\kappa := 9, 7, 6, 6, 5$ for $E_{3,p}$, $Z_{1,p}$, $Q_{2,p}$, $W_{1,p}$, $S_{1,p}$, respectively, and

$$\gamma = \frac{p}{48 \cdot \kappa} \cdot \left(1 + \frac{1}{p + 2\kappa}\right) \geq 0 \tag{14.53}$$

with $\kappa := 6, 5, \frac{9}{2}$ for $W^\sharp_{1,p}$, $S^\sharp_{1,p}$, $U_{1,p}$, respectively.

(e) At the summer school in Cambridge in August 2000 Conjecture 14.8 was confirmed for many other singularities using the computer algebra system Singular and especially the program of M. Schulze for computing spectral numbers, which is presented in [**SchuSt**].

(f) In [**SK8**] K. Saito studied the distribution of the spectral numbers and their characteristic function χ_f heuristically and formulated several questions about them. The G-function might help these problems to be continued.

(g) One can speculate that Conjecture 14.8, if it is true, comes from a more profound hidden interrelation between the Gauß–Manin connection and polarized mixed Hodge structures. The existence of Frobenius manifolds and G-functions alone is not sufficient, as the following shows.

In Remark 11.7 (b) an example of M. Saito [**SM3**, 4.4] is sketched which leads for the semiquasihomogeneous singularity $f = x^6 + y^6 + x^4y^4$ to Frobenius manifold structures with $\{d_1, .., d_\mu\} \neq \{1+\alpha_1-\alpha_i \mid i = 1, \dots, \mu\}$. The number γ in that case is $\gamma = -\frac{1}{144} < 0$.

(h) In the case of the simple singularities A_k, D_k, E_6, E_7, E_8, the parameters $t_1, \dots, t_\mu$ of a suitably chosen unfolding are weighted homogeneous with positive degrees with respect to the Euler field. Therefore $G = 0$ in these cases (cf. [**Gi7**]).

Lemma 14.11 *The number γ of the sum $f(x_0, \dots, x_n) + g(y_0, \dots, y_m)$ of two singularities f and g satisfies*

$$\gamma(f+g) = \mu(f) \cdot \gamma(g) + \mu(g) \cdot \gamma(f). \tag{14.54}$$

Proof: Let $\alpha_1, \dots, \alpha_{\mu(f)}$ and $\beta_1, \dots, \beta_{\mu(g)}$ denote the spectral numbers of f and g. Then the spectrum of $f + g$ as an unordered tuple is [**AGV2**][**SchSt**]

$$(\alpha_i + \beta_j + 1 \mid i = 1, \dots, \mu(f),\ j = 1, \dots, \mu(g)). \tag{14.55}$$

This and the symmetry of the spectra yields (14.54). □

Remarks 14.12 For any Frobenius manifold the variance $\frac{1}{m}\sum_{i=1}^{m}(d_i - \frac{D}{2})^2$ of the eigenvalues $d_1, \dots, d_m$ of ∇E is interesting. It turns up not only as in Corollary 14.5 related to the G-function in the semisimple case, but also in the operator L_0 of the Virasoro constraints in [**DuZ2**, (2.30)] for any Frobenius manifold.

Prior to [**DuZ2**] the Virasoro constraints were postulated in the case of quantum cohomology of projective manifolds with $h^{p,q} = 0$ for $p \neq q$ in [**EHX**]. There a formula for the variance was considered which turned out to be a special case of the following formula from [**LiW**] (cf. also [**Bori**]), which is valid for any projective manifold:

$$\sum_{p,q}(-1)^{p+q}h^{p,q}\left(p - \frac{n}{2}\right)^2 = \frac{n}{12}c_n + \frac{1}{6}c_1c_{n-1}. \tag{14.56}$$

Here c_l is the lth Chern class of the manifold, n is its dimension. The proof uses the Hirzebruch–Riemann–Roch theorem. The formula is generalized to projective varieties with at most Gorenstein canonical singularities in [**Bat**]. Comparing the right hand side with the singularity case, one can speculate $n \sim \alpha_\mu - \alpha_1$, $c_n \sim \mu$ and ask about $\frac{1}{6}c_1c_{n-1} \sim ?$.

Bibliography

[AB] Anosov, D. V., A. A. Bolibruch: The Riemann-Hilbert problem. Aspects of mathematics, vol. **22**, Vieweg, Braunschweig Wiesbaden 1994.

[Ar1] Arnold, V. I.: Critical points of functions and the classification of caustics (in Russian). Uspekhi Mat. Nauk **29.3** (1974), 243–244.

[Ar2] Arnold, V. I.: Critical points of functions on a manifold with boundary, the simple Lie groups B_k, C_k, and F_4 and singularities of evolutes. Russian Math. Surveys **33.5** (1978), 99–116.

[Ar3] Arnold, V. I.: Indices of singular points of 1–forms on a manifold with boundary, convolution of invariants of reflection groups, and singular projections of smooth surfaces, Russian Math. Surveys **34.2** (1979), 1–42.

[AGV1] Arnold, V. I., S. M. Gusein-Zade, A. N. Varchenko: Singularities of differentiable maps, volume I. Birkhäuser, Boston 1985.

[AGV2] Arnold, V. I., S. M. Gusein-Zade, A. N. Varchenko: Singularities of differentiable maps, volume II. Birkhäuser, Boston 1988.

[Au] Audin, M.: Symplectic geometry in Frobenius manifolds and quantum cohomology. J. of Geometry and Physics **25** (1998), 183–204.

[Ba1] Barannikov, S.: Generalized periods and mirror symmetry in dimensions $n > 3$. Preprint, math.AG/9903124.

[Ba2] Barannikov, S.: Quantum periods – I. Semi-infinite variations of Hodge structures. Preprint, math.AG/0006193.

[Ba3] Barannikov, S.: Semi-infinite Hodge structures and mirror symmetry for projective spaces. Preprint, math.AG/0010157.

[BaK] Barannikov, S., M. Kontsevich: Frobenius manifolds and formality of Lie algebras of polyvector fields. Int. Math. Res. Notices **1998–4**, 201–215.

[Bat] Batyrev, V. V.: Stringy Hodge numbers and Virasoro algebra. Preprint, alg-geom/9711019.

[Bj] Björk, J.-E.: Analytic $\mathcal{D}$-modules and applications. Kluwer Academic Publishers, Mathematics and its applications vol. **247**, Dordrecht Boston London 1993.

[Bore] Borel, A.: Introduction aux groupes arithmétiques. Hermann, Paris 1969.

[Bori] Borisov, L. A.: On the Betti numbers and Chern classes of varieties with trivial odd cohomology groups. Preprint, alg-geom/9703023.

[Bou] Bourbaki, N.: Groupes et algèbres de Lie, chapitres 4, 5 et 6. Hermann, Paris 1968.

[Bre] Brélivet, T.: Variance of the spectral numbers and Newton polygons. Preprint, May 2001.

[Bri1] Brieskorn, E.: Singular elements of semisimple algebraic groups. In: Actes Congres Int. Math. **2**, Nice (1970), 279–284.

[Bri2] Brieskorn, E.: Die Monodromie der isolierten Singularitäten von Hyperflächen. Manuscripta Math. **2** (1970), 103–161.

[BrL] Bröcker, Th., L. Lander: Differentiable germs and catastrophes. London Math. Soc. Lecture Note Series **17**, Cambridge University Press 1975.

[Bru] Bruce, J. W.: Vector fields on discriminants and bifurcation varieties. Bull. London Math. Soc. **17** (1985), 257–262.

[CDGP] Candelas, P., X. C. de la Ossa, P. S. Green, L. Parkes: A pair of Calabi-Yau maifolds as an exactly soluble superconformal field theory. Nucl. Phys. B**359** (1991), 21–74.

[Ca] Cartan, H.: Quotient d'un espace analytique par un groupe d'automorphismes. In: Algebraic geometry and topology, a symposium in honour of S. Lefschetz. Princeton Univ. Press, 1957, 90–102.

[CaK] Cattani, E., A. Kaplan: Polarized mixed Hodge structures and the local monodromy of a variation of Hodge structure. Invent. Math. **67** (1982), 101–115.

[ChV] Chmutov, S. V., A. N. Varchenko: Tangent cone to the $\mu =$ constant stratum in terms of mixed Hodge structure. Moscow Univ. Math. Bull. **40** (1985), 7–12.

[Co] Coxeter, H. S. M.: Discrete groups generated by reflections. Ann. Math. **35** (1934), 588–621.

[DD] Nguyen tien Dai, Nguyen huu Duc: Stabilité de l'interaction géométrique entre deux composantes holonomes simples. C. R. Acad. Sc. Paris, Série A, **291** (1980), 113–116.

[De1] Deligne, P.: Equations différentielles à points singuliers réguliers. Lecture Notes in Math. vol. **163**, Springer, Berlin Heidelberg New York 1970.

[De2] Deligne, P.: Théorie de Hodge, II. Publ. Math. I.H.E.S. **40** (1971), 5–57.

[DVV] Dijkgraaf, R., E. Verlinde, H. Verlinde: Topological strings in $d < 1$. Nucl. Phys. B **352** (1991), 59–86.

[Di] Dimca, A.: Monodromy and Hodge theory of regular functions. Proceedings of the summer school on singularities, Newton Institute, Cambridge 2000, to appear.

[Du1] Dubrovin, B.: Integrable systems in topological field theory. Nucl. Phys. **B 379** (1992), 627–689.

[Du2] Dubrovin, B.: Differential geometry of the space of orbits of a Coxeter group. In: Surveys in differential geometry: integral systems [integrable systems]. Surveys in Diff. Geom. IV (C. L. Terng, K. Uhlenbeck, eds.). Int. Press, Boston MA 1998, 181–211.

[Du3] Dubrovin, B.: Geometry of 2D topological field theories. In: Integrable sytems and quantum groups. Montecatini, Terme 1993 (M Francoviglia, S. Greco, eds.). Lecture Notes in Math. 1620, Springer-Verlag 1996, 120–348.

[Du4] Dubrovin, B.: Painlevé transcendents in two-dimensional topological field theory. In: The Painlevé property. One century later (ed. Roberto Conte.). CRM Ser. Math. Phys.. Springer, New York 1999, 287–412.

[Du5] Dubrovin, B.: Geometry and analytic theory of Frobenius manifolds. In: Proceedings of the ICM 1998, Vol II, pp. 315–326.

[DuZ1] Dubrovin, B., Y. Zhang: Bihamiltonian hierarchies in 2D topological field theory at one-loop approximation. Commun. Math. Phys. **198** (1998), 311–361.

[DuZ2] Dubrovin, B., Y. Zhang: Frobenius manifolds and Virasoro constraints. Sel. math., New ser. **5** (1999), 423–466.

[EHX] Eguchi, T., K. Hori, Ch.-Sh. Xiong: Quantum cohomology and Virasoro algebra. Phys. Lett. **B402** (1997), 71–80.

[EV] Esnault, H., Viehweg, E.: Logarithmic de Rham complexes and vanishing theorems. Invent. Math. **86** (1986), 161–194.

[Fi] Fischer, G.: Complex analytic geometry. Lecture Notes in Math. vol. **538**, Springer, Berlin Heidelberg New York 1976.

[Ga] Gabrielov, A. M.: Bifurcations, Dynkin diagrams and modality of isolated singularities. Funct. Anal. **8** (1974), 94–98.

[Ge] Getzler, E.: Intersection theory on $\bar{\mathcal{M}}_{1,4}$ and elliptic Gromov-Witten invariants. J. Amer. Math. Soc. **10** (1997), 973–998.

[Gi1] Givental, A. B.: Convolution of invariants of groups generated by reflections, and connections with simple singularities of functions. Funct. Anal. **14** (1980), 81–89.

[Gi2] Givental, A. B.: Singular Lagrangian manifolds and their Lagrangian maps. J. Soviet Math. **52.4** (1988), 3246–3278.

[Gi3] Givental, A. B.: Homological geometry and mirror symmetry. In: Proceedings of the ICM–94 Zürich. Birkhäuser, Basel 1995, 472–480.

[Gi4] Givental, A. B.: Equivariant Gromov-Witten invariants. Int. Math. Res. Notices **1996–13**, 613–663.

[Gi5] Givental, A. B.: A mirror theorem for toric complete intersections. In: Topological field theory, primitive forms and related topics (Kyoto, 1996). Progr. in Math. **160**, Birkhäuser, Boston MA 1998, 141–175.

[Gi6] Givental, A. B.: Stationary phase integrals, quantum Toda lattices, flag manifolds and the mirror conjecture. In: Topics in singularity theory. Amer. Math. Soc. Transl. Ser. 2, **180**, Amer. Math. Soc., Providence RI 1997, 103–115.

[Gi7] Givental, A. B.: Elliptic Gromov-Witten invariants and the generalized mirror conjecture. In: Integrable systems and algebraic geometry. Proceedings of the Taniguchi Symposium 1997 (M.-H. Saito, Y. Shimizu, K. Ueno, eds.). World Scientific, River Edge NJ 1998, 107–155.

[Gi8] Givental, A. B.: Semisimple Frobenius structures at higher genus. Preprint, math.AG/0008067.

[Gi9] Givental, A. B.: Gromov–Witten invariants and quantization of quadratic hamiltonians. Preprint, math.AG/0108100.

[Gre] Greuel, G.-M.: Der Gauß-Manin-Zusammenhang isolierter Singularitäten von vollständigen Durchschnitten. Math. Ann. **214** (1975), 235–266.

[GrHP] Greuel, G.-M., C. Hertling, G. Pfister: Moduli spaces of semiquasihomogeneous singularities with fixed principal part. J. Algebraic Geometry **6** (1997), 169–199.

[Gri] Griffiths, P.: Periods of integrals on algebraic manifolds: summary of main results and open problems. Bull. Amer. Math. Soc. **76** (1970), 228–296.

[GrH] Griffiths, P., J. Harris: Principles of algebraic geometry. John Wiley and sons, New York 1978.

[He1] Hertling, C.: Analytische Invarianten bei den unimodularen und bimodularen Hyperflächensingularitäten. Dissertation. Bonner Math. Schriften **250** , Bonn 1992.

[He2] Hertling, C.: Ein Torellisatz für die unimodalen und bimodularen Hyperflächensingularitäten. Math. Ann. **302** (1995), 359–394.

[He3] Hertling, C.: Brieskorn lattices and Torelli type theorems for cubics in $\mathbb{P}^3$ and for Brieskorn-Pham singularities with coprime exponents. In: Singularities, the Brieskorn anniversary volume. Progress in Mathematics 162. Birkhäuser Verlag, Basel Boston Berlin 1998, 167–194.

[He4] Hertling, C.: Classifying spaces and moduli spaces for polarized mixed Hodge structures and for Brieskorn lattices. Compositio Math. **116** (1999), 1–37.

[HM] Hertling, C., Yu. Manin: Weak Frobenius manifolds. Int. Math. Res. Notices **1999–6**, 277–286.

[Hi] Hitchin, N. J.: Frobenius manifolds (notes by D. Calderbank). In: Gauge Theory and symplectic geometry, Montreal 1995, J. Hurtubise and F. Lalonde eds., Kluwer Academic Publishers, Netherlands 1997, 69–112.

[Hoe] Hörmander, L.: Fourier integral operators I. Acta. Math. **127** (1971), 71–183.

[Hol] Holmann, H.: Komplexe Räume mit komplexen Transformationsgruppen. Math. Ann. **150** (1963), 327–360.

[Hou] Houzel, C.: Géométrie analytique locale, II, théorie des morphismes finis. Séminaire Henri Cartan 13e année, no. **19**, 1960/61.

[JM] Jimbo, M., T. Miwa: Monodromy preserving deformations of linear ordinary differential equations with rational coefficients II. Physica **2D** (1981), 407–448.

[JMMS] Jimbo, M., T. Miwa, Y. Mori, M. Sato: Density matrix of impenetrable Bose gas and the fifth Painlevé transcendent. Physica **1D** (1980), 80–158.

[JMU] Jimbo, M., T. Miwa, K. Ueno: Monodromy preserving deformations of linear ordinary differential equations with rational coefficients I. Physica **2D** (1981), 306–352.

[Kar] Karpishpan, Y.: Torelli theorems for singularities. Invent. Math. **100** (1990), 97–141. Erratum. Invent. Math. **108** (1992), 667–668.

[Kas1] Kashiwara, M.: Quasi-unipotent constructible sheaves. J. Fac. Sci. Univ. Tokyo, Sec. IA **28** (1982), 757–773.

[Kas2] Kashiwara, M.: Introduction to microlocal analysis. Enseign. Math. **32** (1986), 227–259.

[KK] Kashiwara, M., T. Kawai: On holonomic systems of microdifferential equations III – systems with regular singularities. Publ. RIMS, Kyoto Univ. **17** (1981) 813–979.

[Kaup] Kaup, L., B. Kaup: Holomorphic functions of several variables. De Gruyter, studies in mathematics **3**, Berlin New York 1983.

[KM] Kontsevich, M., Yu. Manin: Gromov-Witten classes, quantum cohomology, and enumerative geometry. Commun. Math. Phys. **164** (1994), 525–562.

[Ku] Kulikov, Va.S.: Mixed Hodge structures and singularities. Cambridge tracts in mathematics **132**, Cambridge University Press, 1998.

[Kun] Kunz, E.: Komplementärmodul und kanonischer Modul. In: Der kanonische Modul eines Cohen-Macaulay-Rings (Herzog, J., Kunz, E., eds.). Lecture Notes in Math. vol. **238**, Springer, Berlin Heidelberg New York 1973, 85–102.

[La] Lazzeri, F.: A theorem on the monodromy of isolated singularities. Singularités a Cargèse, Astérisque **7–8** (1973), 269–275.

[Le] Lê Dũng Tráng: Une application d'un théoréme d'A'Campo a l'equisingularité. Preprint, Centre de Math. de l'Ecole Polytechnique, Paris, 1973.

[LeR] Lê Dũng Tráng, C. P. Ramanujam: The invariance of Milnor's number implies the invariance of the topological type. Amer. J. of Math. **98** (1973), 67–78.

[LiW] Libgober, A. S., J. W. Wood: Uniqueness of the complex structure on Kähler manifolds of certain homotopy types. J. Diff. Geom. **32** (1990), 139–154.

[Lo1] Looijenga, E. J. N: The complement of the bifurcation variety of a simple singularity. Invent Math. **23** (1974), 105–116.

[Lo2] Looijenga, E. J. N.: Isolated singular points on complete intersections. London Math. Soc. Lecture Note Series **77**, Cambridge University Press 1984.

[Lu] Luengo, I.: The μ-constant stratum is not smooth. Invent. Math. **90** (1987), 139–152.

[Ly1] Lyashko, O. V.: The geometry of bifurcation diagrams. Russian Math. Surveys **34.3** (1979), 209–210.

[Ly2] Lyashko, O. V.: Classification of critical points of functions on a manifold with singular boundary. Funct. Anal. **17.3** (1983), 187–193.

[Ly3] Lyashko, O. V.: Geometry of bifurcation diagrams. J. Soviet Math. **27** (1984), 2736–2759.

[Mal1] Malgrange, B.: Intégrales asymptotiques et monodromie. Ann. Sci. École Norm. Sup. **7** (1974), 405–430.

[Mal2] Malgrange, B.: Modules microdifférentiels et classes de Gevrey. In: Mathematical analysis and applications, essays dedicated to L. Schwartz on the occasion of his 65th birthday. Adv. in Math. Suppl. Studies **7B**, Academic Press Inc. 1981, 515–530.

[Mal3] Malgrange, B.: Déformations de systèmes différentiels et microdifférentiels. In: Séminaire de'l ENS, Mathématique et Physique, 1979–1982, Progress in Mathematics vol. 37, Birkhäuser, Boston 1983, pp. 353–379.

[Mal4] Malgrange, B.: Sur les déformations isomonodromiques, I, II. In: Séminaire de'l ENS, Mathématique et Physique, 1979–1982, Progress in Mathematics vol. 37, Birkhäuser, Boston 1983, pp. 401–438.

[Mal5] Malgrange, B.: Deformations of differential systems, II. J. Ramanujan Math. Soc. **1** (1986), 3–15.

[Mal6] Malgrange, B.: Connexions méromorphes. In: Congrés Singularités, Lille 1991 (P. Brasselet ed.), Cambridge University Press, 1994, pp. 251–261.

[Mal7] Malgrange, B.: Connexions méromorphes II: le réseau canonique. Invent. Math. **124** (1996), 367–387.

[Man1] Manin, Yu.: Moduli Fuchsiani. Annali Scuola Normale Sup. di Pisa Ser. III **19** (1965), 113–126.

[Man2] Manin, Yu.: Frobenius manifolds, quantum cohomology, and moduli spaces. American Math. Society, Colloquium Publ. v. **47**, 1999.

[MaM] Manin, Yu., S. A. Merkulov: Semisimple Frobenius (super)manifolds and quantum cohomology of $\mathbb{P}^r$. Topological methods in nonlinear analysis, Journal of the Juliusz Schauder Center **9** (1997), 107–161.

[Math1] Mather, J.: Stability of C^∞-maps III. Finitely determined map germs. Publ. Math. I.H.E.S. **35** (1968), 127–156.

[Math2] Mather, J.: Stability of C^∞-maps IV. Classification of stable germs by $\mathbb{R}$-algebras. Publ. Math. I.H.E.S. **37** (1969), 223–248.

[Matt] Mattei, J.-F.: Modules de feuilletages holomorphes singuliers: I équisingularité. Invent. Math. **103** (1991), 297–325.

[Mi] Milnor, J.: Singular points of complex hypersurfaces. Ann. Math. Stud. vol. 61, Princeton University Press 1968.

[Mue1] Müller, G.: Reduktive Automorphismengruppen analytischer ℂ-Algebren. J. Reine angew. Math. **364** (1986), 26–34.

[Mue2] Müller, G.: Symmetry groups of (semi-)weighted homogeneous singularities. Preprint 1996.

[Nam] Namikawa, Y.: Higher residues associated with an isolated hypersurface singularity. In: Algebraic varieties and analytic varieties, Advanced Studies in Pure Math. 1, North-Holland Publ. Company, 1983, pp. 181–193.

[Nav] Navarro-Aznar, V.: Sobre la invariància topològica de la multiplicitat. Publ. Sec. Mat. Univ. Autònoma Barcelona **20** (1980), 261–262.

[NS] Némethi, A., C. Sabbah: Semicontinuity of the spectrum at infinity. Abh. math. Sem. Univ. Hamburg **69** (1999), 25–35.

[NZ] Némethi, A., A. Zaharia: On the bifurcation set of a polynomial function and Newton boundary. Publ. RIMS, Kyoto Univ. **26** (1990), 681–689.

[No] Noumi, M.: Regular holonomic systems and their minimal extensions I. In: Group representations and systems of differential equations, Advanced Studies in Pure Math. 4, North-Holland Publ. Company, 1984, 209–221.

[Od1] Oda, T.: Introduction to algebraic analysis on complex manifolds. In: Algebraic varieties and analytic varieties, Advanced Studies in Pure Math. 1, North-Holland Publ. Company, 1983, 29–54.

[Od2] Oda, T.: K. Saito's period map for holomorphic functions with isolated singularities. In: Algebraic geometry, Sendai 1985, Advanced Studies in Pure Math. 10, 1987, 591–648.

[Ph1] Pham, F.: Singularités des systèmes différentiels de Gauss-Manin. Progress in Math. **2**, Birkhäuser, Boston 1979.

[Ph2] Pham, F.: Remarque sur l'équivalence des fonctions de phase. C. R. Acad. Sc. Paris, Série A **290** (1980), 1095–1097.

[Ph3] Pham, F.: Structures de Hodge mixtes associées à un germe de fonction à point critique isolé. Astérisque **101–102** (1983), 268–285.

[Ph4] Pham, F.: Vanishing homologies and the n variable saddlepoint method. In: Singularities, Proc. of symp. in pure math. **40.2** (1983), 319–333.

[Ph5] Pham, F.: La descente des cols par les onglets de Lefschetz, avec vues sur Gauss-Manin. Systèmes différentiels et singularités. Asterisques **130** (1985), 11–47.

[Sab1] Sabbah, C.: Monodromy at infinity and Fourier transform. Publ. RIMS, Kyoto Univ. **33** (1997), 643–685.

[Sab2] Sabbah, C.: Frobenius manifolds: isomonodromic deformations and infinitesimal period mappings. Expo. Math. **16** (1998), 1–58.

[Sab3] Sabbah, C.: Hypergeometric period for a tame polynomial. C. R. Acad. Sci. Paris Sér. I Math. **328** (1999), 603–608, and (42 pages) math.AG/9805077.

[Sab4] Sabbah, C.: Déformations isomonodromiques et variétés de Frobenius, une introduction. Centre de Mathematiques, Ecole Polytechnique, U.M.R. 7640 du C.N.R.S., no. 2000–05, 251 pages.

[SK1] Saito, K.: Quasihomogene isolierte Singularitäten von Hyperflächen. Invent. Math. **14** (1971), 123–142.

[SK2] Saito, K.: Regularity of Gauss-Manin connection of flat family of isolated singularities. In: Quelques journeés singulières, Ecole Polytechnique, Paris, 1973.

[SK3] Saito, K.: On a linear structure of the quotient variety by a finite reflexion group. Preprint RIMS–288 (1979), Publ. RIMS, Kyoto Univ. **29** (1993), 535–579.

[SK4] Saito, K.: Theory of logarithmic differential forms and logarithmic vector fields, J. Fac. Sci. Univ. Tokyo, Sec. IA **27** (1981), 265–291.

[SK5] Saito, K.: A characterisation of the intersection form of a Milnor's fiber for a function with an isolated critical point.

[SK6] Saito, K.: Primitive forms for a universal unfolding of a function with an isolated critical point, J. Fac. Sci. Univ. Tokyo, Sect. IA Math. **28** (1982), 775–792.

[SK7] Saito, K.: The higher residue pairings $K_F^{(k)}$ for a family of hypersurface singular points. In: Singularities, Proc. of symp. in pure math. **40.2** (1983), 441–463.

[SK8] Saito, K.: The zeroes of characteristic function χ_f for the exponents of a hypersurface isolated singular point. In: Algebraic varieties and analytic varieties, Advanced Studies in Pure Math. 1, North-Holland Publ. Company, 1983, 195–217.

[SK9] Saito, K.: Period mapping associated to a primitive form. Publ. RIMS, Kyoto Univ. **19** (1983), 1231–1264.

[SM1] Saito, M.: Gauss-Manin systems and mixed Hodge structure, Proc. Japan Acad. **58** (1982), 29–32. Supplément in Astérisque **101–102** (1983), 320–321.

[SM2] Saito, M.: On the structure of Brieskorn lattices. Preprint at Grenoble, September 1983.

[SM3] Saito, M.: On the structure of Brieskorn lattices. Ann. Inst. Fourier Grenoble **39** (1989), 27–72.

[SM4] Saito, M.: Period mapping via Brieskorn modules. Bull. Soc. math. France **119** (1991), 141–171.

[SM5] Saito, M.: Exponents of an irreducible plane curve singularity. Preprint, math. AG/0009133.

[Sche1] Scherk, J.: On the monodromy theorem for isolated hypersurface singularities. Invent. Math. **58** (1980), 289–301.

[Sche2] Scherk, J.: A propos d'un théorème de Mather et Yau. C. R. Acad. Sci. Paris, Série I, **296** (1983), 513–515.

[SchSt] Scherk, J., J. H. M. Steenbrink: On the mixed Hodge structure on the cohomology of the Milnor fibre. Math. Ann. **271** (1985), 641–665.

[Schm] Schmid, W.: Variation of Hodge structure: The singularities of the period mapping. Invent. Math. **22** (1973), 211–319.

[SchuSt] Schulze, M., J. H. M Steenbrink: Computing Hodge-theoretic invariants of singularities. Proceedings of the summer school on singularities, Newton Institute, Cambridge 2000, to appear.

[Seb] Sebastiani, M.: Preuve d'une conjecture de Brieskorn. Manuscripta Math. **2** (1970), 301–308.

[Ser] Serre, J.-P.: Prolongement de faisceaux analytiques cohérents. Ann. Inst. Fourier, Grenoble **16** (1966), 363–374.

[ShS1] Shcherbak, I., A. Szpirglas: Boundary singularities: Topology and duality. In: Advances in Soviet Math. **21**, 1994, 213–223.

[ShS2] Shcherbak, I., A. Szpirglas: Boundary singularities: double coverings and Picard-Lefschetz formulas. C. R. Acad. S. Paris, Série I, **322** (1996), 557–562.

[ShO] Shcherbak, O. P.: Wavefronts and reflection groups. Russian Math. Surveys **43.3** (1988), 149–194.

[Sl] Slodowy, P.: Einige Bemerkungen zur Entfaltung symmetrischer Funktionen. Math. Z. **158** (1978), 157–170.

[Stn] Steenbrink, J. H. M.: Mixed Hodge structure on the vanishing cohomology. In: Real and complex singularities, Oslo 1976, P. Holm (ed.). Alphen aan den Rijn: Sijthoff and Noordhoff 1977, 525–562.

[Stv] Stevens, J.: On the μ-constant stratum and the V-filtration: an example. Math. Z. **201** (1989), 139–144.

[Sz] Szpirglas, A.: Singularités de bord: dualité, formules de Picard Lefschetz relatives et diagrammes de Dynkin. Bull. Soc. math. France **118** (1990), 451–486.

[Te1] Teissier, B.: Déformations à type topologique constant. Séminaire de géometrie analytique 1971–1972, Astérisque **16**, 215–249.

[Te2] Teissier, B.: The hunting of invariants in the geometry of discriminants. In: Real and complex singularities, Oslo 1976, P. Holm ed., Alphen a/d Rijn: Sijthoff & Noordhoff, 565–677.

[Ter] Terao, H.: The bifurcation set and logarithmic vector fields. Math. Ann. **263** (1983), 313–321.

[Va1] Varchenko, A. N.: The asymptotics of holomorphic forms determine a mixed Hodge structure. Sov. Math. Dokl. **22** (1980), 772–775.

[Va2] Varchenko, A. N.: Asymptotic mixed Hodge structure in vanishing cohomologies. Math. USSR Izvestija **18** (1982), 469–512.

[Va3] Varchenko, A. N.: The complex singular index does not change along the stratum $\mu =$ constant. Functional Anal. Appl. **16** (1982), 1–9.

[Va4] Varchenko, A. N.: A lower bound for the codimension of the stratum $\mu =$ constant in terms of the mixed Hodge structure. Moscow Univ. Math. Bull. **37** (1982), 30–33.

[Va5] Varchenko, A. N.: Local classification of volume forms in the presence of a hypersurface. Funct. Anal. **19.4** (1985), 269–276.

[Va6] Varchenko, A. N.: On the local residue and the intersection form on the vanishing cohomology. Math. USSR Izvestija **26** (1986), 31–52.

[VaS] Vassiliev, V. A., V. V. Serganova: On the number of real and complex moduli of singularities of smooth functions and realizations of matroids. Math. Notes **49** (1991), 15–20.

[Vo] Voisin, C.: Symétrie miroir. Panoramas et synthèses **2**, Soc. Math. de France 1996.

[Wah] Wahl, J.: Equisingular deformations of plane algebroid curves. Trans. Amer. Math. Soc. **193** (1974), 143–170.

[Wal1] Wall, C. T. C.: A note on symmetry of singularities. Bull. London Math. Soc. **12** (1980), 169–175.

[Wal2] Wall, C. T. C.: A second note on symmetry of singularities. Bull. London Math. Soc. **12** (1980), 347–354.

[War] Warner, F. W.: Foundations of differentiable manifolds and Lie groups. Graduate texts in mathematics **94**. Springer-Verlag 1983.

[Was] Wassermann, G.: Stability of unfoldings. Lecture Notes in Math. **393**, Springer-Verlag, 1974.

[Wir] Wirthmüller, K.: Singularities determined by their discriminant. Math. Ann. **252** (1980), 237–245.

[Wit] Witten, E.: Two-dimensional gravity and intersection theory on moduli spaces. Surveys in Diff. Geom. **1** (1991), 243–310.

[Za] Zakalyukin, V. M.: Reconstruction of wave fronts depending on one parameter. Funct. Anal. **10.2** (1976), 139–140.

Index